A TEXTBOOK OF ORGANIC CHEMISTRY

Volume I

Mandeep Dalal

DALAL INSTITUTE

Publisher:

Dalal Institute, Main Market, Sector 14, Rohtak, Haryana 124001, India

(info@dalalinstitute.com, +91-9802825820)

www.dalalinstitute.com

Dedicated to my mother "Darshana Devi"

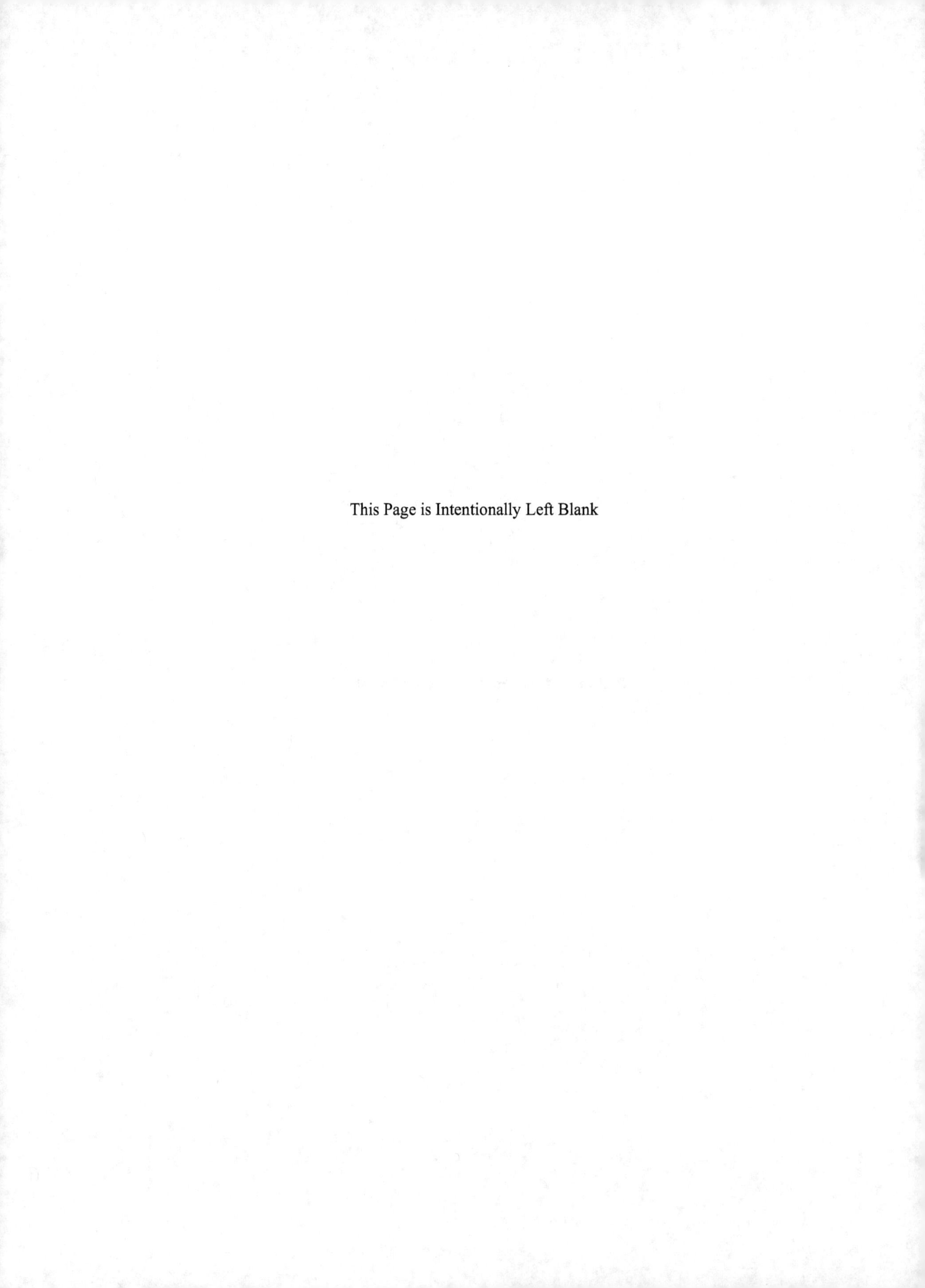

PREFACE

The preface writing has always been a wonderful feeling that cannot be expressed in words as it relates you to your audience through your work. I conceived the idea of writing a new advanced-level textbook in organic chemistry during my Ph.D pursuit when I saw post-graduate chemistry students who were tired in search of the syllabus topics because of their ill-resourced university or college library. I also decided to write the textbooks of physical and organic chemistry because I think that someone who wants to teach or text one stream must have the core conceptual understanding of all the three streams of chemical science otherwise one would not be able to connect and explain the interdisciplinary topics in a comprehensive manner.

Out of the series of three textbooks, the present book, entitled "A Textbook of Organic Chemistry – Volume 1", is the first installment of "A Textbook of Organic Chemistry", which is a four-volume set in all. All the students and teachers are advised to read and consult all the four volumes in a subsequent pattern for a more efficient and thorough understanding of the subject of organic chemistry.

I also celebrate this opportunity for expressing the bottom-hearted gratitude towards the people who supported me at all stages of my work. First of all, I would like to express my sincere gratitude to my doctoral supervisors, Prof. S. P. Khatkar and Prof. V.B. Taxak for their continuous support and guidance from day one. Then I would like to record appreciation to my lovely sister, Jyoti Dalal, for her unconditional love, support and for being the guiding light when the life threw me in the darkest of corners. I am very much thankful to my beautiful wife, Anita Sangwan, who always stands shoulder to shoulder with me in my good and bad times. I especially want to thank my brother Sandeep Dalal for his positive criticism, encouragement, motivation and truly selfless support. A special thanks to my dearest sister Garima Sheoran for her love, care, and all-time encouragement. I also wish to thank my entire family, friends, and teachers for providing a loving environment for me.

Lastly, and most importantly, I wish to thank my mother, Darshana Devi, who bore me, raised me, supported me, taught me, and loved me.

Mandeep Dalal

This Page is Intentionally Left Blank

Table of Contents

CHAPTER 1

Nature of Bonding in Organic Molecules

❖ Delocalized Chemical Bonding

The bonding in most of the organic compounds can be described by using a single Lewis structure (localized bond), and the orbital picture of such chemical bonds is treated either by the valence bond approach or by the molecular orbital theory. However, in some organic compounds, the electron density is not confined between two nuclei only, and therefore, cannot be depicted by a single Lewis structure. These types of molecules are said to possess "delocalized bonds"; and like localized bonds, the wave mechanical picture of such molecules can also be obtained by valence bond theory, as well as from the molecular orbital approach.

The valence bond approach is to consider many possible Lewis structures and then averaging out them all. These possible Lewis structures are generally called as canonical or resonating structures, and the phenomenon is labeled as "resonance". For instance, let us consider the case of the benzene molecule. If we treat it by the first extension of valence bond theory, three carbon-carbon bonds should be shorter and three carbon-carbon bonds should be longer due to their double and single bond order character.

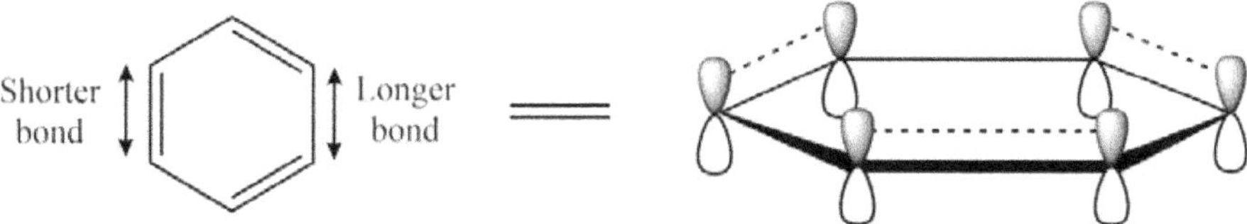

Figure 1. The expected structure of benzene molecule without resonance.

However, in the actual molecule, all the bond lengths and bond angles were found equal. All this can be explained only after considering the phenomenon of resonance in which all the possible resonating structures are given below.

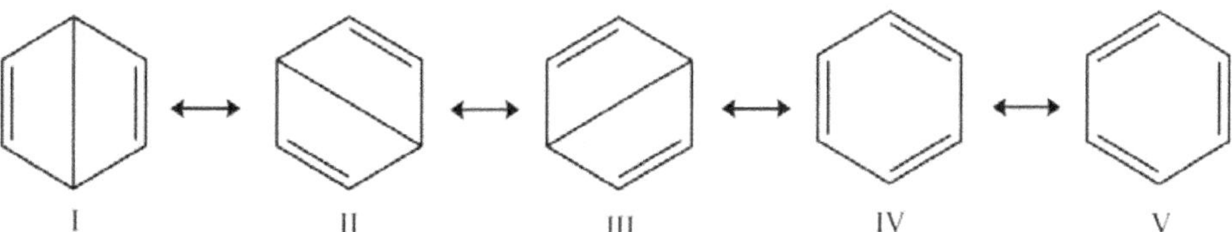

Figure 2. Different canonical structures of benzene molecule from resonance.

It has been found that after quantum mechanical calculations that Dewar structures (I–III) contribute only 7.33% each whereas Kekule structures (IV and V) contribute around 39% each to the resultant structure of benzene in which all the bond lengths and bond angles are equal. The bond-order of carbon-carbon bonds is greater than one but less than two showing an averaged strength for the resonance hybrid. It is also worthy to note that the energy of resonance hybrid is always less than the most stable resonating structure, and the energy difference between the two is simply called as resonance energy.

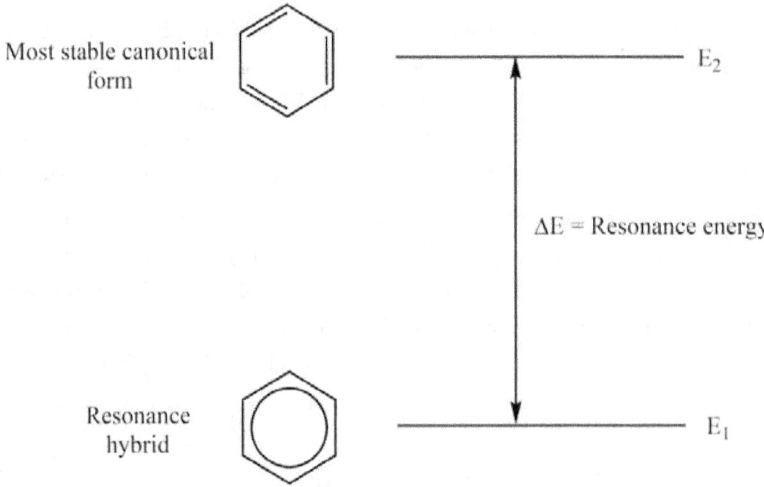

Figure 3. The depiction of resonance hybrid, most stable canonical form, and resonance energy in benzene molecule.

Now although the introduction of the concept of resonance in valence bond theory is quite useful in explaining the nature of "delocalized bond" in a lot of organic compounds, serious complications arise during the similar treatment of some other molecules like pyridine or the naphthalene. So, the molecular orbital theory, which is a more sophisticated approach, must be used to rationalize the nature of delocalization in such cases. However, for a comparative purpose, we will the discuss superficial molecular orbital treatment of benzene here. In this case, σ-bonding is treated in a valence bond framework and only π-bonding is examined in molecular orbital theory. In this method, the mixing of six atomic p_z-orbitals from six sp^2-hybridized carbons gives rise to six molecular orbitals, in which three are bonding and three are non-bonding. The number of nodes in the most bonding molecular orbital is zero whereas two other degenerate bonding molecular orbitals are with one node. The degenerate set of antibonding atomic orbitals, ψ_4 and ψ_5, are having two mutually perpendicular nodes; whereas three nodes are present in the most antibonding molecular orbital in benzene. All the six π-electrons are filled in bonding molecular orbitals leaving antibonding orbitals empty. The pictorial representation of various molecular orbitals in the benzene molecule is given below.

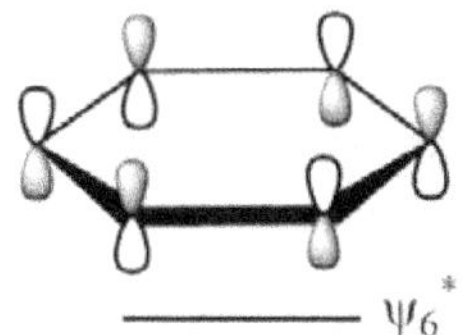

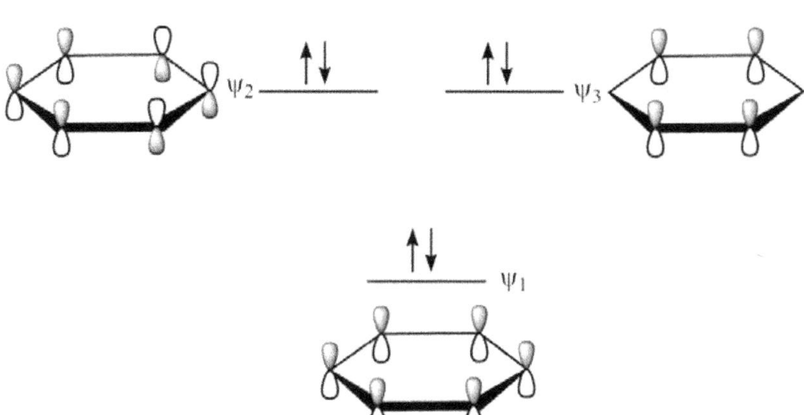

Figure 4. The depiction of molecular orbitals in benzene molecule.

Now, after having the general discussion on delocalized chemical bonds, it's time to discuss some phenomena and effects that arise due to delocalized bonds like conjugation, cross conjugation, resonance, hyperconjugation, and tautomerism.

❖ Conjugation

The phenomenon of conjugation in organic compounds is typically characterized by the presence of alternate single and double bonds which are perpendicular to the molecular plane.

Typically, the orbitals required for the perpendicular π-bonding are supplied by the sp^2-hybridized carbon atom; nevertheless, it is not the hard-and-fast condition for the conjugation to take place because as long as the adjacent atoms are in a chain and have *p*-orbitals, the system can be treated as conjugated in nature. For instance, the furan molecule is a 5-membered cyclic system with two alternating double bonds neighboring oxygen in a five-membered ring. The oxygen atom has two lone pairs of electrons, one pair fills a *p* orbital perpendicular to the molecular plane, and therefore, sustaining the conjugation of that 5-membered cyclic system by overlap with the perpendicular *p* orbital on every adjacent C atoms. The 2nd lone pair remains in the molecular plane and does not contribute to the phenomenon of conjugation. The two main types of conjugations are given below.

➤ *Double Bond in Conjugation with a Double Bond*

In these compounds, a double or triple bond is in conjugation with another double or triple bond. Some of the common examples of these types of molecules are given below.

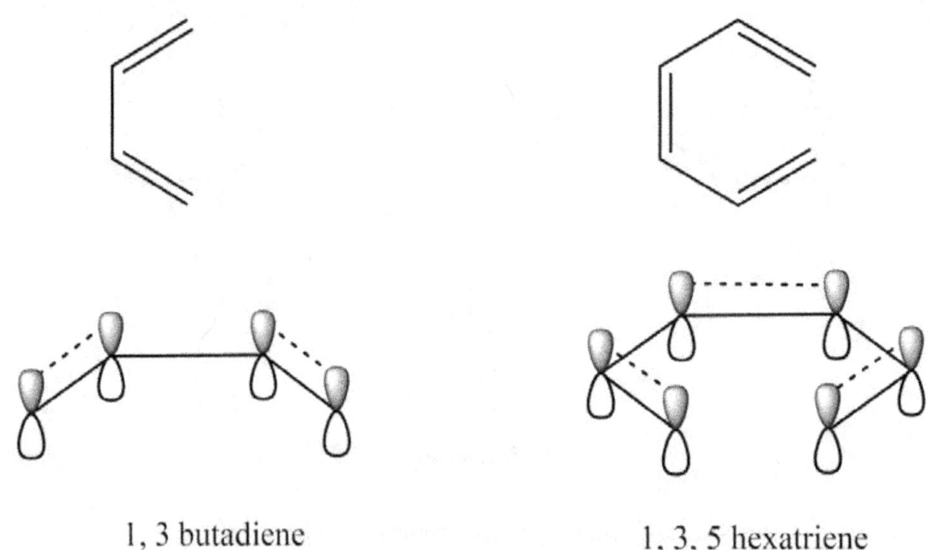

1, 3 butadiene   1, 3, 5 hexatriene

Figure 5. Some of the most common examples of acyclic molecular geometries bond-conjugated systems in them.

It should also be noted that the carbon that possesses the conjugated orbital is almost always got sp^2 hybridization even if contains a lone pair of electron.

> ### *Double Bond in Conjugation with p-Orbital on the Neighboring Atom*

In these compounds, a double or triple bond is in conjugation with the atomic p-orbital of an adjacent carbon or heteroatom. Some of the common examples of these types of molecules are given below.

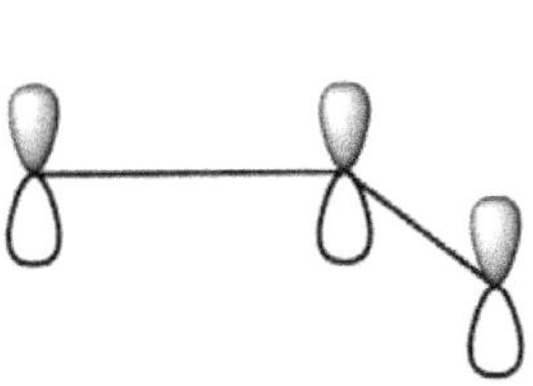

Allyl radical

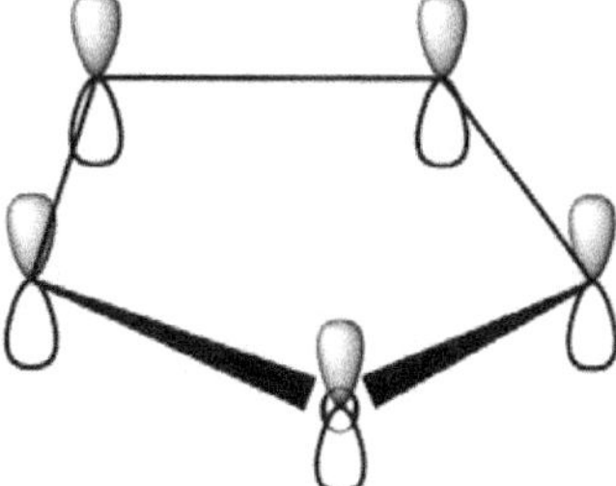

Furan

Figure 6. Common examples of some bond-orbital conjugated molecules.

It should also be noted that the carbon that possesses the conjugated orbital is almost always got sp^2 hybridization even if contains a lone pair of electrons.

❖ Cross Conjugation

The phenomenon of cross-conjugation is a special type of conjugation in a molecule when in a set of three π-bonds only two pi-bonds interact with each other by conjugation, the third one is excluded from the interaction.

Alternatively, we can also say that a system with cross-conjugation has an alkene fragment bonded to one of the central atoms of a 2nd conjugated chain via a single bond. It is different from normal conjugation in which a polyene typically has alternating single and double bonds along with consecutive atoms. Some of the common examples of these types of molecules are given below.

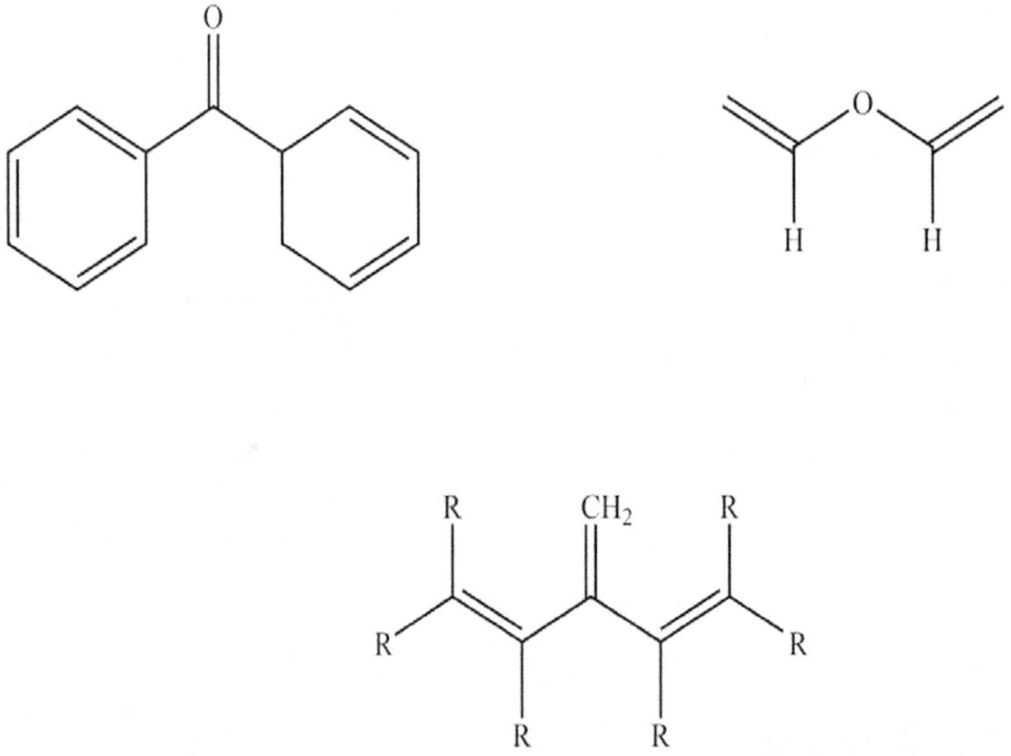

Figure 7. Common examples of molecules with cross-conjugation.

Classically speaking, the branching of double-bonds happens instead of continuous conjugation, and the fragment of that same primary chain is in conjugation with the side group, but all portions are not conjugated with each other very firmly. Common examples of the cross-conjugation effect can be found in molecules such as divinylketones, benzophenone, dendralenes, p-quinones, radialenes, Indigo dye, and fullerene.

Furthermore, it should also be mentioned that the phenomenon of cross conjugation affects the reactivity and molecular electronic transitions in a significant manner. For instance, the lone pair on nitrogen activates the ring at *o*- and *p*-position in aniline (I), supporting the formation of tribromo aniline for bromination reaction. On the other hand, the bromination leads to only monosubstituted acetanilide due to cross conjugation in acetanilide.

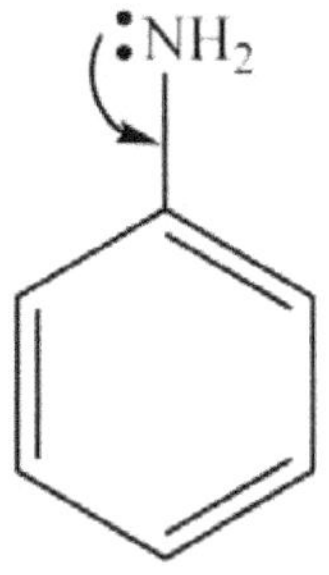

aniline

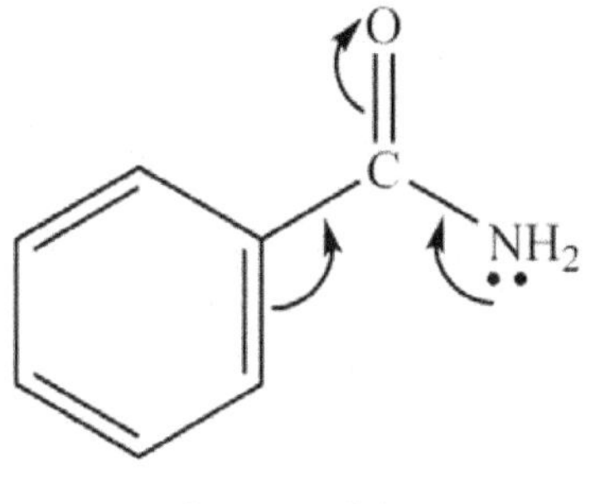

benzamide

Figure 8. The effect of cross-conjugation upon reactivity.

It is also worthy to note that the phenomenon of cross conjugation is different than homoconjugation-which arises when two π-systems are separated by one non-conjugating group.

❖ Resonance

The concept of resonance is an extension of valence bond theory to explain the nature of the "delocalized bond" in a lot of organic compounds which cannot be described by a single Lewis structure.

The valence bond approach to delocalized bonds is to consider many possible Lewis structures and then averaging out them all. These possible Lewis structures are generally called as canonical or resonating structures, and this phenomenon is labeled as "resonance". For instance, let us consider the case of the benzene molecule. If we treat it by the first extension of valence bond theory, three carbon-carbon bonds should be shorter and three carbon-carbon bonds should be longer due to their double and single bond order character.

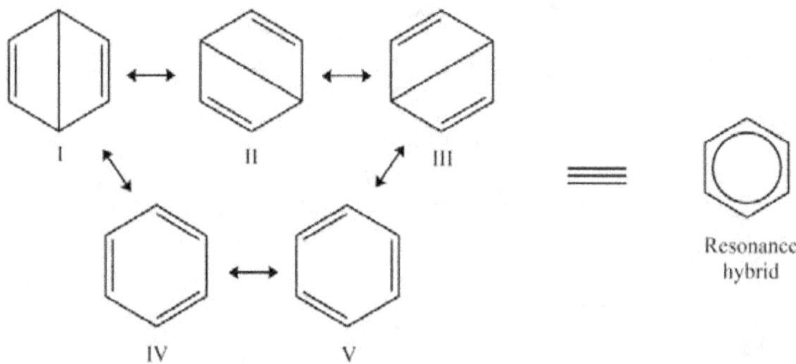

Figure 9. Different resonating structures of the benzene molecule.

However, in the actual molecule, all the bond lengths and bond angles were found equal. All this can be explained only after considering the phenomenon of resonance in which all the possible resonating structures are given in 'Figure 8'. It has been found that after quantum mechanical calculations that Dewar structures (I–III) contribute only 7.33% each whereas Kekule structures (IV and V) contribute approx. 39% each to the resultant structure of benzene in which all the bond lengths and bond angles are equal. The bond order of carbon-carbon bonds is greater than one but less than two showing an averaged strength for the resonance hybrid. The resonance phenomena in some other molecules is given below.

Figure 10. Resonance phenomena in carbon dioxide and acetic acid.

Calculation of resonance energy: It is also worthy to note that the energy of resonance hybrid is always less than the most stable resonating structure, and the energy difference between the two is simply called as resonance energy.

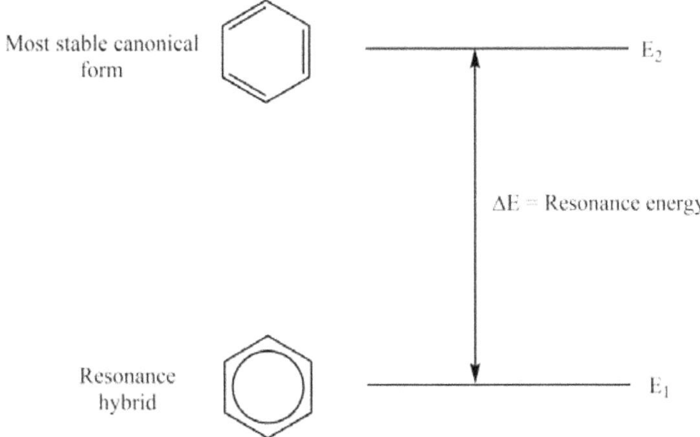

Figure 11. The depiction of resonance hybrid, most stable canonical form, and resonance energy in benzene molecule.

Since the Lewis structures are not real ones, the resonance energy for the benzene molecule can be obtained via two routes as discussed below.

i) From heat of hydrogenation: The hypothetical six-membered cyclic ring system with no resonance would be cyclohexatriene whose heat of hydrogenation should be three times of cyclohexene molecule.

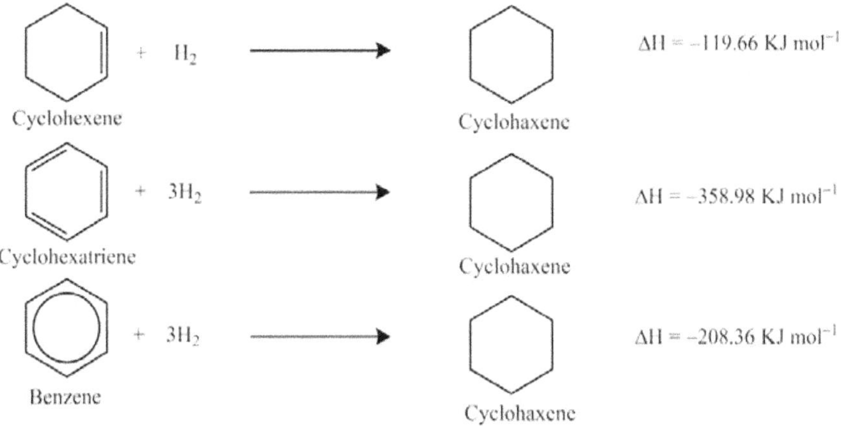

Now since the enthalpy of hydrogenation of the actual benzene molecule is less than its hypothetical localized version (358.98 – 208.36 = 150.62 kJ mol^{-1}), the extra stabilization can be attributed to the phenomena of delocalization easily.

ii) From heat of formation: The hypothetical six-membered cyclic ring system with no resonance would be cyclohexatriene (Kekule structure) whose heat of formation should be the sum of the bond energies of different carbon-carbon and carbon-hydrogen bonds.

$$\Delta H_{calculated} = 6 \times Bond\ dissociation\ energy\ of\ C-H\ bond \tag{1}$$
$$+ 3 \times Bond\ dissociation\ energy\ of\ C-C\ bond$$
$$+ 3 \times Bond\ dissociation\ energy\ of\ C=C\ bond$$

$$= 6 \times 413.3\ KJ\ mol^{-1} + 3 \times 347.7\ KJ\ mol^{-1} + 606.7\ KJ\ mol^{-1} \tag{2}$$
$$= 5343\ KJ\ mol^{-1}$$

Now since the experimental enthalpy of formation of the actual benzene molecule is −5494.84 kJ mol^{-1} which is more than its hypothetical localized version (−5343 kJ mol^{-1}), the extra stabilization of −151.84 kJ mol^{-1} can be attributed to the phenomena of delocalization easily.

Conditions for resonance: In order to show the phenomenon of resonance, the following condition must be satisfied by the molecule under consideration.

1. Different canonical forms of the molecule should differ only with respect to the position of electrons and not the nuclei.

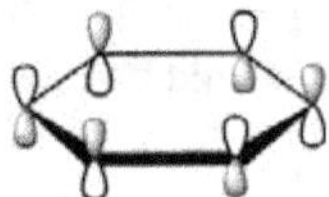

2. All the participating atoms must lie in the same plane.

3. All resonating structures must possess the same number of unpaired and paired electrons.

4. If two resonating structures are having the same energy, they will contribute equally to the resonance hybrid.

5. More stable resonating structure will contribute higher to the resonance hybrid.

6. Greater is the number of canonical structures, higher will be the stability of corresponding resonance hybrid.

7. The following factors must be taken into consideration before we decide the relative stability of participating canonical structure:

i) Resonating structures with a greater number of covalent bonds will be more stable, and therefore, will show more contribution towards resonance hybrid.

More stable (more contribution)   Less stable (less contribution)

ii) Structures with charge separation will be less stable, and therefore, will contribute less towards the resonance hybrid.

More stable (more contribution)   Less stable (less contribution)

iii) Structures violating the octet rule should not be considered (involving second-period elements).

Stable (will show contribution)   Highly unstable (should be discarded)

iv) Resonating structures with the negative charge on a more electronegative atom will be less stable, and therefore, will contribute less to the resonance hybrid.

Most stable (most contribution)   Least stable (least contribution)   Less stable (less contribution)

v) If a structure helps to delocalize the positive charge will contribute significantly toward the resonance hybrid even if the positive charge is on the electronegative atom.

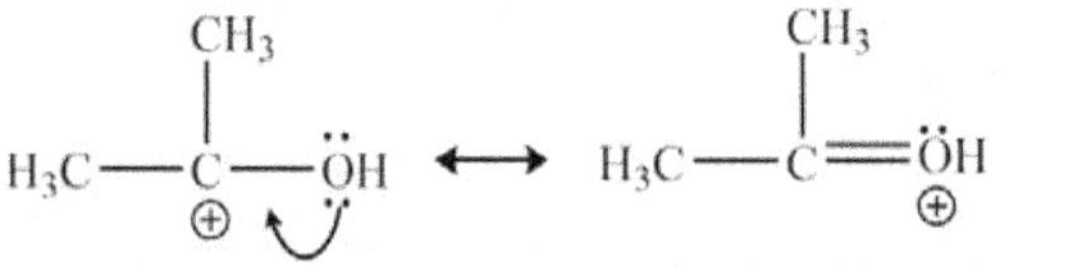

Most stable   Significantly stable
(more contribution)   (significant contribution)

vi) Resonating structures with like charges on adjacent atoms in space will be less stable sterics, and therefore, will contribute less towards the resonance hybrid.

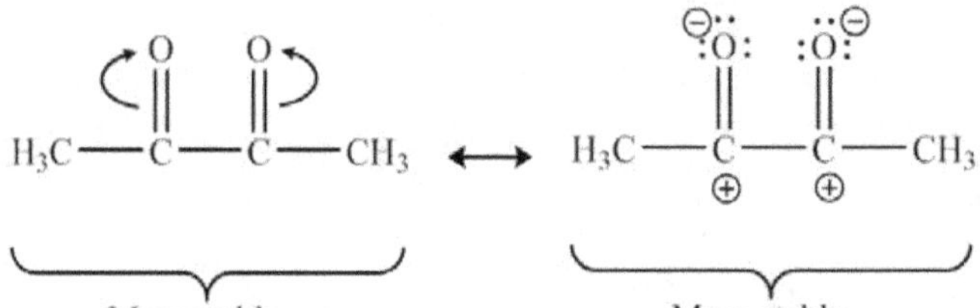

More stable   More stable
(more contribution)   (more contribution)

Resonance effect:

The resonance or mesomeric effect in organic compounds may simply be defined as the polarity produced in the molecule due to resonance phenomena.

The effect is used in a qualitative way and describes the electron-withdrawing or releasing properties of substituents based on relevant resonance structures and is symbolized by the letter R or M. The resonance effect is negative (–R) when the substituent is an electron-withdrawing group and the effect is positive (+R) when the substituent is an electron releasing group. A detailed explanation of the two effects with some typical examples is given below.

i) *+R effect:* Some groups release the electron density to the multiple bonds through resonance and are said to have a +M or +R effect in general.

7. The following factors must be taken into consideration before we decide the relative stability of participating canonical structure:

i) Resonating structures with a greater number of covalent bonds will be more stable, and therefore, will show more contribution towards resonance hybrid.

$$H_2C=C-C=CH_2 \longleftrightarrow H_2C-C=C-CH_2 \longleftrightarrow H_2C-C=C-CH_2$$

More stable
(more contribution)

Less stable
(less contribution)

ii) Structures with charge separation will be less stable, and therefore, will contribute less towards the resonance hybrid.

$$H_3C-C-OH \longleftrightarrow H_3C-C-OH$$

More stable
(more contribution)

Less stable
(less contribution)

iii) Structures violating the octet rule should not be considered (involving second-period elements).

$$H_2C=C-NH_3 \longleftrightarrow H_2C-C=NH_3$$
$$8\ e^- \qquad\qquad 10\ e^-$$

Stable
(will show contribution)

Highly unstable
(should be discarded)

iv) Resonating structures with the negative charge on a more electronegative atom will be less stable, and therefore, will contribute less to the resonance hybrid.

$$H_3C\diagdown C=O \longleftrightarrow H_3C\diagdown C-O \longleftrightarrow H_3C\diagdown C-O$$

Most stable
(most contribution)

Least stable
(least contribution)

Less stable
(less contribution)

v) If a structure helps to delocalize the positive charge will contribute significantly toward the resonance hybrid even if the positive charge is on the electronegative atom.

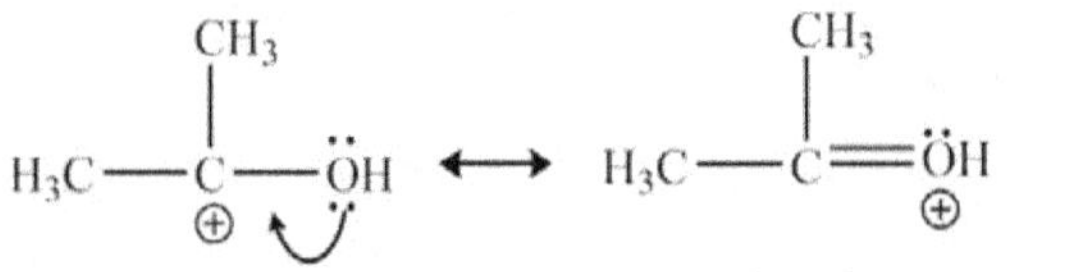

$$H_3C - \underset{\oplus}{\overset{CH_3}{C}} - \overset{..}{O}H \quad \longleftrightarrow \quad H_3C - \overset{CH_3}{C} = \underset{\oplus}{\overset{..}{O}}H$$

Most stable   Significantly stable
(more contribution)   (significant contribution)

vi) Resonating structures with like charges on adjacent atoms in space will be less stable sterics, and therefore, will contribute less towards the resonance hybrid.

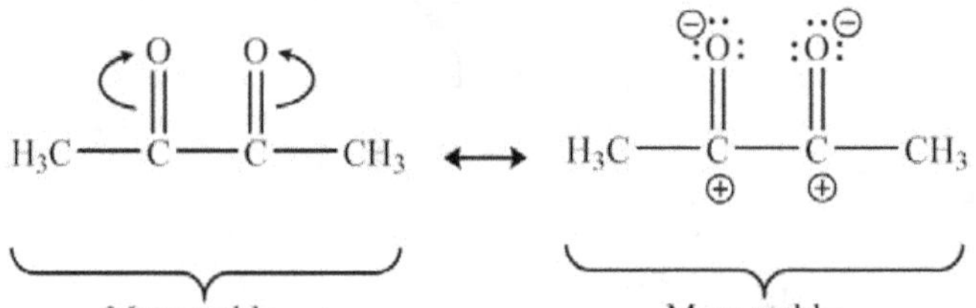

$$H_3C - \overset{O}{\overset{\|}{C}} - \overset{O}{\overset{\|}{C}} - CH_3 \quad \longleftrightarrow \quad H_3C - \underset{\oplus}{\overset{\overset{\ominus}{:}\overset{..}{O}:}{C}} - \underset{\oplus}{\overset{:\overset{..}{O}:\overset{\ominus}{}}{C}} - CH_3$$

More stable   More stable
(more contribution)   (more contribution)

Resonance effect:

The resonance or mesomeric effect in organic compounds may simply be defined as the polarity produced in the molecule due to resonance phenomena.

The effect is used in a qualitative way and describes the electron-withdrawing or releasing properties of substituents based on relevant resonance structures and is symbolized by the letter R or M. The resonance effect is negative (–R) when the substituent is an electron-withdrawing group and the effect is positive (+R) when the substituent is an electron releasing group. A detailed explanation of the two effects with some typical examples is given below.

i) +R effect: Some groups release the electron density to the multiple bonds through resonance and are said to have a +M or +R effect in general.

$$H_2C = \underset{H}{C} - \overset{..}{Br}: \quad \longleftrightarrow \quad \overset{\ominus}{H_2C} - \underset{H}{C} = \overset{\oplus}{Br}:$$

$$H_2C = \underset{H}{C} - \overset{..}{O}H \quad \longleftrightarrow \quad \overset{\ominus}{H_2C} - \underset{H}{C} = \overset{\oplus}{O}H$$

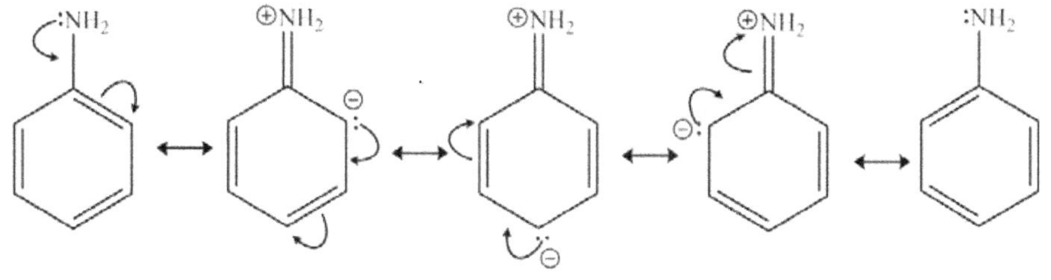

It is also worthy to mention that groups showing the +R effect are having lone pair of the electron that can be put into conjugation with the double of the chain or ring to which it gets attached with. Some of the typical groups showing the +R effect are given below.

$$-O^- > -NH_2 > -NHR > -OR > -NHCOR > -OCOR > -Ph > -F > -Cl > -Br > -I$$

ii) −R effect: Some groups withdraw the electron density to the multiple bond through resonance, and are said to have −M or −R effect in general.

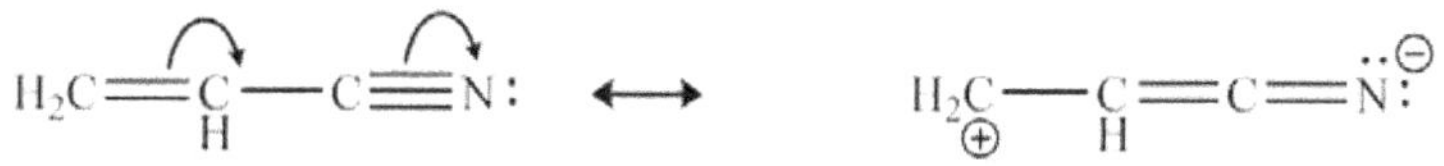

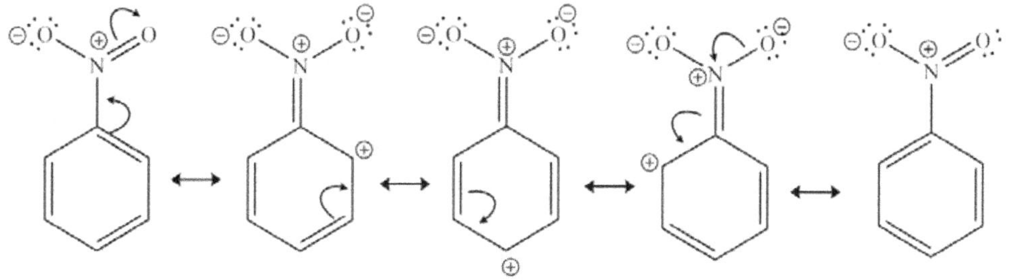

It is also worthy to mention that groups showing the −R effect are having a double bond that can be put into conjugation with the double of the chain or ring to which it gets attached with. Some of the typical groups showing the −R effect are given below.

$$-NO_2 > -CN > -S(=O)_2-OH > -CHO > -C{=}O > -COOCOR > -COOR > -COOH > -CONH_2 > -COO^-$$

It should also be noted that the resonance is different from the inductive effect and can exert its electron-withdrawing or electron releasing effect in a direction along or opposite to the inductive effect. Unlike inductive effect which can also act in saturated molecules, the resonance effect is observed only in conjugated systems. Furthermore, the inductive effect involves σ-electrons and is operative only up to four carbon atoms, whereas the resonance effect involves unpaired or π-electrons which can get delocalized over longer distances.

Applications of resonance effect: Some of the important applications of the resonance or mesomeric effect that affect the properties to a great extent are given below.

i) Low reactivity of vinyl and aryl halides: The existence of the +R effect in vinyl and aryl halides makes them less reactive towards the nucleophilic substitution reaction due to the partial double character of the carbon-halogen bond as shown below.

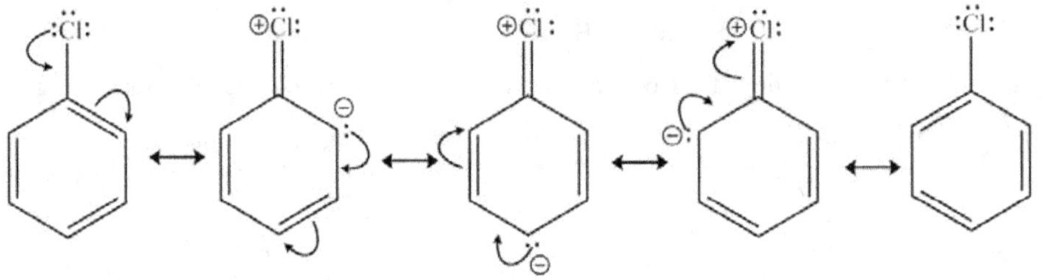

ii) High reactivity of allyl and benzyl halides: The existence of the −R effect in allyl and benzyl halides makes them more reactive towards the nucleophilic substitution reaction due to the formation resonance stabilized cation bond as shown below.

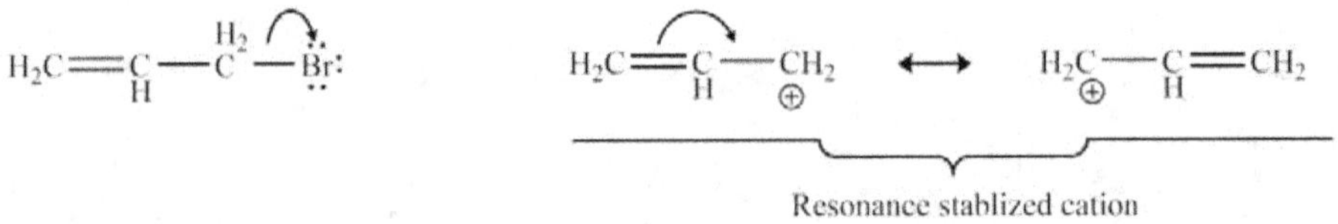

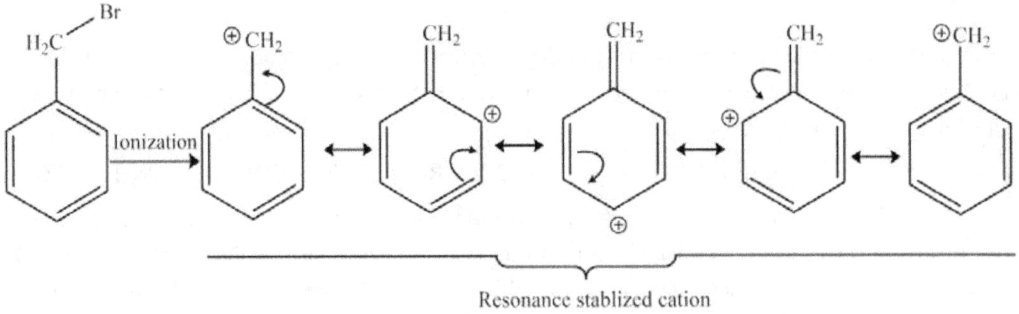

Resonance stablized cation

Resonance stablized cation

iii) Acidic character of carboxylic acid: The existence of charge separation in one of the resonating structures of carboxylic acid makes it less stable than the carboxylate anion which is formed by the loss of a proton.

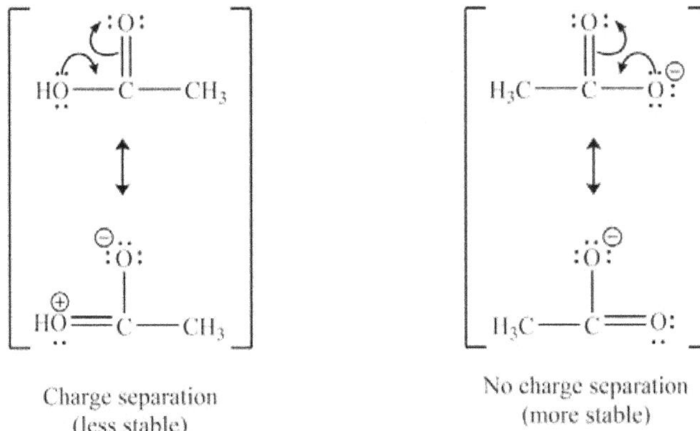

Charge separation
(less stable)
                               No charge separation
(more stable)

iv) Acidic strength of alcohols and phenols: Alcohols are less acidic than phenols because, unlike alkoxide ion, the phenoxide (formed by the loss of proton) resonance stabilized.

No resonance (less stable)

Charge separation (less stable)

No charge separation (more stable)

v) Basic strength of ethylamine and aniline: Ethylamine is more basic than aniline because the electron pair in aniline is less available for donation due to resonance. On the other hand, the +I effect of the ethyl group increases the electron density on the nitrogen atom making it more basic.

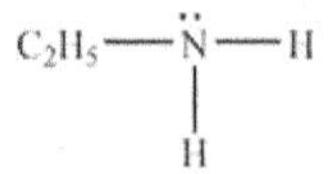

$$C_2H_5 — \overset{\displaystyle ..}{N} — H$$
$$|$$
$$H$$

+I effect of -C_2H_5 (more basic)

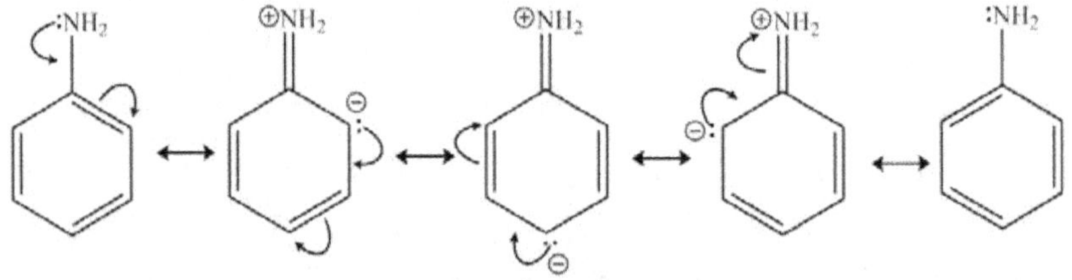

+R effect (less basic)

vi) Basic strength of ethylamine and acetamide: Ethylamine is more basic than acetamide because the electron pair in acetamide is less available for donation due to resonance. On the other hand, the +I effect of the ethyl group in ethylamine increases the electron density on the nitrogen atom making it more basic.

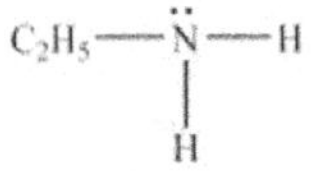

$$C_2H_5 — \overset{\displaystyle ..}{N} — H$$
$$|$$
$$H$$

+I effect of -C_2H_5 (more basic)    +R effect (less basic)

vii) Effect on dipole moments: The dipole moments of different organic molecules are also affected by the resonance effect to a very large extent. For instance, the dipole moment of vinyl chloride is found to be only 1.69D which is much less than the value expected from a large inductive effect (2.05D).

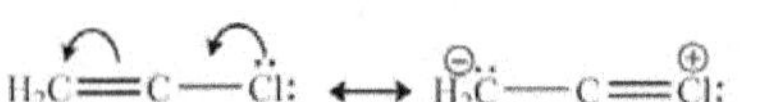

Direction of dipole due to –I effect of Cl ⟶
Direction of dipole due to +R effect of Cl ⟵

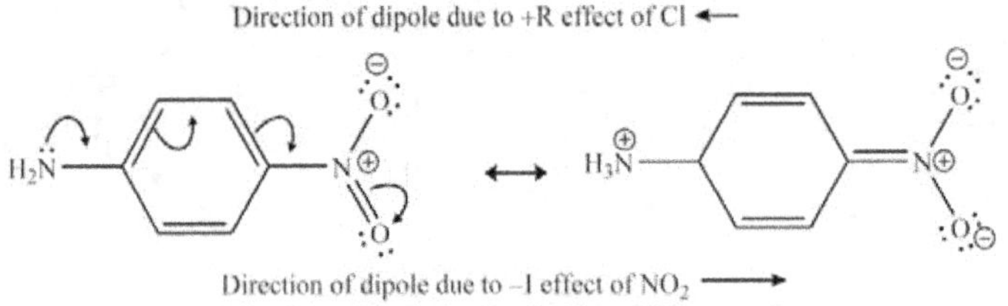

Direction of dipole due to –I effect of NO_2 ⟶
Direction of dipole due to –R effect of NO_2 ⟶

❖ Hyperconjugation

The phenomenon of hyperconjugation (or σ-conjugation) may simply be defined as the delocalization of electrons with the participation of bonds of primarily σ-character.

Typically, the phenomenon of hyperconjugation involves the interaction of the σ-electrons from the C–H bond with an adjacent empty nonbonding p or antibonding σ^* or π^* orbitals to give a pair of extended molecular orbitals. Nevertheless, antibonding σ^* orbitals, which are low-lying energetically, may also interact with fully-filled orbitals of lone pair character (n), which in that case is called as 'negative hyperconjugation'. Increased delocalization of electron density associated with the phenomenon of hyperconjugation raises the stability of the system. Particularly, the new bonding orbital is stabilized, resulting in high overall stabilization of the molecular system. The Baker-Nathan effect, which is also sometimes synonymously used for hyperconjugation, is an explicit application employed to some reactions or structure types.

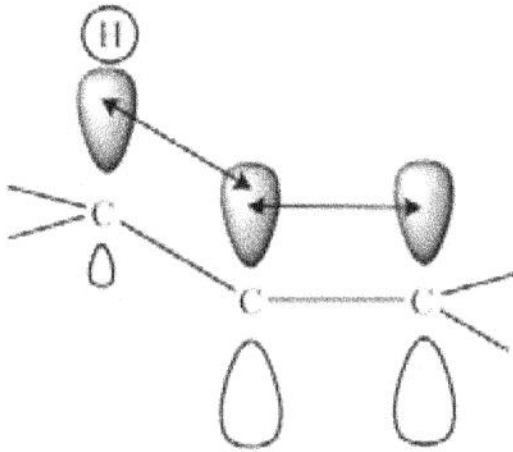

Figure 12. The orbital picture of hyperconjugation.

This effect also reverses the order of the +I effect of alkyl groups if they are attached to an unsaturated carbon. The general order of +I effect is $(CH_3)_3C- > (CH_3)_2CH- > CH_3-CH_2- > -CH_3$. However, consider the hyperconjugation after attaching them to unsaturated carbon i.e.

Therefore, the greater the number of hydrogen at the carbon attached to unsaturated carbon, the more will be the number of possible hyper conjugative structures resulting in a larger electron-donating effect. The number of α-hydrogens in $(CH_3)_3C-$, $(CH_3)_2CH-$, CH_3-CH_2- and $-CH_3$ are 0, 1, 2 and 3, respectively; and thus, the new order of electron donating strength becomes is $(CH_3)_3C- < (CH_3)_2CH- < CH_3-CH_2- < -CH_3$.

Applications of hyperconjugation effect: Some of the important applications of the hyperconjugation effect that affect the properties to a great extent are given below.

i) Directive influence of alkyl groups: The existence of the hyperconjugation effect can be used to rationalize the *o*- and *p*-directing influence of alkyl groups as shown below.

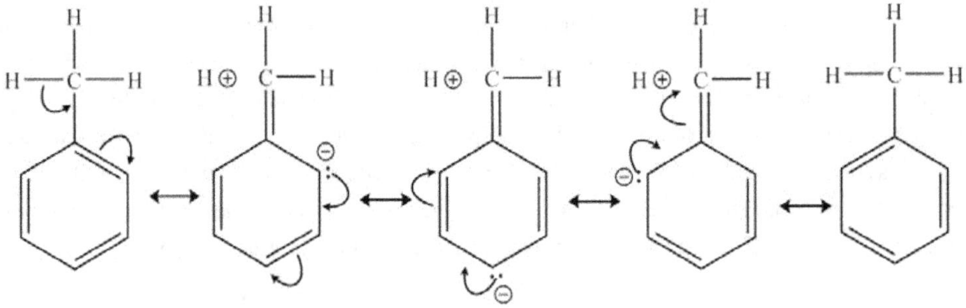

Similarly, six more hyperconjugative structures can also be written for two other H-atoms; and it is obvious that the electron density is increased at *o*- and *p*-sites.

ii) Shortening of C–C bond adjacent to multiple bonds: The existence of hyperconjugation effect can be used to rationalize the shortening of C–C bond adjacent to multiple bonds as shown below.

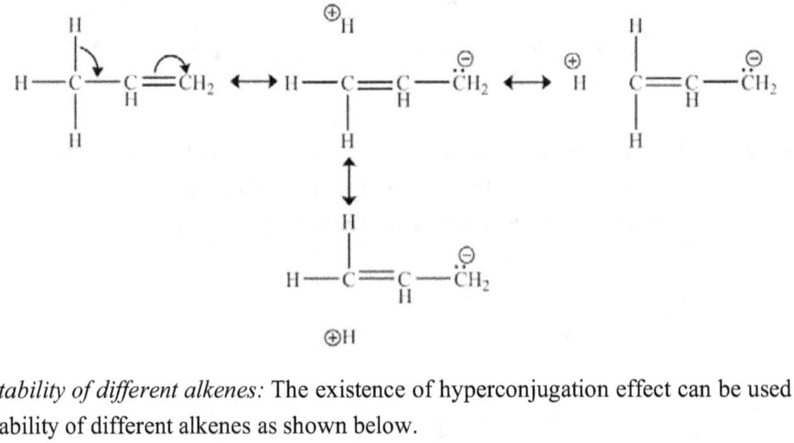

iii) Relative stability of different alkenes: The existence of hyperconjugation effect can be used to rationalize the relative stability of different alkenes as shown below.

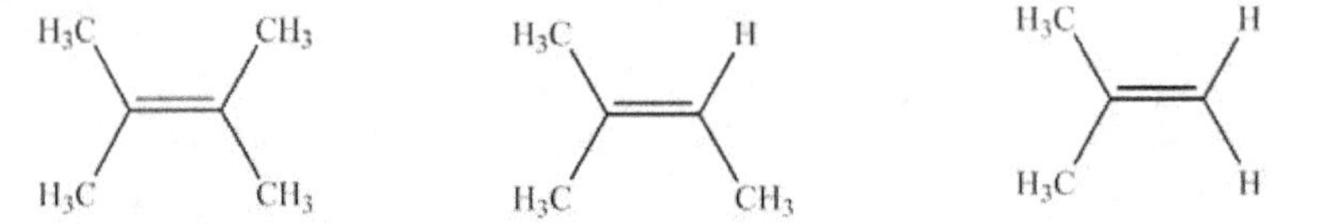

2,3-dimethylbut-2-ene

$$\left(\begin{array}{c} 12\ \alpha\ \text{hydrogens} \\ \Delta_{H2} = 26.6\ \text{Kcal mol}^{-1} \end{array} \right)$$

2-methylbut-2-ene

$$\left(\begin{array}{c} 9\ \alpha\ \text{hydrogens} \\ \Delta_{H2} = 26.9\ \text{Kcal mol}^{-1} \end{array} \right)$$

2-methylprop-1-ene

$$\left(\begin{array}{c} 6\ \alpha\ \text{hydrogens} \\ \Delta_{H2} = 28.6\ \text{Kcal mol}^{-1} \end{array} \right)$$

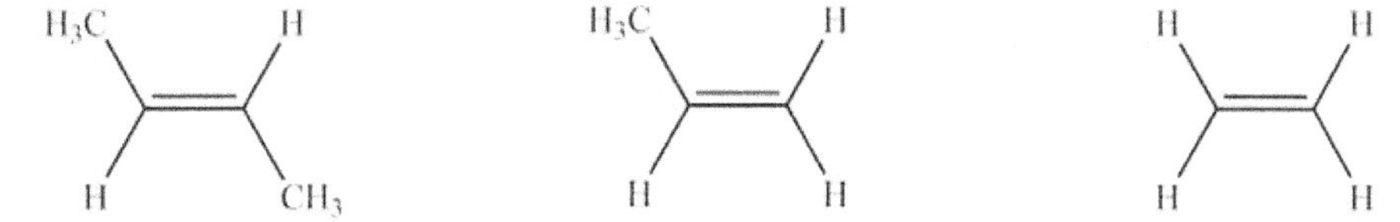

(*E*)-but-2-ene

$$\left(\begin{array}{c} 6\ \alpha\ \text{hydrogens} \\ \Delta_{H2} = 27.6\ \text{Kcal mol}^{-1} \end{array}\right)$$

prop-1-ene

$$\left(\begin{array}{c} 3\ \alpha\ \text{hydrogens} \\ \Delta_{H2} = 30.1\ \text{Kcal mol}^{-1} \end{array}\right)$$

ethene

$$\left(\begin{array}{c} 0\ \alpha\ \text{hydrogens} \\ \Delta_{H2} = 32.8\ \text{Kcal mol}^{-1} \end{array}\right)$$

iv) Relative stability of different carbocations: The existence of hyperconjugation effect can be used to rationalize the relative stability of different carbocations as shown below.

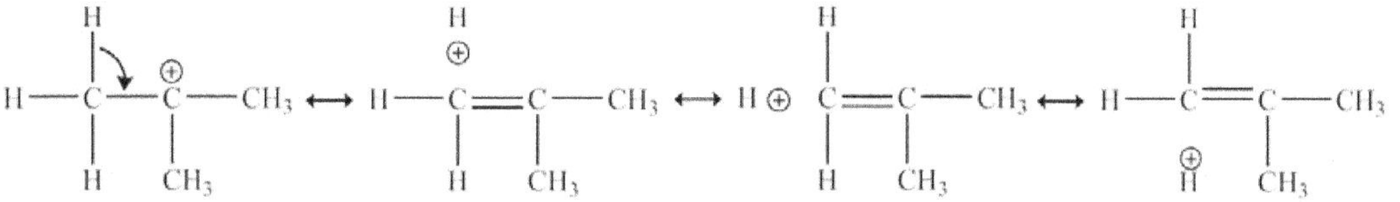

tert-butyl carbocation

$$\left(\begin{array}{c} 3\ \text{hyperconjugative structures for one methyl group} \\ \text{Total hyperconjugative structures} = 9 \end{array}\right)$$

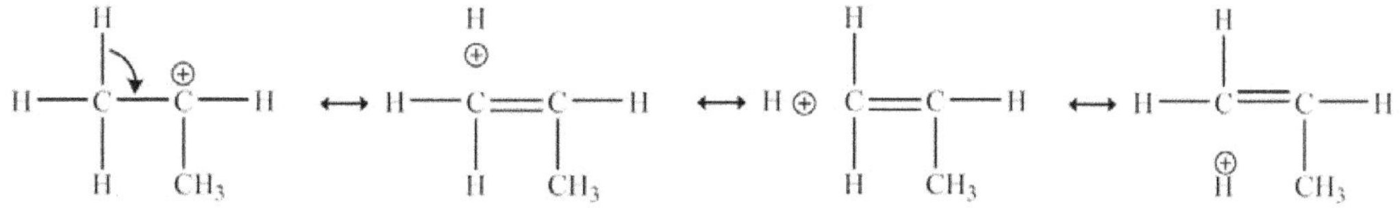

iso-propyl carbocation

$$\left(\begin{array}{c} 3\ \text{hyperconjugative structures for one methyl group} \\ \text{Total hyperconjugative structures} = 6 \end{array}\right)$$

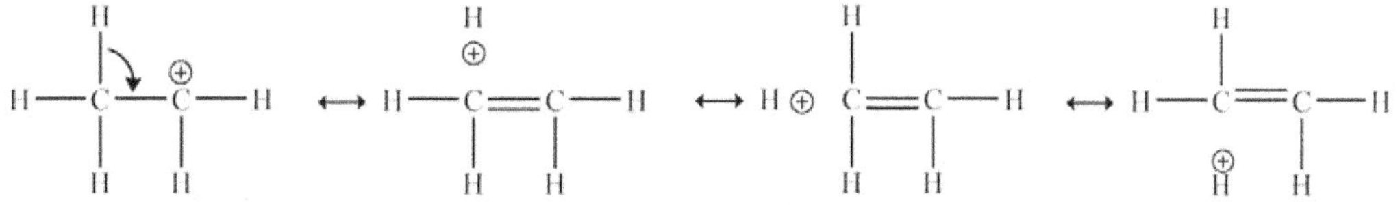

ethyl carbocation

$$\left(\begin{array}{c} 3\ \text{hyperconjugative structures for one methyl group} \\ \text{Total hyperconjugative structures} = 3 \end{array}\right)$$

Hence, as far as the number of possible hyperconjugative structures possible is concerned, tertiary carbocation should be more stable than secondary, which in turn should be more stable than primary.

v) Relative stability of different free radicals: The existence of hyperconjugation effect can be used to rationalize the relative stability of different free radicals as shown below.

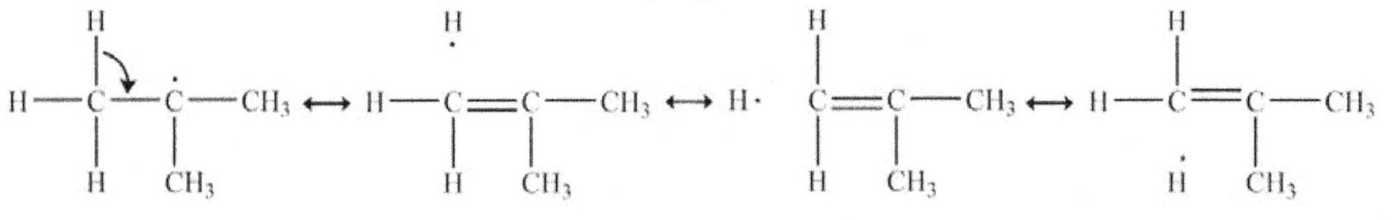

tert-butyl free radical

$$\left(\begin{array}{c}\text{3 hyperconjugative structures for one methyl group}\\ \text{Total hyperconjugative structures} = 9\end{array}\right)$$

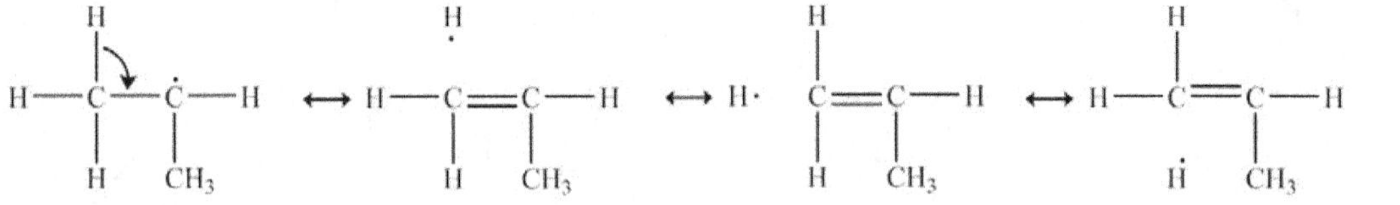

iso-propyl free radical

$$\left(\begin{array}{c}\text{3 hyperconjugative structures for one methyl group}\\ \text{Total hyperconjugative structures} = 6\end{array}\right)$$

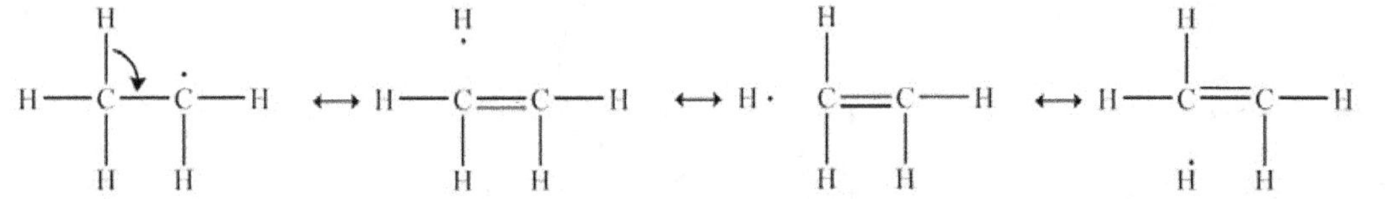

ethyl free radical

$$\left(\begin{array}{c}\text{3 hyperconjugative structures for one methyl group}\\ \text{Total hyperconjugative structures} = 3\end{array}\right)$$

Hence, as far as the number of possible hyperconjugative structures possible is concerned, tertiary free radicals should be more stable than secondary, which in turn should be more stable than primary.

The phenomenon of hyperconjugation can also clarify numerous other effects where explanations may also not be as clear as that for the rotational barrier of C_2H_6. For instance, the Lewis structure for an ammonium ion shows a positive charge on the N atom; though, the hydrogen atoms are more electropositive than is nitrogen atoms, and therefore, are the more genuine carriers of the positive charge. We have known this fact since bases eliminate the protons instead of the nitrogen atom. The matter of the ethane's rotational barrier is not settled by the scientific community till now. An investigation of quantitative molecular orbital theory demonstrates that 2-orbital-4-electron (steric) repulsions are much more dominant over the phenomenon of hyperconjugation. The valence bond theory also put a strong emphasis on the steric significance.

❖ Tautomerism

The phenomenon of tautomerism may simply be defined as the fast interconversion of structural isomers (constitutional isomers) of chemical compounds.

The phenomenon of tautomerism is also called as "desmotropism". The chemical reaction interconverting the two is called tautomerization. One should not be confused with tautomers at the depictions of 'contributing structures' in typical resonance phenomenon. Tautomers are absolutely different chemical species and can be recognized as such by their distinguished spectroscopic profile, whereas resonating structures are nothing but convenient portrayals and do not have physical meaning. One of the most common examples of tautomerism is the interconversion of the keto- and enol forms of an organic compound.

$$H_3C - \underset{\underset{O}{\|}}{C} - CH_3 \quad \rightleftharpoons \quad H_2C = \underset{\underset{OH}{|}}{C} - CH_3$$

propan-2-one   prop-1-en-2-ol

It is also worthy to mention before we discuss the mechanism involved that the phenomenon of tautomerism commonly results in the relocation of a proton.

Mechanism: The phenomenon of tautomerism can occur in the acidic as well as in the basic medium as discussed below.

i) In acidic medium: The most positive species in acidic medium is H^+ ions and the most negative species is the water molecule itself. Therefore, the following mechanism is proposed for keto- to enol conversion.

Similarly, the mechanism responsible for enol- to keto conversion is can also be proposed as given below.

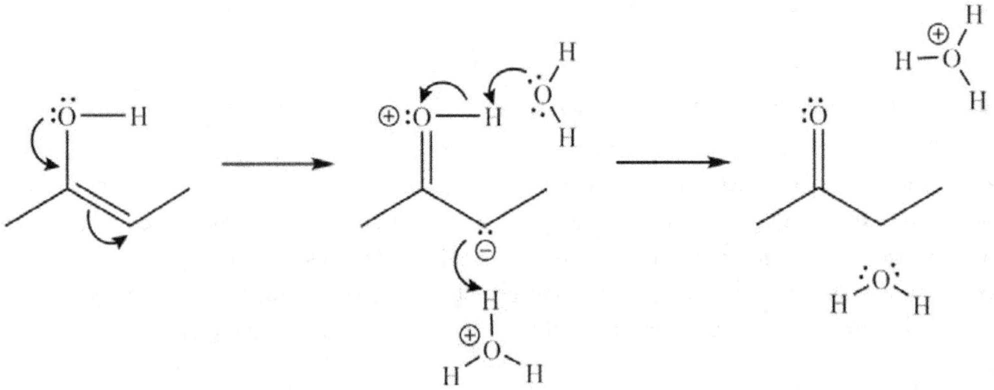

ii) In basic medium: The most positive species in a basic medium is H_2O ions and the most negative species is the hydroxyl ion. Therefore, the following mechanism is proposed for keto- to enol conversion.

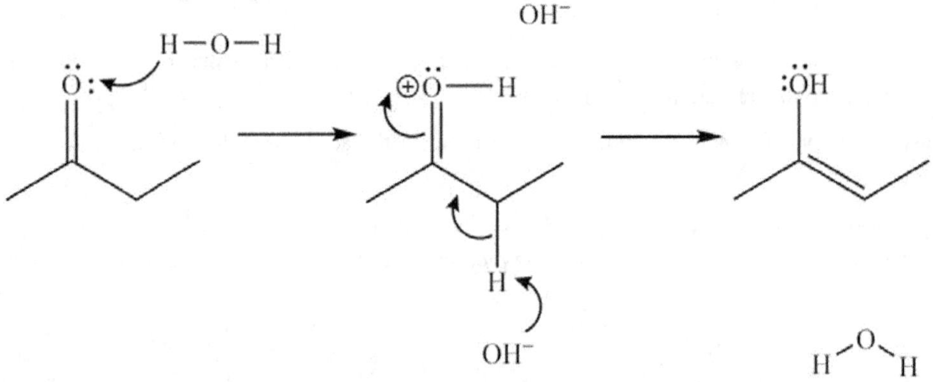

Similarly, the mechanism responsible for enol- to keto conversion is can also be proposed as given below.

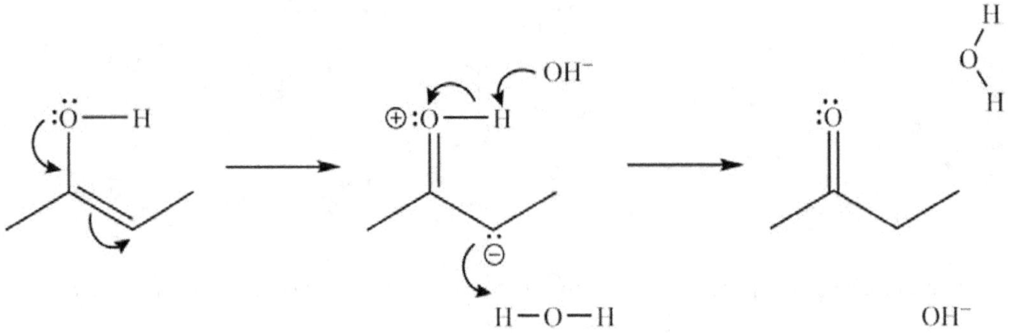

❖ Aromaticity in Benzenoid and Nonbenzenoid Compounds

In order to understand the aromatic character of benzenoid and non-benzenoid compounds, we need to understand the aromaticity first.

The phenomenon of aromaticity in organic chemistry may simply be defined as a special property of planar cyclic geometries with a ring of resonating bonds that gives enhanced stability relative to other connective or geometric arrangements with the same atomic set.

Aromatic compounds are quite stable and don't get fragmented easily to react with other chemical compounds. Organic compounds that don't have aromatic character are categorized as aliphatic compounds, which might be cyclic, but only aromatic rings have low reactivity character. Now because most common aromatic compounds are derived from benzene, the word aromatic generally refers to benzene derivatives. Nonetheless, many non-benzenoids also exist. Alternatively saying, these compounds can mainly be classified into two categories as given below.

➤ *Aromaticity in Benzenoid Compounds*

Benzenoid aromatic compounds are the organic molecular species either with isolated benzene rings or with multiple benzene rings which fused to form a more complex structure. Therefore, these compounds can further be classified into monocyclic aromatic compounds and polycyclic aromatic compounds.

i) Monocyclic aromatic compounds:

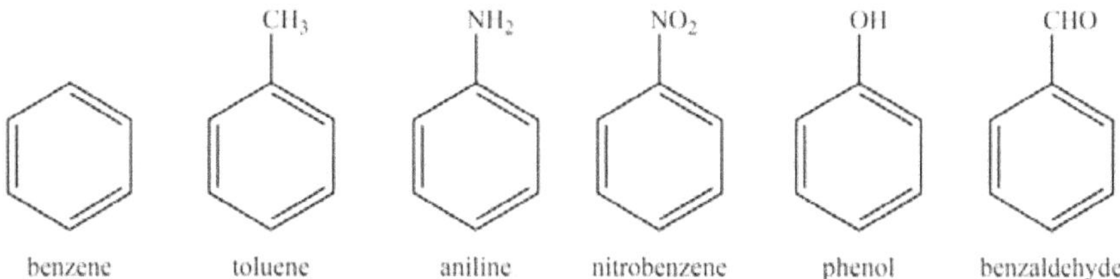

ii) Polycyclic aromatic compounds:

> ➤ *Aromaticity in Nonbenzenoid Compounds*

Nonbenzenoid aromatic compounds are the organic molecular species either with all carbons in the cycle or with one or more heteroatoms in the rings. Therefore, these compounds can further be classified into homocyclic (carbocyclic) aromatic compounds and heterocyclic aromatic compounds.

i) Homocyclic aromatic compounds:

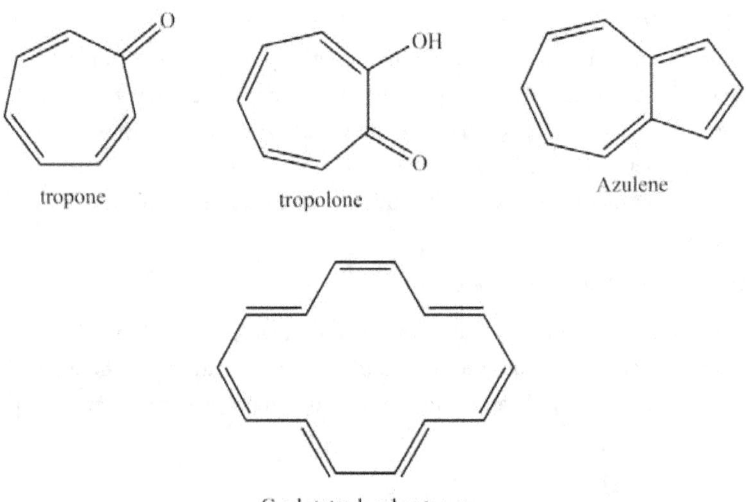

tropone   tropolone   Azulene

Cyclotetradecaheptaene

ii) Heterocyclic aromatic compounds:

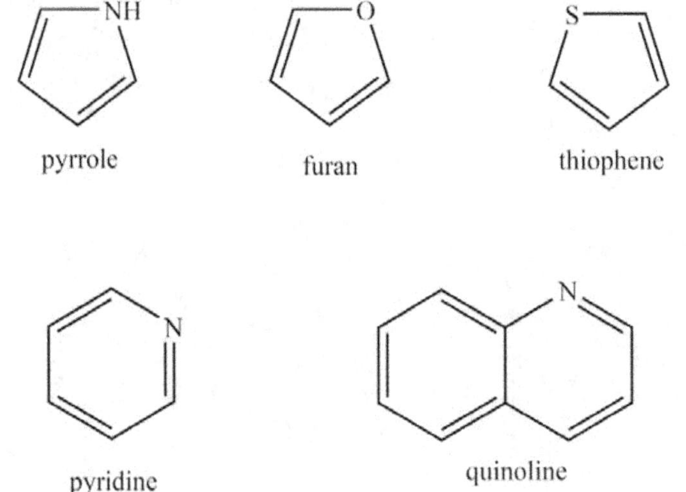

pyrrole   furan   thiophene

pyridine   quinoline

❖ Alternant and Non-Alternant Hydrocarbons

Besides benzenoid and non-benzenoid grouping, all aromatic hydrocarbons can also be classified as alternant and non-alternant hydrocarbons. A general discussion on the two types is given below.

➤ *Alternant Hydrocarbons*

An alternant hydrocarbon is any conjugated hydrocarbon system that doesn't possess an odd-membered cycle. In such cases, it is quite plausible to follow a starring route, where all the C atoms are divided into two groups; in one group, all the carbons are marked with a star in such a way that no two-starred or unstarred atoms are mutually bonded.

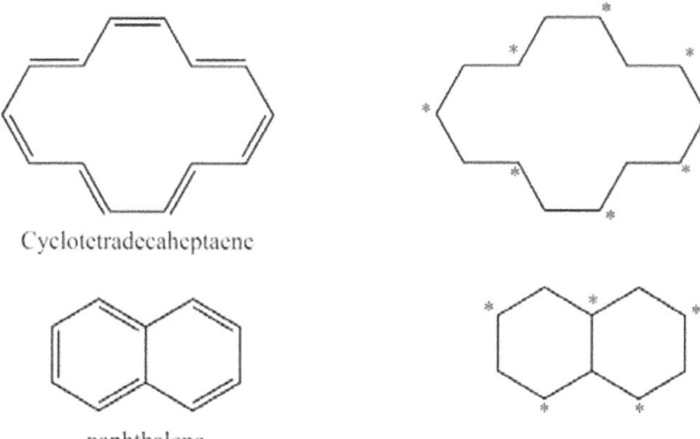

Cyclotetradecaheptaene

naphthalene

In alternate hydrocarbons, the secular determinant in the Hückel approximation has a simpler form, since cross-diagonal elements between atoms in the same set are necessarily 0.

Important properties of alternant hydrocarbons:

i) The coefficients of two paired MOs are the same at the same position, except for a change of sign in the unstarred group.

ii) The electron density or population at all sites is equal to 1 in the ground electronic state, so that the distribution of π-electron density is uniform across the entire molecular structure.

iii) The energies of molecular orbitals for the π system are paired, that is for an orbital of energy $E = \alpha + x\beta$ there is one of energy $E = \alpha - x\beta$.

Furthermore, if the alternant hydrocarbon has an odd number of carbon atoms then there must be an unpaired orbital with no bonding energy (i.e., a non-bonding wave function). For this wavefunction, the coefficients on the atomic sites can be given without calculation; the coefficients on all the orbitals belonging to the smaller set are 0, and the sum of the coefficients of the orbitals around them should also be zero. Modest calculations allow the assignment of different coefficients with normalization. This route allows the prediction of patterns of reactivity and can be employed to find Dewar's reactivity numbers for all positions.

> ### *Non-Alternant Hydrocarbons*

A non-alternant hydrocarbon is any conjugated hydrocarbon system in which it is not possible to mark all the carbon atoms in a complete alternate fashion.

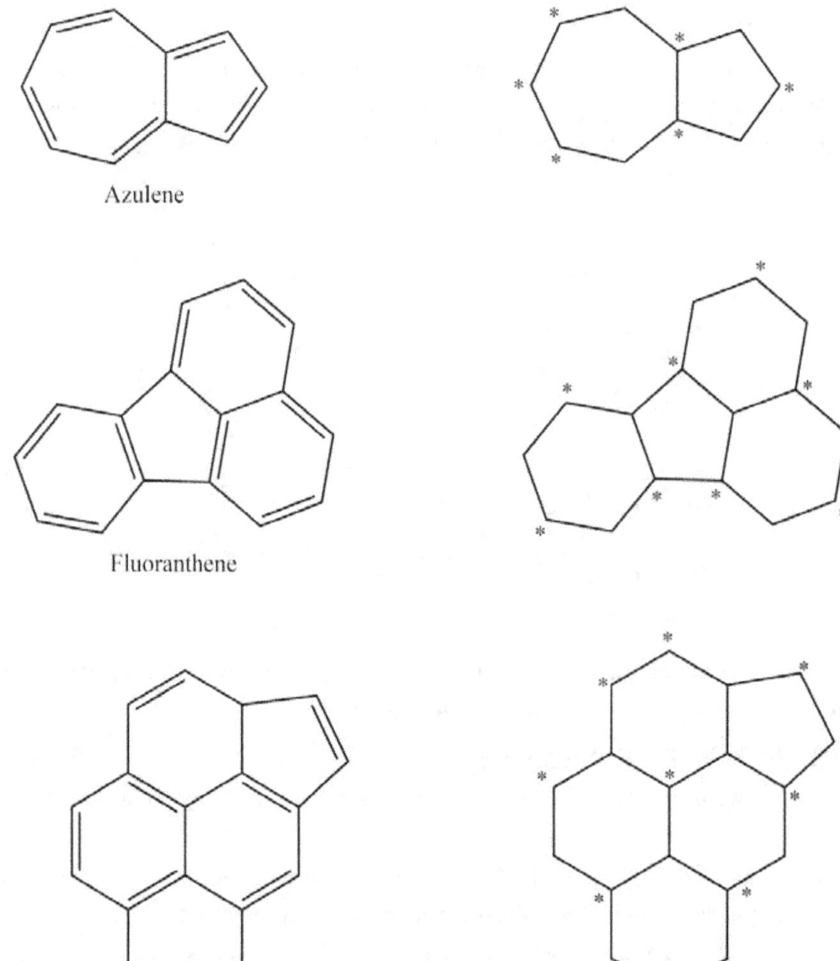

Azulene

Fluoranthene

2a,6-dihydrocyclopenta[*cd*]pyrene

In non-alternate hydrocarbons, the energies of the antibonding and bonding molecular orbitals are not the same and opposite. Furthermore, is also worthy to note that the charge distributions in non-alternate hydrocarbons also are not the same in the anions cations and in radical species.

❖ Huckel's Rule: Energy Level of π-Molecular Orbitals

Huckel's rule states that all the planar and conjugated systems with 4n + 2 π-electrons will be aromatic, and therefore, will be more stable than their acyclic conjugates.

The theoretical basis for Hückel's rule is the Hückel molecular orbital theory (HMO), which was proposed by Erich Hückel in 1930. This theory considers a very simple linear combination of atomic orbitals molecular orbitals method for the determination of energies of molecular orbitals of π-electrons in π-delocalized molecules, such as ethylene, benzene, butadiene, and pyridine.

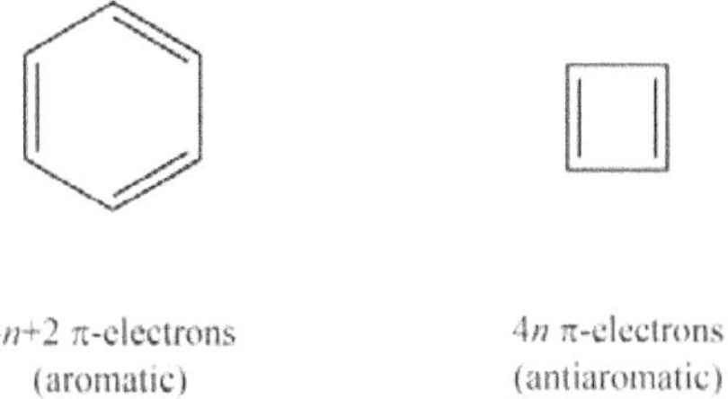

4n+2 π-electrons   4n π-electrons
(aromatic)   (antiaromatic)

> ### *Main Features of Hückel Molecular Orbital Theory*

1. It is limited to conjugated hydrocarbons only.

2. Only π-electron density of MOs is included because these determine much of the spectral and chemical properties of these compounds. The σ-electron density is thought to form the framework of the molecule and σ-connectivity is employed to find whether two π-orbitals interact or not. Nevertheless, the orbitals generated by σ-orbitals are overlooked and presumed not to interact with π-electrons, which is called as σ-π separability. It is defensible by the orthogonality of π and σ orbitals in planar molecular geometries. That's why the Hückel method is only limited to systems that are nearly or completely planar.

3. The theory is based on the variational method to the linear combination of atomic orbitals and makes simple assumptions about the resonance, overlap, and Coulomb integrals of these atomic orbitals. Also, it doesn't try to solve the Schrödinger equation, and neither the functional form of the atomic basis nor particulars of the Hamiltonian involved.

4. For different hydrocarbons, the theory takes atomic connectivity as the only input value; empirical parameters are only required after the incorporation of heteroatoms.

5. The theory successfully predicts the number of energy levels for a given molecule, their degeneracy and it expresses the energies molecular orbital in terms of two parameters, called α, the energy of an electron in an atomic $2p$ orbital, and β which represents the energy of interaction between two $2p$ orbitals. The typical convention makes both α and β be as negative quantities. In order to fully understand and equate systems in a qualitative or semi-quantitative manner, clear numerical values for these parameters are typically not needed.

6. This theory also enables the calculation of charge density for each atom in the π framework, the fractional bond order between any two carbon atoms, and the overall dipole moment of the molecule.

> ### *Energy Level of Some Typical π-Molecular Orbitals*

The exact procedure to determine the energy levels of conjugated π-bonded systems is beyond the scope of this book. However, the results from theory can be concluded in some simple mathematical expressions which can be used to yield the required energy levels. For linear and cyclic systems (with N atoms), general solutions are given below.

1. Linear system (polyene/polyenyl): The energy of different molecular orbitals for these types of systems can be obtained from the expression given below.

$$E_k = \alpha + 2\beta \; Cos \frac{(k+1)\pi}{(N+1)} \tag{1}$$

Where $k = 0, 1, 2, 3 \ldots N - 1$ and N is the number conjugated carbons. Once the energies of different molecular orbitals are obtained, they should be arranged in accenting order on the energy level diagram.

i) Ethylene: In case of ethylene, $N = 2$, we get

for $k = 0$
$$E_0 = \alpha + 2\beta \; Cos \frac{(0+1)\pi}{(2+1)} = \alpha + \beta \tag{2}$$

for $k = 1$
$$E_1 = \alpha + 2\beta \; Cos \frac{(1+1)\pi}{(2+1)} = \alpha - \beta \tag{3}$$

The corresponding energy level diagram is given below.

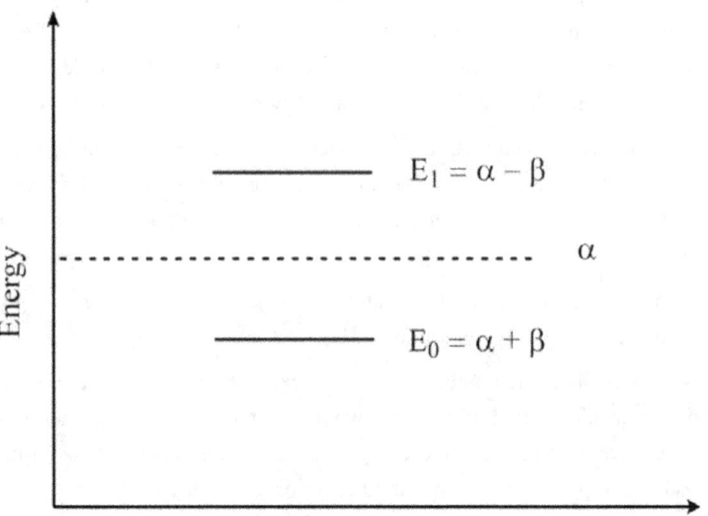

Figure 13. The π-molecular orbital energy level diagram of ethylene.

ii) Allyl: In case of allyl, $N = 3$, we get

for $k = 0$
$$E_0 = \alpha + 2\beta\, Cos\frac{(0+1)\pi}{(3+1)} = \alpha + \sqrt{2}\beta \tag{4}$$

for $k = 1$
$$E_1 = \alpha + 2\beta\, Cos\frac{(1+1)\pi}{(3+1)} = \alpha \tag{5}$$

for $k = 2$
$$E_2 = \alpha + 2\beta\, Cos\frac{(2+1)\pi}{(3+1)} = \alpha - \sqrt{2}\beta \tag{6}$$

The corresponding energy level diagram is given below.

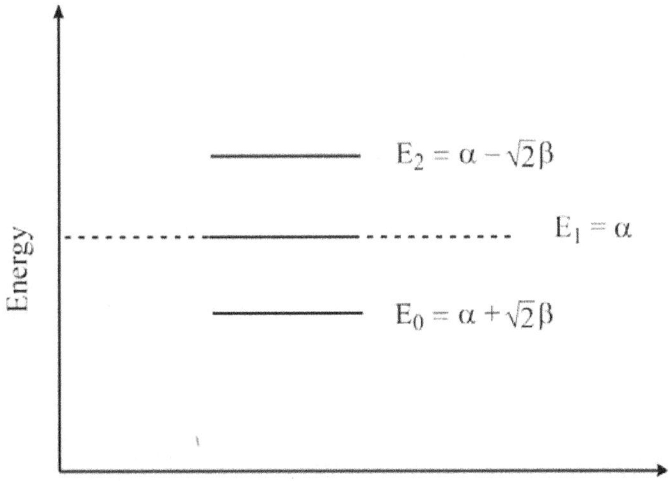

Figure 14. The π-molecular orbital energy level diagram of allyl.

iii) Butadiene: In case of butadiene, $N = 4$, we get

for $k = 0$
$$E_0 = \alpha + 2\beta\, Cos\frac{(0+1)\pi}{(4+1)} = \alpha + 1.618\,\beta \tag{7}$$

for $k = 1$
$$E_1 = \alpha + 2\beta\, Cos\frac{(1+1)\pi}{(4+1)} = \alpha + 0.618\,\beta \tag{8}$$

for $k = 2$
$$E_2 = \alpha + 2\beta\, Cos\frac{(2+1)\pi}{(4+1)} = \alpha - 0.618\,\beta \tag{9}$$

for $k = 3$
$$E_3 = \alpha + 2\beta\, Cos\frac{(3+1)\pi}{(4+1)} = \alpha - 1.618\,\beta \tag{10}$$

The corresponding energy level diagram is given below.

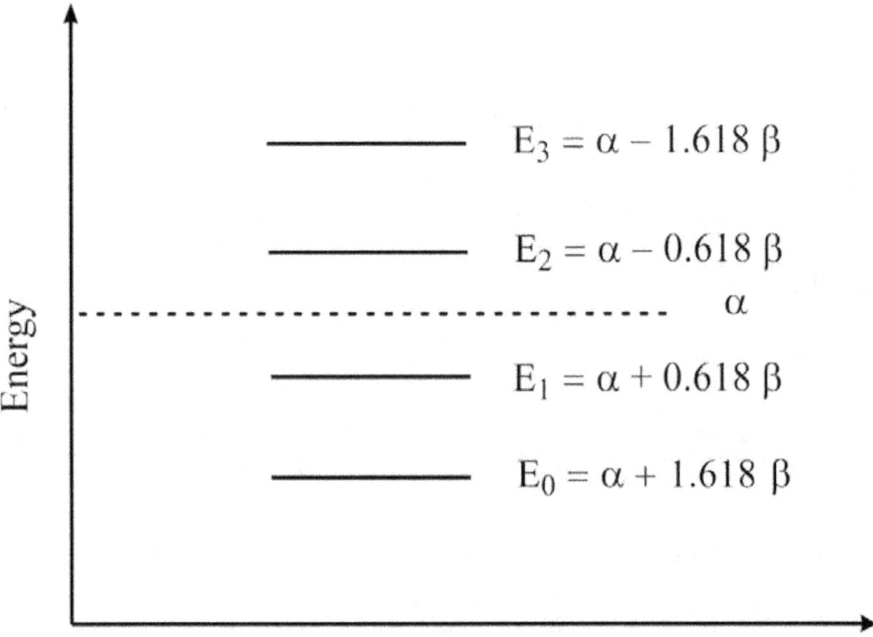

Figure 15. The π-molecular orbital energy level diagram of butadiene.

2. Cyclic system, Hückel topology (annulene/annulenyl): The energy of different molecular orbitals for these types of systems can be obtained from the expression given below.

$$E_k = \alpha + 2\beta\,Cos\,\frac{2k\pi}{N} \tag{11}$$

Where $k = 0, 1, 2, 3 \dots N - 1$ and N is the number conjugated carbons. Once the energies of different molecular orbitals are obtained, they should be arranged in accenting order on the energy level diagram.

i) Cyclopropenyl: In case of cyclopropenyl, $N = 3$, we get

for $k = 0$
$$E_0 = \alpha + 2\beta\,Cos\,\frac{2(0)\pi}{3} = \alpha + 2\beta \tag{12}$$

for $k = 1$
$$E_1 = \alpha + 2\beta\,Cos\,\frac{2(1)\pi}{3} = \alpha - \beta \tag{13}$$

for $k = 1$
$$E_2 = \alpha + 2\beta\,Cos\,\frac{2(2)\pi}{3} = \alpha - \beta \tag{14}$$

The corresponding energy level diagram is given below.

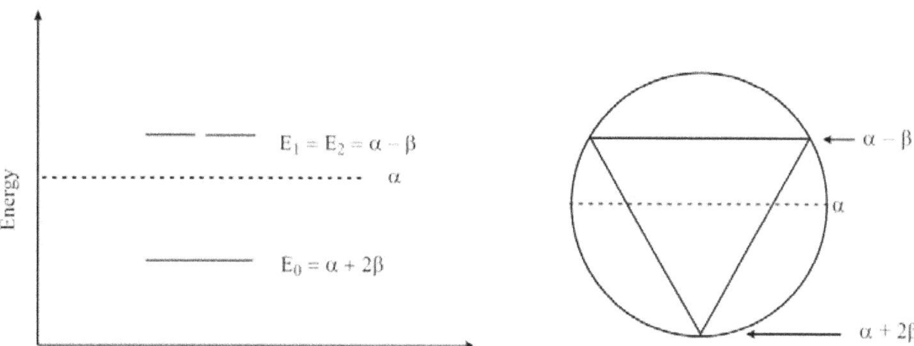

Figure 16. The π-molecular orbital energy level diagram of cyclopropenyl.

ii) Cyclobutadiene: In case of cyclobutadiene, $N = 4$, we get

for $k = 0$
$$E_0 = \alpha + 2\beta \, Cos\frac{2(0)\pi}{4} = \alpha + 2\beta \qquad (12)$$

for $k = 1$
$$E_1 = \alpha + 2\beta \, Cos\frac{2(1)\pi}{4} = \alpha \qquad (13)$$

for $k = 2$
$$E_2 = \alpha + 2\beta \, Cos\frac{2(2)\pi}{4} = \alpha - 2\beta \qquad (14)$$

for $k = 3$
$$E_3 = \alpha + 2\beta \, Cos\frac{2(3)\pi}{4} = \alpha \qquad (15)$$

The corresponding energy level diagram is given below.

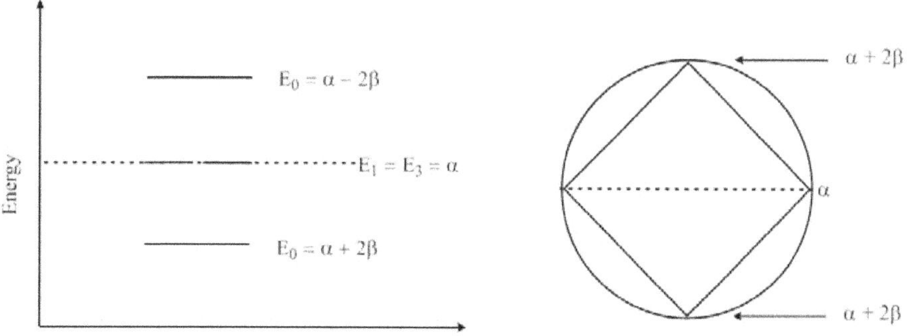

Figure 17. The π-molecular orbital energy level diagram of cyclobutadiene.

iii) Benzene: In case of butadiene, $N = 6$, we get

for $k = 0$

$$E_0 = \alpha + 2\beta \, Cos \frac{2(0)\pi}{6} = \alpha + 2\beta \qquad (16)$$

for $k = 1$

$$E_1 = \alpha + 2\beta \, Cos \frac{2(1)\pi}{6} = \alpha + \beta \qquad (17)$$

for $k = 2$

$$E_2 = \alpha + 2\beta \, Cos \frac{2(2)\pi}{6} = \alpha - \beta \qquad (18)$$

for $k = 3$

$$E_3 = \alpha + 2\beta \, Cos \frac{2(3)\pi}{6} = \alpha - 2\beta \qquad (19)$$

for $k = 4$

$$E_4 = \alpha + 2\beta \, Cos \frac{2(4)\pi}{6} = \alpha - \beta \qquad (20)$$

for $k = 5$

$$E_5 = \alpha + 2\beta \, Cos \frac{2(1)\pi}{6} = \alpha + \beta \qquad (21)$$

The corresponding energy level diagram is given below.

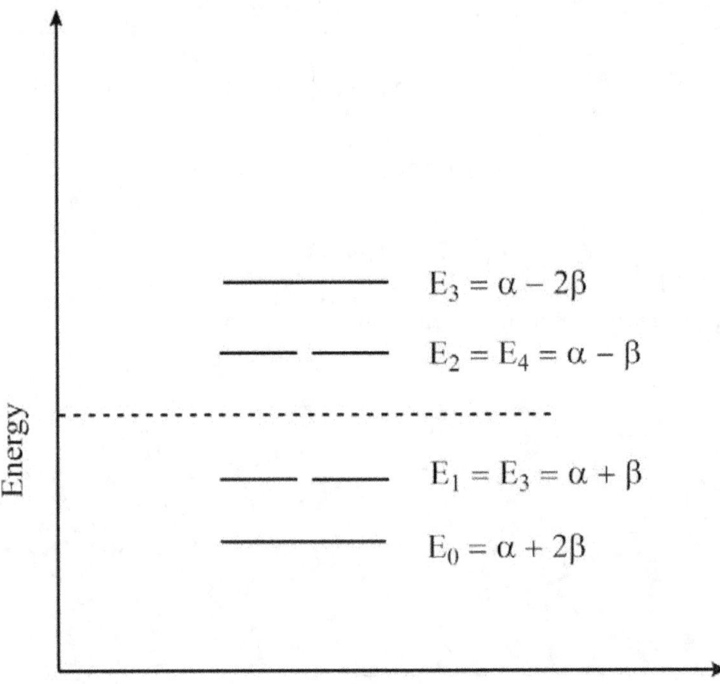

Figure 18. The typical π-molecular orbital energy level diagram of the benzene molecule.

The corresponding Frost circle is given below.

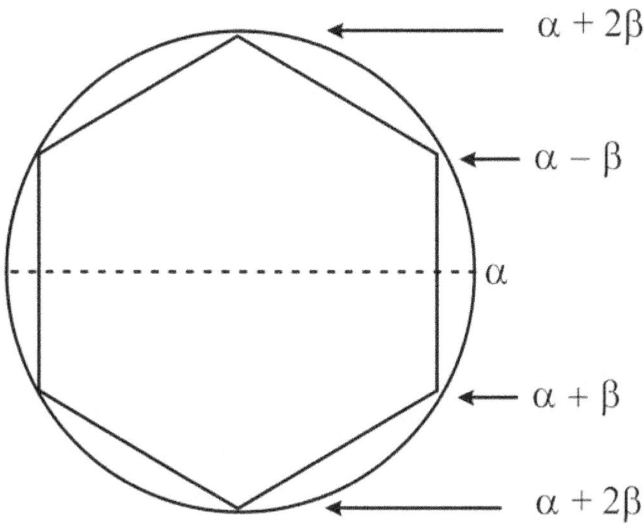

The energy levels for cyclic systems can be obtained using the Frost circle method (named after the American chemist A. A. Frost). A circle with a center at α with a radius of 2β is inscribed with a regular polygon with N sides with one vertex pointing downward; the y-coordinate of the vertices of the polygon then shows the energies of different orbital of the [N]annulene/annulenyl system. Corresponding mnemonics exist for Möbius and linear systems. The energy level diagrams of different cyclic systems are assisted with the corresponding Frost cycle.

In the later period, it was extended to conjugated molecules with heteroatoms like pyrrole, pyridine, and furan. A more dramatic extension of the method (Huckel molecular orbital method) to include σ-electrons, known as the extended Hückel method (EHM), was developed by Roald Hoffmann.

The Hückel method's extension provides some degree of quantitative accuracy for organic compounds in general and was employed to give computational justification for the famous Woodward-Hoffmann rules. In order to differentiate the original route from Hoffmann's extension, the Hückel method is also well-known as the 'simple Hückel method' (SHM). Even though the method is quite simple, the Hückel route in its original form makes chemically useful and qualitatively accurate predictions for numerous common molecular geometries, and therefore, is a very powerful and widely popular educational toolkit. It is also included in textbooks of preliminary quantum chemistry and physical organic chemistry. Furthermore, many organic chemists, in specific, still apply Hückel theory routinely to find back-of-the-envelope, the very approximate profile of π-interactions.

❖ Annulenes

Annulenes are the monocyclic conjugated polyenes that contain the maximum number of non-cumulated double bonds.

Annulenes have the general formula of C_nH_n (when n represents an even number) or C_nH_{n+1} (when n represents an odd number). The IUPAC nomenclature of annulenes with 7 or more carbon atoms follows as [n]-annulene, where n represents the number of C atoms in their cycle, though the smaller annulenes are referred to using the same notation sometimes (benzene is also an annulene). The first 3 even annulenes include cyclobutadiene, benzene, and cyclooctatetraene. Many annulenes like cyclobutadiene i.e. [4]-annulene, cyclodecapentaene i.e. [10]-annulene, cyclododecahexaene i.e. [12]-annulene and cyclotetradecaheptaene i.e., [14]-annulene, are not stable, with cyclobutadiene at extreme position. Annulenes may be aromatic (benzene, [6]-annulene and [18]-annulene), non-aromatic ([8]- and [10]-annulene), or anti-aromatic (cyclobutadiene, [4]-annulene). Cyclobutadiene or [4]-annulene is the only annulene with considerable antiaromatic character due to unavoidable planarity. With [8]-annulene takes on a tub shape that permits it to evade conjugation of double bonds making it less unstable. [10]-Annulene has very strange geometry to be planar structure; in a planar conformation, ring strain arising from either steric hindrance of internal H atoms (when some double bonds are trans) or the distortion of bond angle (if double bonds are all cis) is unavoidable. hence, it doesn't get appreciable aromatic character.

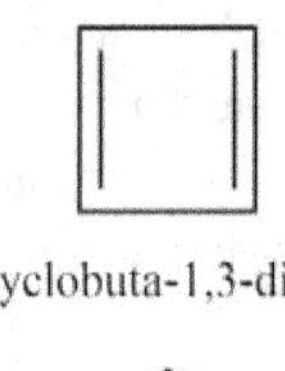

cyclobuta-1,3-diene

[4]-annulene

4 π-electrons, planer

Antiaromatic

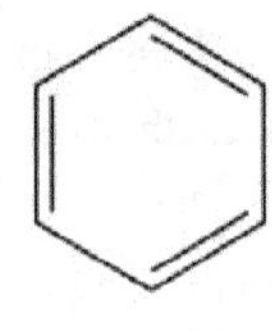

benzene

[6]-annulene

6 π-electrons, planer

Aromatic

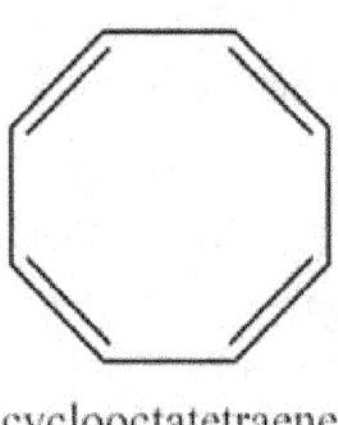

cyclooctatetraene

[8]-annulene

8 π-electrons, non-planer

Nonaromatic

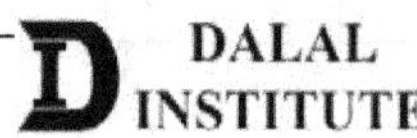

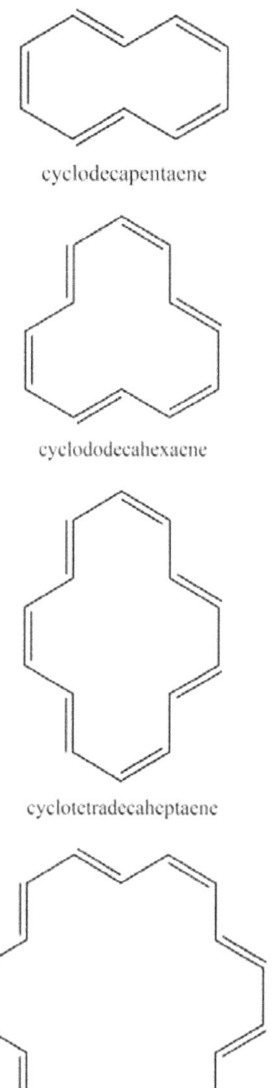

cyclodecapentaene

[10]-annulene

10 π-electrons, non-planer

Nonaromatic

cyclododecahexaene

[12]-annulene

12 π-electrons, planer

Antiaromatic

cyclotetradecaheptaene

[14]-annulene

14 π-electrons, planer

Aromatic

[18]-annulene

18 π-electrons, planer

Aromatic

Cyclooctadecanonaene

When the annulene is large enough, [18]-annulene, for example, there is enough room internally to accommodate hydrogen atoms without significant distortion of bond angles. [18]-Annulene possesses several properties that qualify it as aromatic. Nevertheless, no bigger annulenes are as stable as [6]-annulene i.e. benzene, as their reactivity resembles closely a conjugated polyene than a hydrocarbon of aromatic nature.

❖ Antiaromaticity

The phenomenon of antiaromaticity in organic chemistry may simply be defined as a property of planar conjugated cyclic structures by which it becomes less stable than its acyclic conjugate.

In order to find whether the compound is antiaromatic or not, it is a necessary pre-condition to confirm that its delocalization energy is less than its acyclic conjugate. After studying a lot of conjugated molecules, it was found that in almost all antiaromatic compounds, the total number of the π-electrons can be fitted by the formula $4n$ delocalized (π or lone pair) electrons in it. Unlike aromatic systems, which are governed by Hückel's rule ($4n+2$ π-electrons) and are very stable compounds, these antiaromatic systems are extremely unstable and very reactive. In order to evade this instability due to antiaromaticity, the compound may transform its shape, becoming a non-planar, and therefore, removing some of the π-interactions. In contrast to the diamagnetic ring current of aromatic systems, antiaromatic molecules have a paramagnetic type ring current, which can be detected by the NMR spectroscopic methods. Common examples of antiaromatic systems are. Pentalene, biphenylene, cyclopentadienyl cation. The archetypal case of antiaromaticity, cyclobutadiene molecule, is widely debated with some scientists saying that antiaromaticity is not the primary factor to its large destabilization.

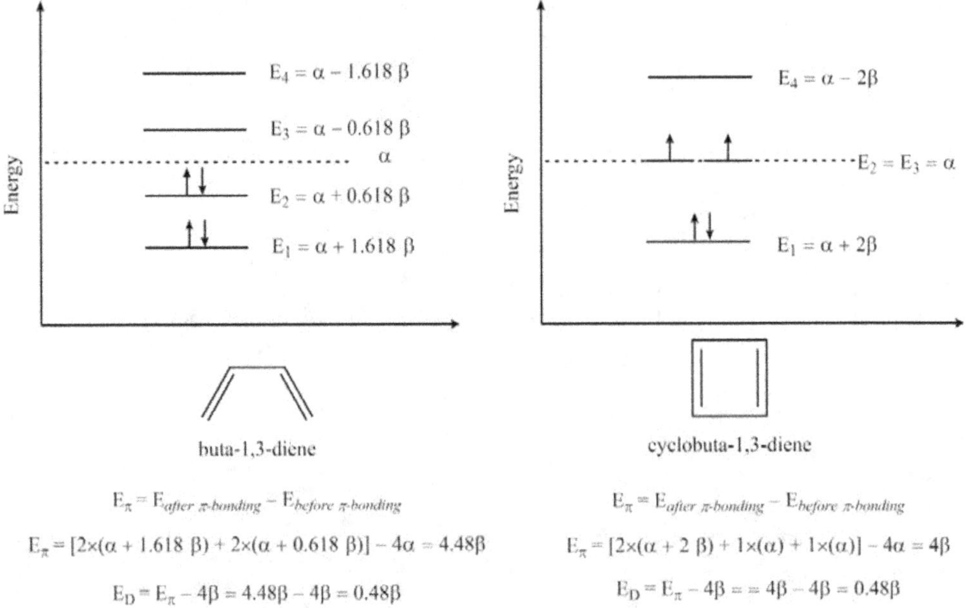

$$E_\pi = E_{after\ \pi\text{-}bonding} - E_{before\ \pi\text{-}bonding}$$

$$E_\pi = [2\times(\alpha + 1.618\ \beta) + 2\times(\alpha + 0.618\ \beta)] - 4\alpha = 4.48\beta$$

$$E_D = E_\pi - 4\beta = 4.48\beta - 4\beta = 0.48\beta$$

$$E_\pi = E_{after\ \pi\text{-}bonding} - E_{before\ \pi\text{-}bonding}$$

$$E_\pi = [2\times(\alpha + 2\ \beta) + 1\times(\alpha) + 1\times(\alpha)] - 4\alpha = 4\beta$$

$$E_D = E_\pi - 4\beta = 4\beta - 4\beta = 0.48\beta$$

Figure 19. The π-molecular orbital energy level diagram of cyclobutadiene.

It is obvious that the delocalization energy (E_D) of cyclobutadiene is less than its acyclic conjugate, and therefore, it is antiaromatic in nature.

Similarly, we can prove the antiaromaticity of cyclopropenyl anion as given below.

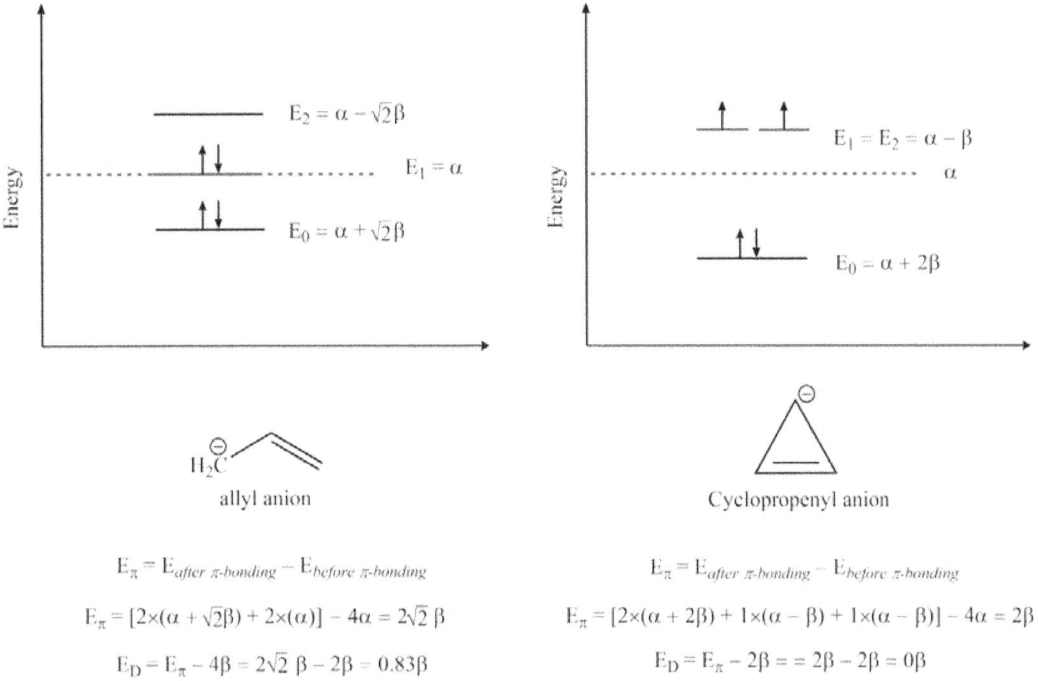

$$E_\pi = E_{after\ \pi\text{-}bonding} - E_{before\ \pi\text{-}bonding}$$

$$E_\pi = [2\times(\alpha + \sqrt{2}\beta) + 2\times(\alpha)] - 4\alpha = 2\sqrt{2}\,\beta$$

$$E_D = E_\pi - 4\beta = 2\sqrt{2}\,\beta - 2\beta = 0.83\beta$$

$$E_\pi = E_{after\ \pi\text{-}bonding} - E_{before\ \pi\text{-}bonding}$$

$$E_\pi = [2\times(\alpha + 2\beta) + 1\times(\alpha - \beta) + 1\times(\alpha - \beta)] - 4\alpha = 2\beta$$

$$E_D = E_\pi - 2\beta = = 2\beta - 2\beta = 0\beta$$

Figure 20. The π-molecular orbital energy level diagram of cyclobutadiene.

Cyclooctatetraene is the most popular case of a molecule becoming a non-planar geometry to avoid antiaromatic destabilization. Had it been planar, it would have a single 8π-electron system around the cycle, nonetheless, it becomes a boat-like shape with four separate π bonds.

Since antiaromatic systems are usually short-lived and difficult to detect experimentally, the antiaromatic destabilization energy is frequently modeled by computer simulation rather than by actual experiment.

❖ Homoaromaticity

The homoaromaticity in organic chemistry may simply be defined as a special case of aromatic behavior where the normal conjugation is interrupted by a single sp³-hybridized carbon atom.

Though this sp^3 center disturbs the continuity of overlap between p-orbitals, conventionally thought to be a primary need for aromaticity, substantial thermodynamic stability, and many of the magnetic, spectroscopic, and chemical properties related to aromatic compounds are still observed for such molecules. This official discontinuity is bypassed by p-orbital overlap, preserving a contiguous cycle of π-electron density that is accountable for its extra stability. The idea of homoaromaticity was proposed by S. Winstein in 1959, encouraged by his analysis of the "tris-homocyclopropenyl" ion. After Weinstein's publication, abundant research has been dedicated to the understanding of these molecules, which embody an additional class of aromatic compounds included broadening the definition of aromatic behavior. Up to the present time, homoaromatic molecules are quite well-known to exist as anionic and cationic species, and some reports support the presence of neutral homoaromatic compounds, nonetheless, these are not so common. It seems that the 'homotropylium' cation ($C_8H_9^+$) is the most studied case of a homoaromatic system.

The label "homoaromaticity" originates from the similar structural profile of homoaromatic compounds with analogous homo-conjugated alkenes observed previously. The IUPAC Gold Book needs that bis-, tris-, etc. type prefixes to be used to define homoaromatic compounds where two, three, etc. sp^3 centers separately interrupt the phenomenon of conjugation of the aromatic molecules.

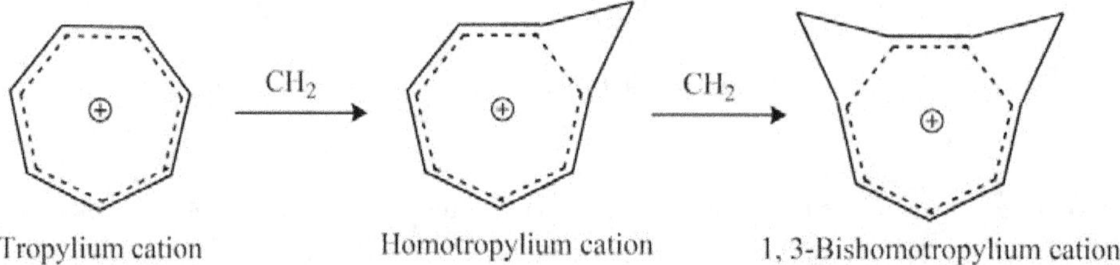

The homoaromatic compounds can simply be divide into three different categories, depending on the nature of the charge on them. A general discussion on all three types is given below.

➤ *Cationic Homoaromatic Compounds*

The most well-known and proven homoaromatic species are cationic entities. As we have earlier mentioned, the homotropenylium cation is the most studied homoaromatic compound. Numerous homoaromatic cationic species act as a basis for a tropylium cation, cyclopropenyl cation, or a cyclobutadiene dication because these species have very strong aromaticity. Additionally, another well-known cationic homoaromatic compound is the norbornen-7-yl cation, which is quite strongly homoaromatic, proven both experimentally and theoretically.

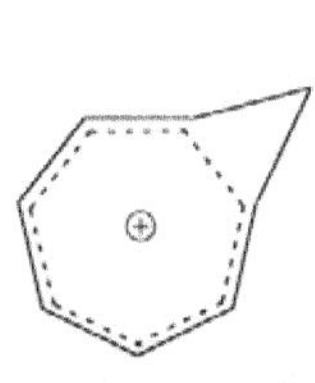

Homotropylium cation

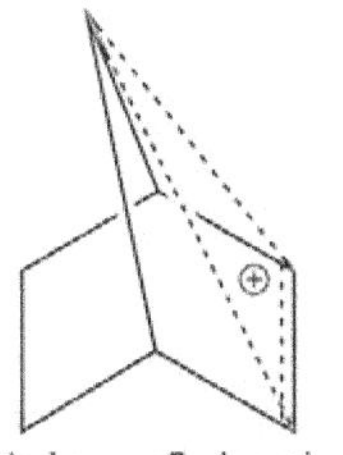

Norbornen 7-yle cation

> ### ➤ *Anionic Homoaromatic Compounds*

The homoaromatic anions are widely accepted to have "true" homoaromaticity. These anionic species are frequently obtained from their neutral parent molecules via Li-based reduction. 1,2-diboretanide derivatives display strong homoaromatic behavior via their 3-atom (B, B, and C), 2-electron bond, which contains shorter carbon-boron bonds than in the neutral analog. These 1,2-diboretanides can be expanded to bigger ring sizes with different groups and all have some degree of homoaromaticity.

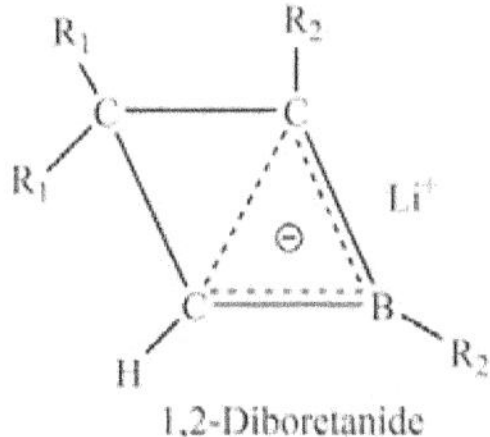

1,2-Diboretanide

> ### ➤ *Neutral Homoaromatic Compounds*

Numerous classes of neutral homoaromatic compounds have been observed though there is a debate whether they are truly homoaromatic or not. One class of the type of neutral homoaromatics is called monohomoaromatics, including cycloheptatriene. One more example is a 60-carbon fulleroid derivative that has only one methylene bridge. The NMR and UV -visible study has proven that the aromatic character of this modified fulleroid is not disturbed by the addition of a homoconjugate bond, and therefore, this molecule is homoaromatic for sure.

1,2-Diboretane

❖ PMO Approach

The PMO (perturbation molecular orbital) theory is nothing but an extension of the general molecular orbital theory. In this approach, the participating atomic orbitals are considered as the initial state, which in turn, is actually perturbed by the interaction to give the final state i.e. molecular orbitals. Furthermore, the PMO approach can also be applied to molecular orbitals to give new molecular orbitals. Now, depending upon the energy of basis orbitals i.e. unperturbed orbitals, the perturbed molecular orbitals can take two routes as discussed below.

➤ *Perturbed Molecular Orbitals from Degenerate Basis*

The perturbed molecular orbitals are formed by the additive and subtractive interaction of the basis orbitals; and if the unperturbed basis orbitals are degenerate, they equally contribute toward the additive and subtractive interaction.

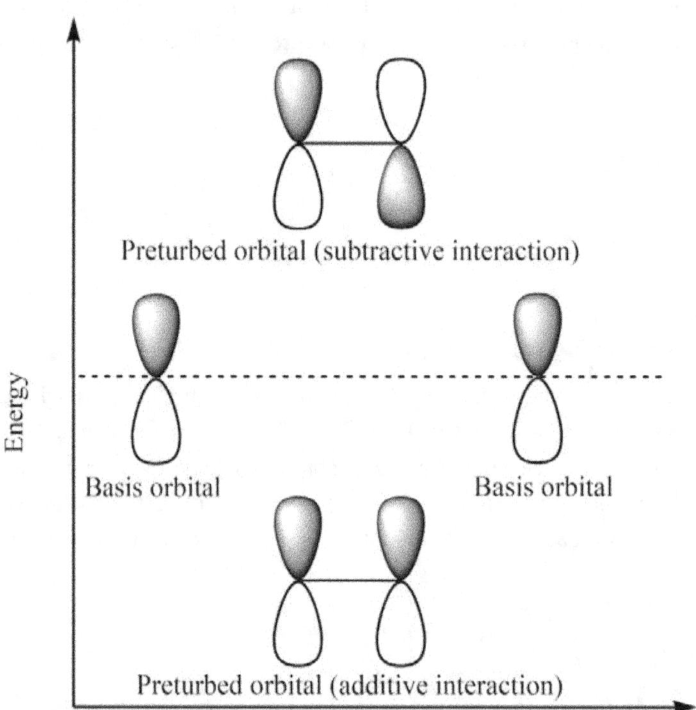

Figure 21. The perturbed molecular orbitals from degenerate basis orbitals.

In other words, the subtractive perturbation increases the energy whereas the additive interaction decreases the total energy of the molecular orbital formed.

> ### *Perturbed Molecular Orbitals from Non-Degenerate Basis*

The perturbed molecular orbitals are formed by the additive and subtractive interaction of the basis orbitals; and if the unperturbed basis orbitals are not degenerate, they do not equally contribute toward the additive and subtractive interaction. Consider the two unperturbed basis orbitals ψ_1 and ψ_2 with different energies.

$$\psi_+ = \psi_1 + \lambda\psi_2 \tag{1}$$

$$\psi_- = \psi_2 - \lambda\psi_1 \tag{2}$$

Where the symbol λ whose value varies from 0 to 1. As the energy difference between ψ_1 and ψ_2 increases, the magnitude of λ approaches to zero.

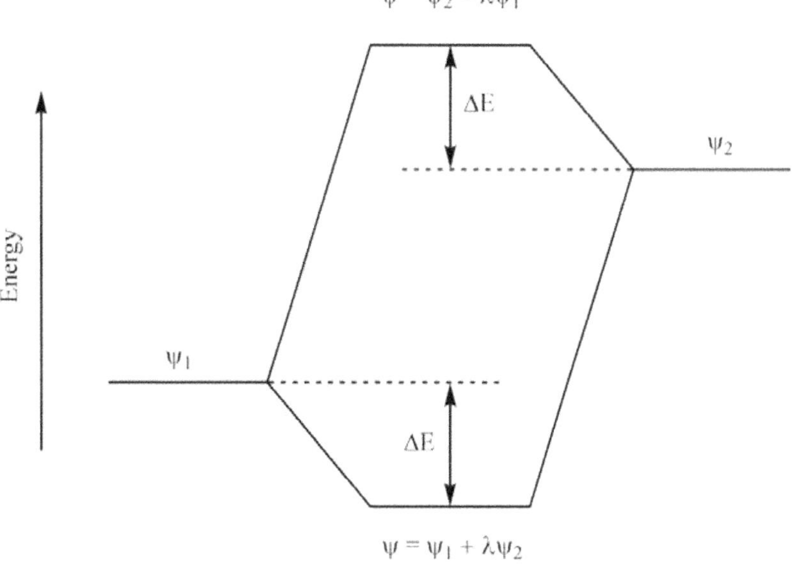

Figure 22. The perturbed molecular orbitals from non-degenerate basis orbitals.

In other words, the subtractive perturbation increases the energy whereas the additive interaction decreases the total energy of the molecular orbital formed. It is obvious from the above energy level diagram that the perturbation made the lower-energy basis orbital to lowered down and higher energy-orbital to rise above in energy further.

➤ *Characteristic Features of PMO Approach*

Before we apply the PMO method to draw all the molecular orbitals of different linear and cyclic systems, it is very important to discuss some characteristic features of perturbed molecular orbitals first as given below.

1. Depiction of perturbed molecular orbitals: Unlike conventional MO theory, the molecular orbitals in the PMO approach are portrayed as in terms of basis orbitals rather than the actual shapes. Therefore, every shape with more than one atomic orbital is actually a molecular orbital.

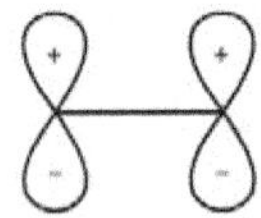

Bonding molecular
orbital (BMO)

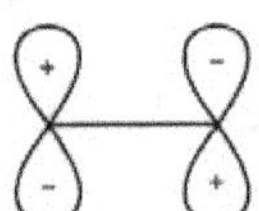

Antibonding molecular
orbital (BMO)

2. Symmetry of perturbed molecular orbitals: If a molecular orbital can be divided into two equal halves, the MO will be said to possess mirror or 'm' symmetry.

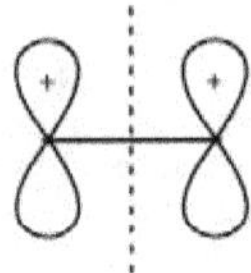

Mirror symmetry
present

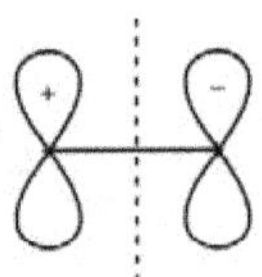

Mirror symmetry
absent

Similarly, a molecular orbital has a two-fold symmetry axis, the MO will be said to possess inversion or 'C_2' symmetry.

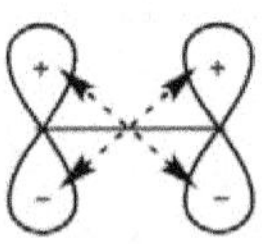

C_2 symmetry
absent

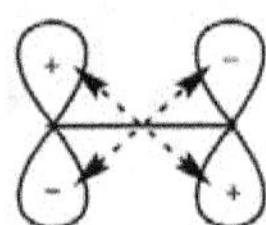

C_2 symmetry
present

3. Representation of basis orbitals: The basis orbitals in PMO approach are portrayed combined number of participating atoms. For instance, if the molecular orbital of ethylene and a p-orbital are basis orbital, both should be depicted three carbon centres.

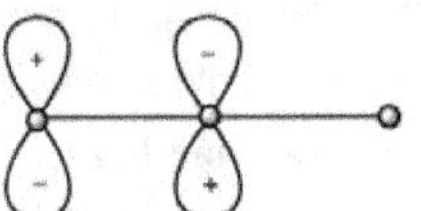

Ethylene

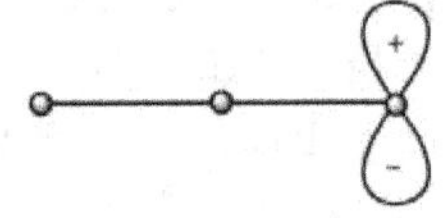

p-orbital

D DALAL
**INSTITUTE

> *Construction of Perturbed Molecular Orbitals of Linear Conjugated Systems*

The π-molecular orbitals of linear conjugated systems can primarily be classified into two categories, even-numbered systems, and odd-numbered conjugated systems.

1. Molecular orbitals of even-numbered linear conjugated systems: If the number of carbon atoms in the conjugated system is even, half of the molecular orbitals will be bonding and half of the molecular orbitals will be antibonding in nature. The number of vertical nodes in the nth molecular orbital will be $n-1$.

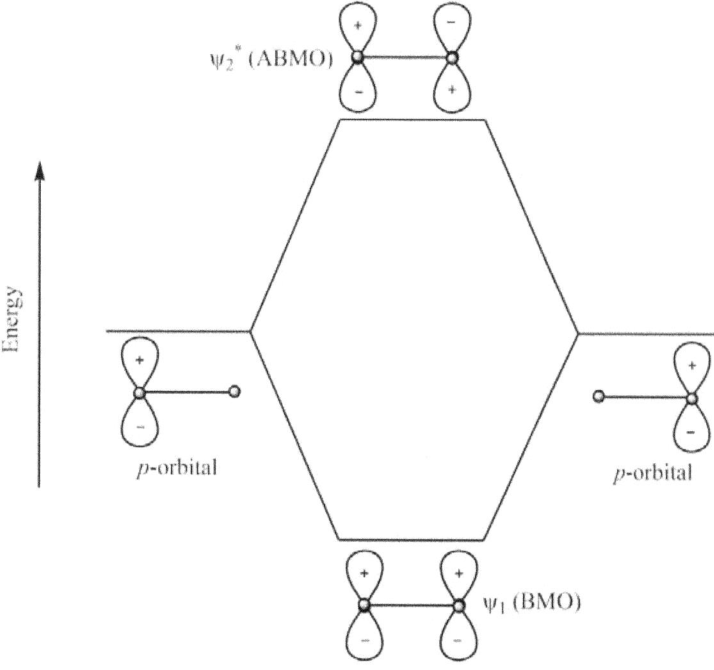

Figure 23. The perturbed molecular orbitals of ethylene.

The nature of the symmetry and number of nodes in the perturbed molecular orbitals (PMO) of ethylene are given below.

Table 1. Symmetry and nodes in different PMOs of ethylene.

PMO	Number of nodes	Mirror symmetry	C_2-symmetry
ψ_1	0	Present	Absent
ψ_2	1	Absent	Present

The nature of the symmetry and number of nodes in the perturbed molecular orbitals (PMO) of other even-numbered π-bonding systems are given below.

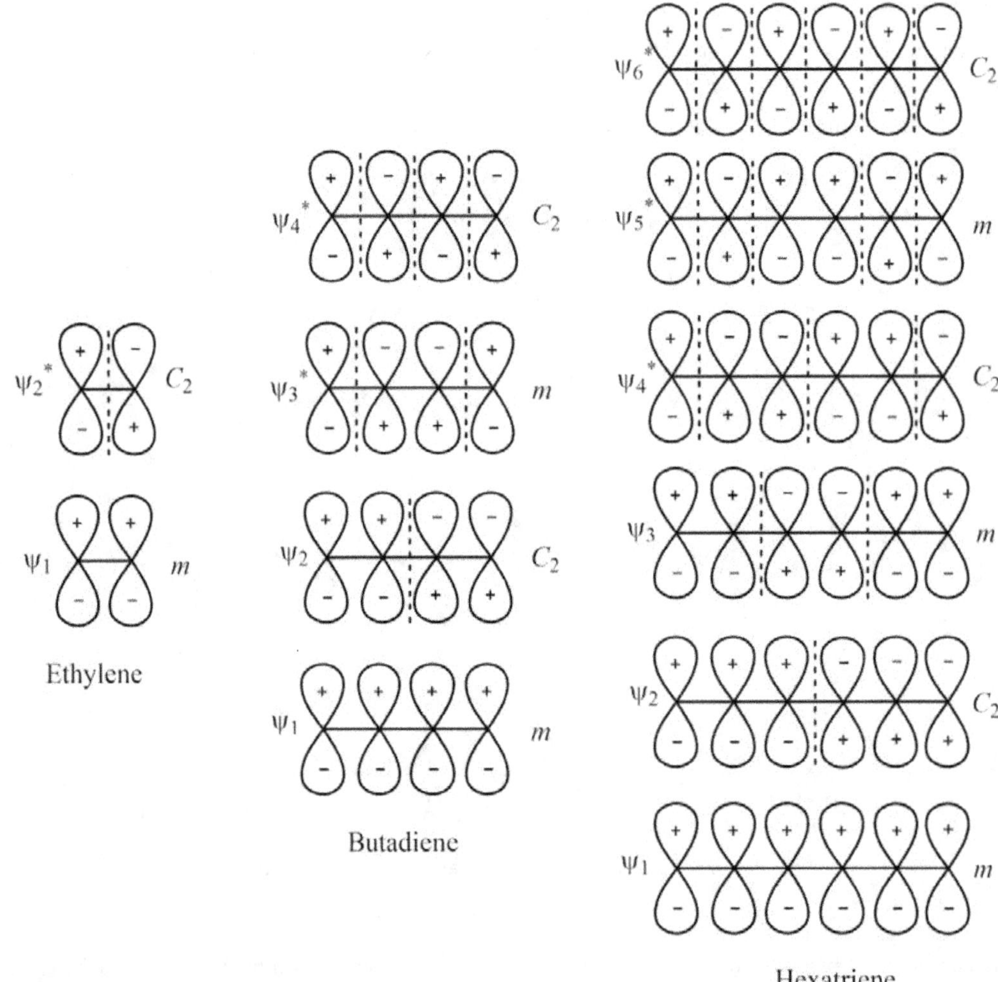

Figure 24. The perturbed molecular orbitals (along with the symmetry) of ethylene, butadiene, and hexatriene.

It is obvious from the above diagram that all the nodes in linear conjugated systems with an even number of carbon atoms pass through in-between the carbon nuclei.

2. Molecular orbitals of odd-numbered linear conjugated systems: If the number of carbon atoms in the conjugated system is odd, say N; $N/2$ of the molecular orbitals will be bonding whereas $N/2$ of the molecular orbitals will be antibonding in nature. One perturbed molecular orbital will be non-bonding in nature. The number of vertical nodes in the nth molecular orbital will be $n-1$.

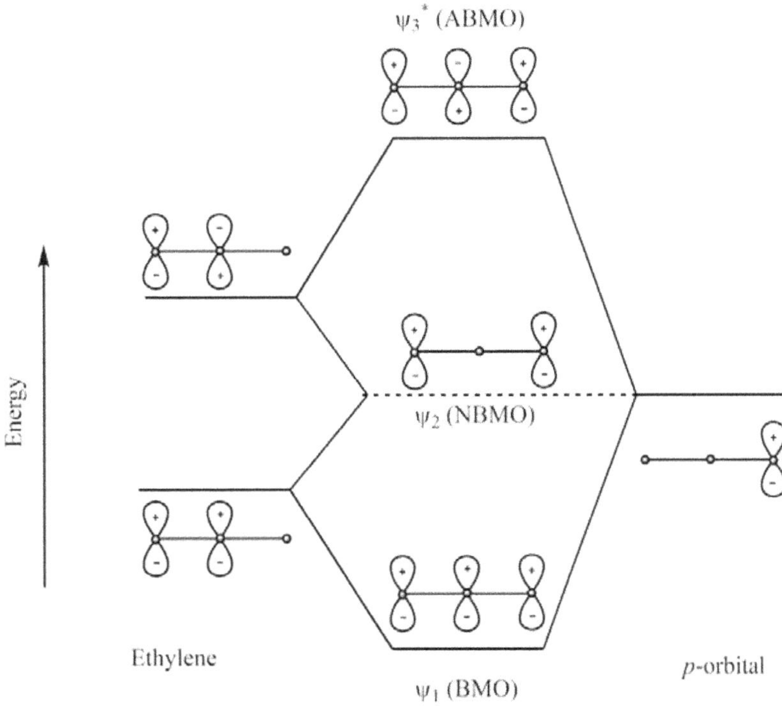

Figure 25. The perturbed molecular orbitals of allyl.

The nature of the symmetry and number of nodes in the perturbed molecular orbitals (PMO) of ethylene are given below.

Table 2. Symmetry and nodes in different PMOs of allyl.

PMO	Number of nodes	Mirror symmetry	C_2-symmetry
ψ_1	0	Present	Absent
ψ_2	1	Absent	Present
ψ_3	2	Present	Absent

The nature of the symmetry and number of nodes in the perturbed molecular orbitals (PMO) of other odd-numbered π-bonding systems are given below.

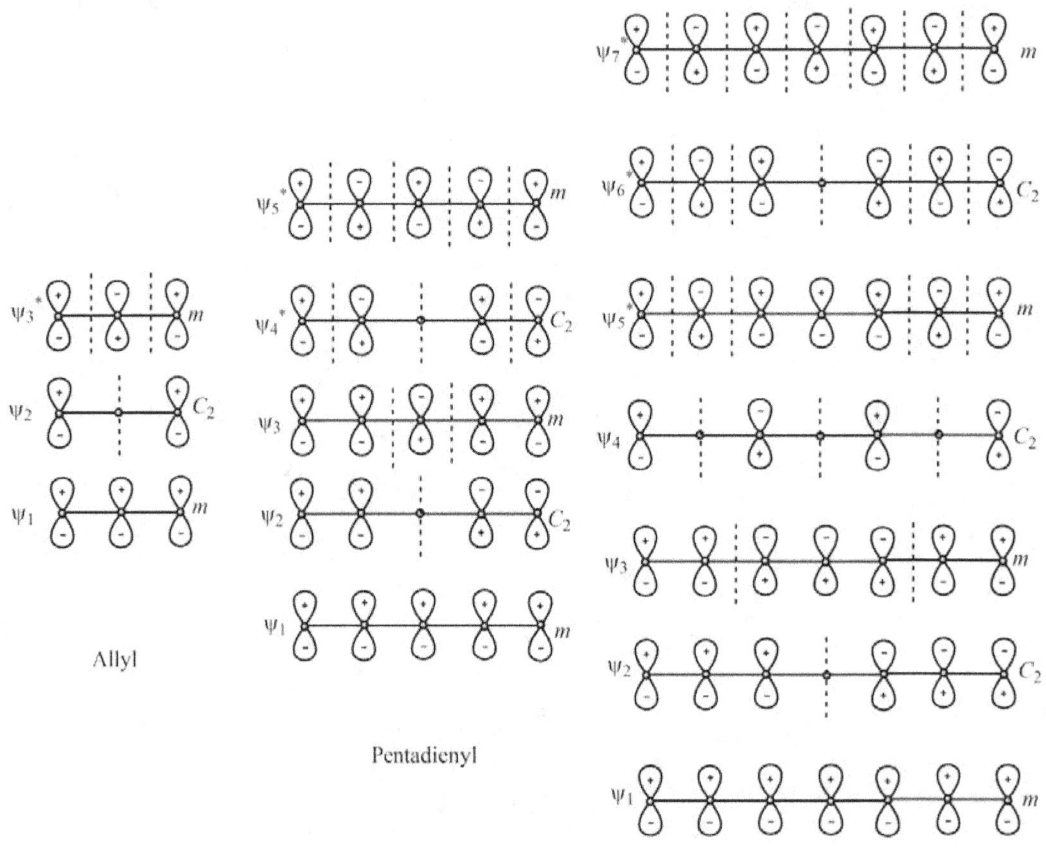

Figure 26. The perturbed molecular orbitals (along with the symmetry) of allyl, pentadienyl, and heptatrienyl.

It is obvious from the above diagram that all the nodes in ψ_{odd} pass through in-between the carbon nuclei whereas all nodes in ψ_{even} pass through in-between the carbon nuclei excepting one that passes through the central carbon nucleus. Furthermore, all the nodes in the case of non-bonding molecular orbitals pass through the carbon nuclei.

> ### *PMO Approach for Aromaticity*

One of the most important applications of the PMO approach is that it can also be used to predict the aromatic character of different organic compounds. According to this method, we need to observe the total energy of the conjugated system when it is molded into a ring. If the energy of the cyclic conjugated system becomes less than the open chain counterpart, it will be labeled as aromatic. Contrariwise, If the energy of the cyclic conjugated system becomes greater than the open chain counterpart, it will be labeled as antiaromatic. M. J. Goldstein and Roald Hoffmann wrote a very important paper in 1971 on symmetry, topology, and aromaticity. In their method, a cyclic conjugated system is considered as a derivative of two open-chain fragments that are joined at the ends, something like a donor-acceptor assembly. Now after taking mirror symmetry into consideration, we should expect a stable association if the symmetry of HOMO of one fragment matches with the symmetry of LUMO of the other fragment and vice-versa.

For simplicity, we will study conjugated systems with an even number of π-electrons only. Goldstein and Hoffmann divided these types of systems into two categories; Mode 2 (HOMO symmetric and LUMO antisymmetric) and Mode 0 (HOMO antisymmetric and LUMO symmetric).

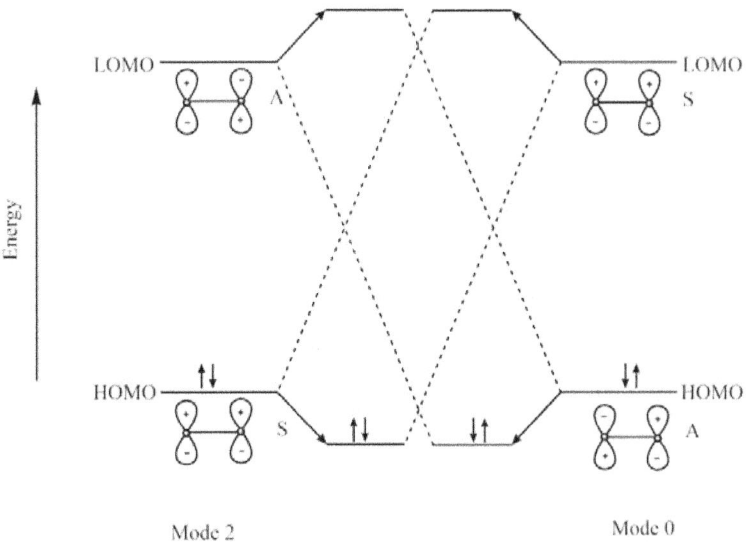

Figure 27. The pictorial depiction of Mode 2 overlap with Mode 0.

Similarly, if we consider Mode 2 interaction with Mode 2 or Mode 0 with Mode 0, we will get an unstable cyclic system owing to a symmetry disparity. Now since Mode 2 and Mode 0 fragments contain $4n + 2$ and $4n$ π-electrons respectively; the combination of Mode 2 and Mode 0 will create a total of $4n + 2$ π-electrons (aromatic), whereas Mode-2–Mode-2 or Mode-0–Mode-0 will create a total of $4n$ π-electrons (antiaromatic).

❖ Bonds Weaker Than Covalent

The bonds we have studied so far in this chapter possess bond dissociation energy in the range of 200–400 kJ mol^{-1} i.e., covalent bonds. However, there are some bonds with very low bond dissociation energies as well (10–40 kJ mol^{-1}). Therefore, in this section, we will discuss different types of bonds that are weaker than the typical covalent bond and results in supramolecular systems.

➤ *Ion-Ion Interactions*

Though ionic bonds are stronger than typical covalent bonds, the ion-ion interaction between a cation and organic anion is somewhat weaker than typical covalent interaction because of the bond energy in the range of 100–350 kJ mol.

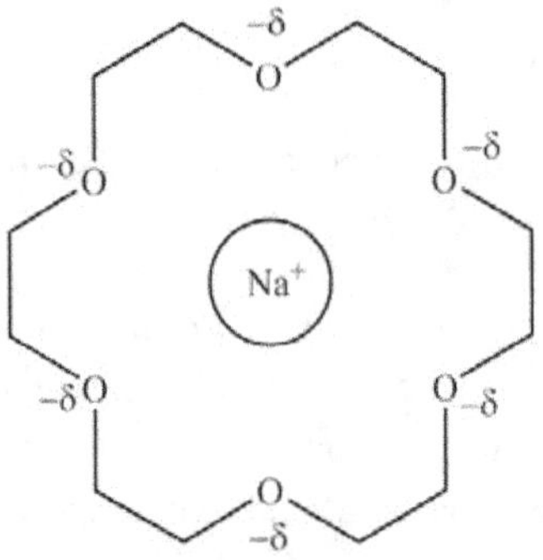

➤ *Ion-Dipole Interactions*

The interaction of an ion (like sodium) with a polar molecule (like water), is the case of ion-dipole interaction, whose bond energy ranges from 50–200 kJ mol^{-1}. This type of interaction is observed both in the solution and in the solid state. The ion-dipole bonding in supramolecular systems of alkali cations' complexes with macrocyclic (large ring) ethers are labeled as crown ethers. The oxygen atoms of the ether play the same role as played by the polar H_2O molecules, though the complex gets its extra stability from the chelate- and macrocyclic effects.

> ### *Dipole-Dipole Interactions*

The alignment of molecular dipoles with each other can produce substantial attractions. This can be of two types; the first is the matching of a single pair of poles on adjacent molecules, and the second one involves the opposing alignment of one dipole with the other. The bond energy of dipole-dipole interactions ranges from 5–50 kJ mol^{-1}. Many carbonyl derivatives show this type of behavior in the solid state, and studies have proposed that the second type of interactions have energy around 20 kJ mol^{-1}, which is quite comparable to a hydrogen bond of intermediate strength. Nevertheless, the lower boiling point of ketones suggests that dipole-dipole interactions of the second type are relatively weak in solution (acetone's boiling point = 56°C).

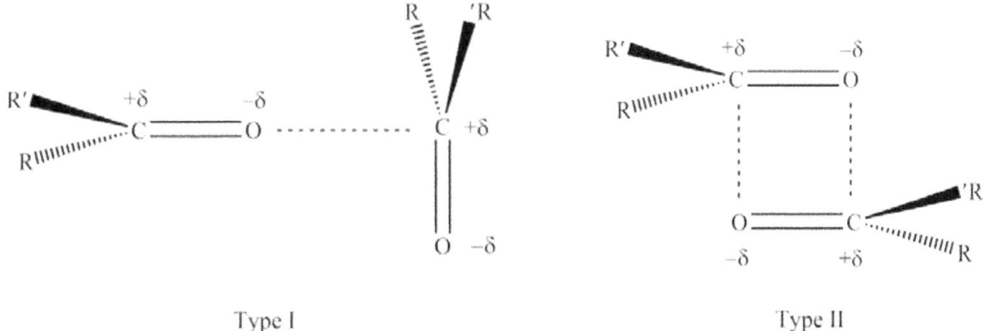

Type I   Type II

> ### *Hydrogen Bonding*

The hydrogen bonding or H-bonding may be considered as a special type of dipole-dipole interaction where an H atom is in quite strong attraction with an electronegative atom of the same or neighboring molecule. If the H-bonding takes place within the molecule itself, it is called as intramolecular H-bonding. On the other hand, if the H-bonding takes place between two molecules, it is called as intermolecular H-bonding. The bond energy for H-interactions ranges from 10–120 kJ mol^{-1}. A typical case of H bonding in supramolecular systems is the generation of carboxylic acid dimers. H-bonds are ubiquitous, and are responsible for the overall shape of numerous proteins, recognition of substrates by many enzymes (along with π-π interactions) and also for DNA's double helix structure.

Intramolecular H-bonding   Intermolecular H-bonding

 DALAL INSTITUTE

> ➤ *π-π Interactions*

Since π-π interactions happen to be present between aromatic rings, they are also known as aromatic π-π stacking interactions, and generally involve one electron-deficient and one electron-rich species. The bond energy of these types of interactions ranges from 50–500 kJ mol^{-1}. These π-interactions are two kinds; one is face-to-face, and the second one is edge-to-face. The face-to-face π-π interactions are accountable for the slippery nature of graphite and its beneficial lubricant feature. Similar kind of π-π interactions between the aryl rings of nucleobase pairs also helps the DNA double helix to get stabilized. The second type (i.e. edge-to-face interactions) may be considered as weak types of H-bonds between the electron-rich π-cloud of one aromatic ring and the slightly electron-deficient H atoms of another aromatic ring. Hence, they must not be called as π-stacking because no stacking of the π-electron surfaces is present.

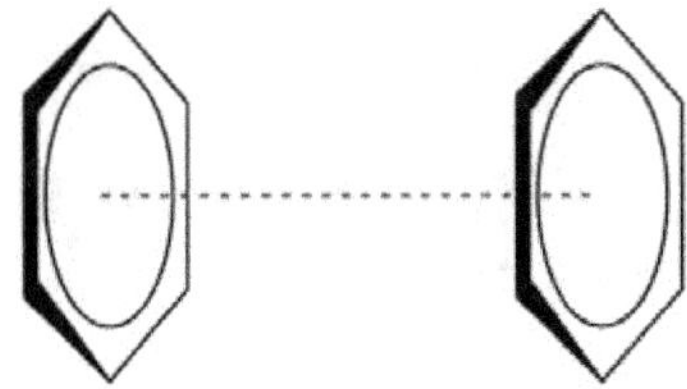

Sandwitch

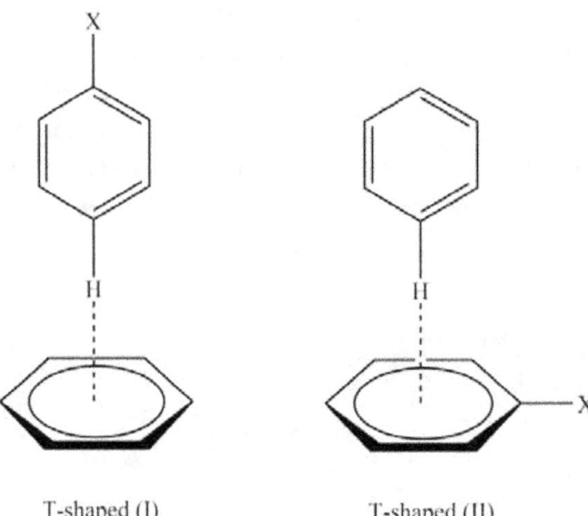

T-shaped (I)   T-shaped (II)

The edge-to-face interactions are also accountable for the characteristic herringbone packing in the crystal lattice of a range of small aromatic hydrocarbons (benzene included).

> ➢ *Cation-π Interactions*

Many cations of d-block metals (like Fe^{2+}, Pd^{2+}, Pt^{2+}) are known to yield complexes with aromatic and olefinic systems such as ferrocene i.e. $[Fe(C_5H_5)_2]$ and Zeise's salt i.e. $[PtCl_3(C_2H_4)]^-$. The nature of bonding in such systems is quite strong and could not be considered non-covalent, because it is closely linked with the partially filled d-orbitals of the metals. For instance, the bond between Ag^+ and C_6H_6 has a significant covalent character. Nevertheless, the bonding of alkali metals' cations or alkaline earth metals' cations with C=C double bonds is quite weak, and therefore, can be considered as non-covalent interaction. These interactions have also been proved to play a significant role in biological systems. For instance, the bond energy of the K^+-benzene bond in the gas phase is around 80 kJ mol^{-1}, which is quite comparable to K^+-water interaction (75 kJ mol^{-1}). This can be attributed to the fact that the K^+ ion is more soluble in water than in C_6H_6 as more water molecules can interact with the potassium ion, but only a few C_6H_6 molecules can fit around it due to bulkier size.

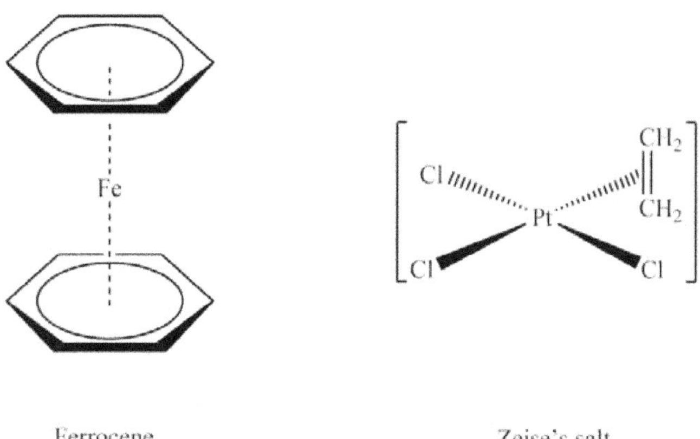

Ferrocene   Zeise's salt

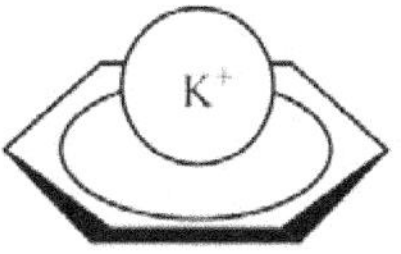

Cation-π interactions in between
K^+ ion and benzene

The bonding of cations of non-metals like RNH_3^+ with double bonds can be regarded as a type of X−H $\cdots$π H-interaction.

> ### *Anion-π Interactions*

It seems that, unlike cation-π interactions, anion-π interactions should not be favorable as it appears to be repulsive; for instance, the affinity of the aromatic ring with cryptand for halides quickly dopes in the order F^- >> Cl^- > Br^- ~ I^- since repulsions arising from anion-π interaction is more for bigger halides. Nevertheless, there is a difference in the charges between an anion and an overall neutral aromatic ring, and hence, there is a possibility of an electrostatic attraction. These kinds of short anion-π interactions are noted for organometallic calixarene derivatives where the aromatic ring has a substantial positive charge.

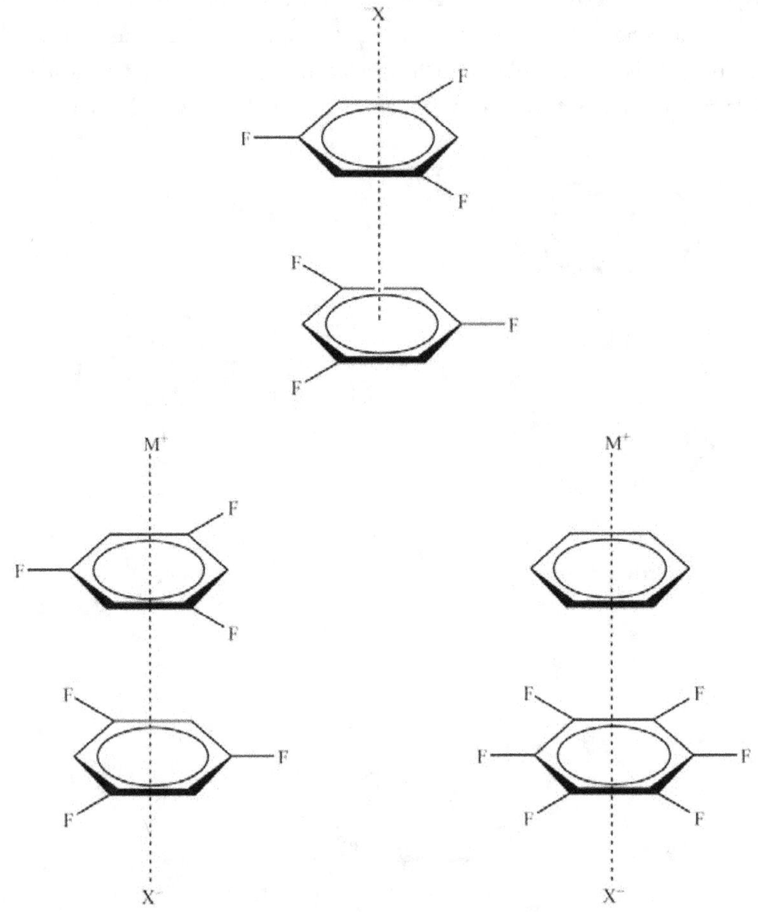

Anion-π interaction

The anion-π interactions are also caught up as controlling factors in self-assembly reactions of monovalent silver complexes with π-acidic aromatic rings.

> ### *Van der Waal's Forces*

Van der Waals forces are attractive and arise from the polarization of an electronic cloud by the closeness of an adjacent nucleus, yielding a weak attraction of electrostatic nature. The characteristic range for Van der Waals forces ranges from 0.50–40 kJ mol^{-1}. The Van der Waals forces are non-directional, and therefore, have only restricted scope in designing specific hosts for choosy complexation of specific guest molecules. Generally speaking, van der Waals forces deliver a generic attraction for most 'easily polarisable' (soft) species with the interaction energy relational to the surface area of direct contact. In the case of supramolecular systems, the van der Waals forces are extremely important in the generation of 'inclusion compounds', where very organic molecules small are loosely united within molecular cavities or crystal lattices; for instance, the inclusion of toluene into the molecular cavity of the *p*-tert-butylphenol-based macrocycle. In the case of molecular assemblies in solid-state, the effects of crystal close packing dictate the overall structural arrangement.

The hydrophobic interactions also perform a vital role in supramolecular systems as these are accountable for displacing the H$_2$O molecules from the host's cavity by an organic molecular guest.

➤ *Closed Shell Interactions*

It is quite a well-known fact that an atom with partially filled electronic shells gives rise to a strong covalent bond. Nevertheless, many ions usually have fully-filled or closed valence electronic shells but feel strong attractions between oppositely charged pairs. Therefore, one might think that closed shell atoms of neutral or like charges should not show any significant interaction; though in some cases, they show significant attraction. These kinds of interactions are labeled as closed shell interactions and involve some secondary bonding interactions; one is halogen bonding and the other is metalophilic interactions. As far as the strength is concerned, these interactions are quite comparable to moderate H-bonds. These interactions are considered to originate from electron correlation effects, which are further strengthened by relativistic effects in the case of heavy metals like gold (i.e., aurophilic interactions). These interactions are very noticeable for heavy metals with electron configurations ranging from d^8 to $d^{10} s^2$.

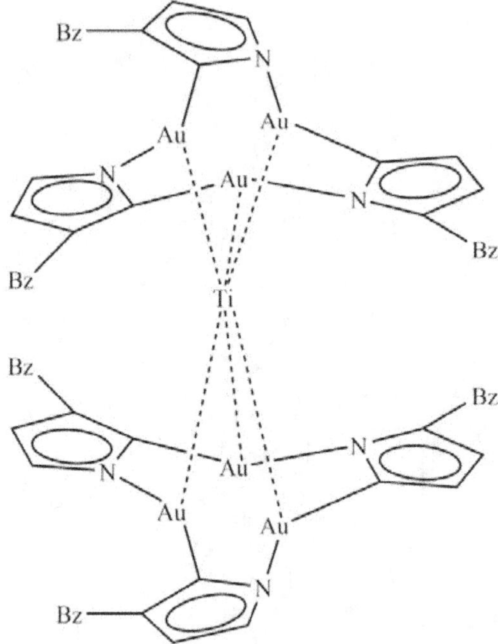

Closed Shell Interactions

Furthermore, it is also worthy to note the fact that the overall strength of halogen bonding decreases in the following order.

$$I > Br > Cl \gg F$$

❖ Addition Compounds: Crown Ether Complexes and Cryptands, Inclusion Compounds, Cyclodextrins

In this section, we will discuss some organic compounds in which the strength of the bonds is less than a typical covalent bond. One of the major types of compounds having weak non-covalent interaction is addition Compounds. An adduct, or addition compound, may simply be defined as the product of the direct addition of two or more distinct molecules, resulting in a single reaction product containing all atoms of all components. Now there are many types of addition compounds but we will confine our discussion only on adducts with bonds weaker than covalent such as electron donor-acceptor complexes.

➢ *Electron Donor-Acceptor Complexes*

An electron donor-acceptor complex, or simply the charge-transfer complex (CT complex), is an association of two or more molecular entities, or of different parts of one big molecule, where a fraction of electronic charge is transferred from one molecular entity to another.

The resulting electrostatic force of attraction delivers a stabilizing drive for the molecular adduct so formed. The source molecule from which the charge is transferred is called the electron donor whereas the charge receiving species is labeled as the electron acceptor. These electron donor-acceptor complexes usually have spectra that different from the sum of the spectra of the individual participating molecules. There are two primary types of electron donor-acceptor complexes.

1. When the acceptor is a metal ion and the donor is an aromatic ring an alkene: Some of the typical examples of charge transfer complexes where the donor is an alkene or an aromatic ring and acceptor is a metal ion are given below.

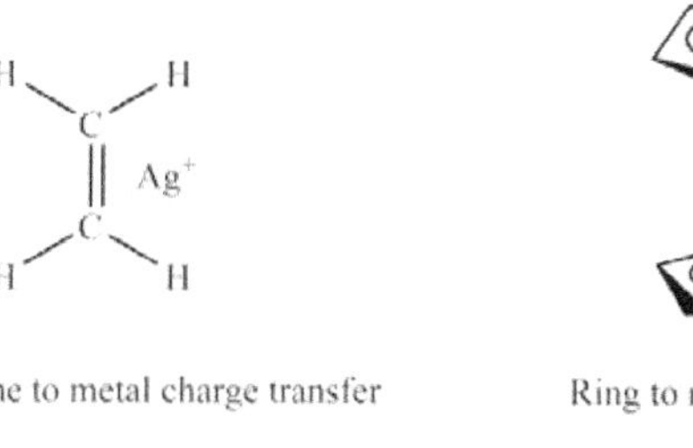

Alkene to metal charge transfer   Ring to metal charge transfer

Allyl to metal charge transfer   Open chain to metal charge transfer

DALAL
INSTITUTE

2. When the Acceptor is an organic molecule: Some of the typical examples of charge transfer complexes where the acceptor entity is an organic molecule are given below.

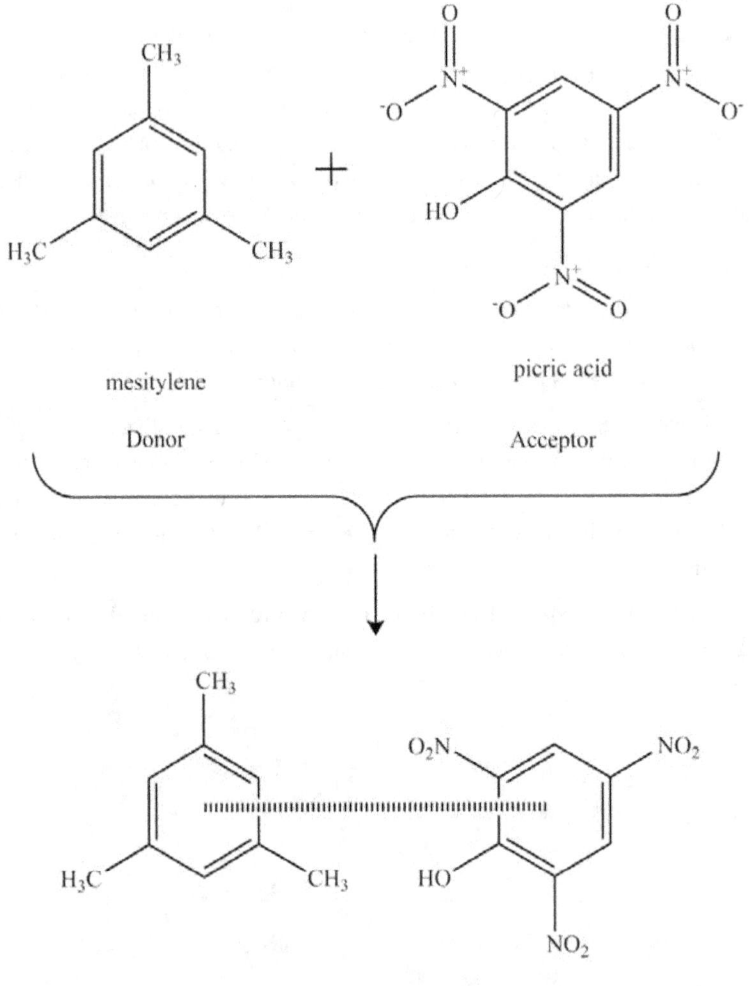

mesitylene   picric acid

Donor   Acceptor

π-complex
(Picrate)

It is also worthy to note that owing to the very small energy gap between the ground and excited states, electron donor-acceptor complexes show absorption peaks in the visible or near-UV region, and therefore, usually show color in sunlight.

> *Crown Ether Complexes and Cryptates*

More typical examples of addition compounds are crown ether complexes and cryptands. A general discussion on both these types is given below.

1. Crown ether complexes: Crown ethers may simply be defined as the cyclic compounds that are consisted of a ring with several ether groups. Some of the most common examples of crown ethers are cyclic oligomers of ethylene oxide, where ethyleneoxy is the repeating unit. Chief members of this Crown ethers' series are the hexamer ($n = 6$), the pentamer ($n = 5$), and the tetramer ($n = 4$).

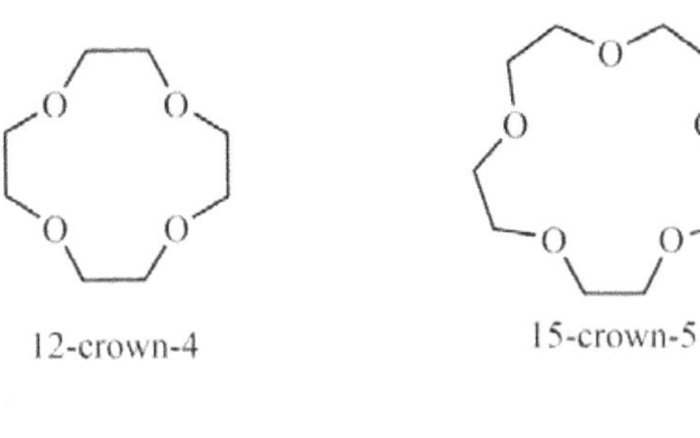

12-crown-4   15-crown-5

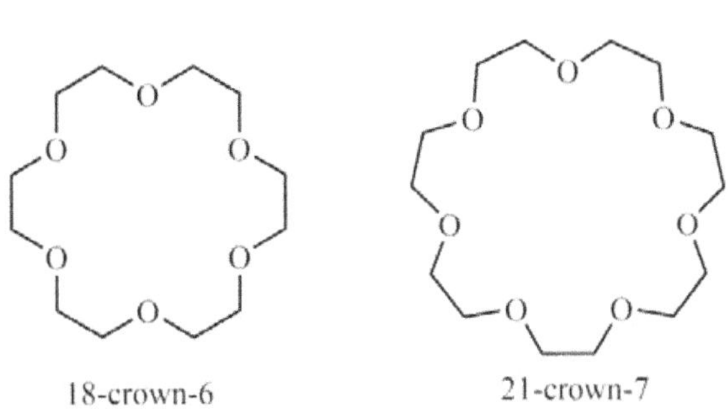

18-crown-6   21-crown-7

Here the label "crown" means the resemblance with the structure of a king's crown. As far as the nomenclature is concerned, the first number means the number of atoms in the cycle, and the 2nd number means the number of oxygen atoms. Crown ethers are much broader than the oligomers of ethylene oxide; a very important class that is catechols' derivative.

These compounds have a very strong tendency to bind with certain cations, and therefore, generating complexes. The O atoms are well placed to bind with a certain cation situated within the ring, while the ring's exterior remains hydrophobic. The cations thus produced generally form salts that are solvable in non-polar solvents, and therefore, crown ethers are very valuable in the application of phase transfer catalysis. The polyether's denticity affects the affinity of the crown ether for different cations. For instance, 18-crown-6 has more affinity for K^+, 15-crown-5 for Na^+, and 12-crown-4 for Li^+ ion. The very high affinity of 18-crown-6 for K^+ cation is primarily responsible for its toxic character.

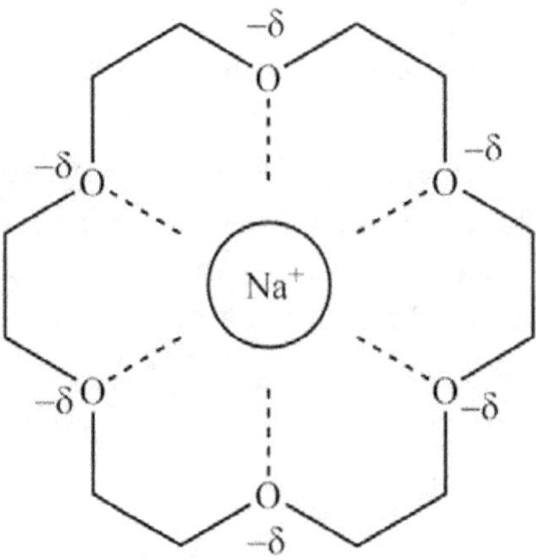

Less stable

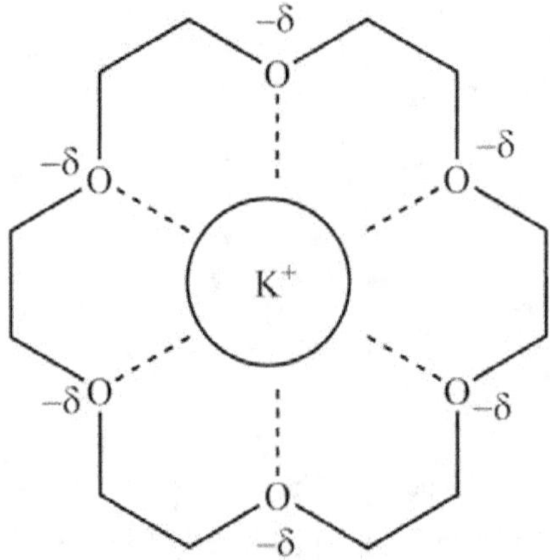

More stable

It is also worthy to note that crown ethers aren't the only macrocyclic ligands that capable of binding with K^+ cation. Some ionophores like valinomycin also show a very strong preference for the K^+ ion over other cationic species. These compounds have also been known to bind with Lewis acids via σ-hole (i.e., halogen bond) interactions, electrostatic interactions; in other words, bonding takes place between the electrophilic Lewis acid center and the Lewis basic oxygen atoms of the crown ether.

2. Cryptands: Cryptands may simply be defined as a family of synthetic bicyclic and polycyclic multidentate ligands that are capable of binding with a range of cationic species. In 1987, Donald J. Cram, Jean-Marie Lehn, and Charles J. Pedersen got the Nobel Prize in chemistry for the discovery and finding uses of cryptands and crown ethers, and therefore, starting a novel field of supramolecular studies. The label cryptand means that this ligand gets attached to the substrates in a crypt, burying the guest. These molecular systems are 3-dimensional analogs of crown ethers, nevertheless, are more choosy and have a stronger tendency for complex formation giving lipophilic assemblies.

One of the most commonly studied and important cryptand is $N[CH_2CH_2OCH_2CH_2OCH_2CH_2]_3N$; whose IUPAC name is 1,10-diaza-4,7,13,16,21,24-hexaoxabicyclo[8.8.8]hexacosane. This cryptand is labeled as [2.2.2]cryptand, with numbers inferring about the number of ether groups (and so the binding positions) in each bridge between the nitrogen sites. Numerous cryptands are available commercially by the tradename of Kryptofix. Furthermore, almost all of the amine cryptands show a very high affinity for alkali metal ions, which in turn, made the isolation of salts of K^- possible.

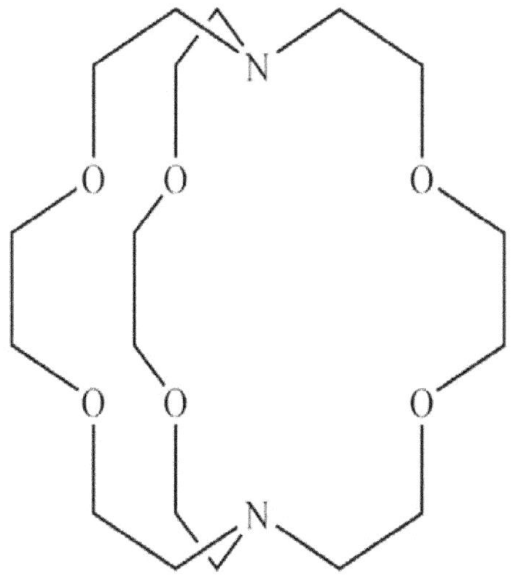

2.2.2]Cryptand

Also, the 3-dimensional void of a cryptand gives a binding site for guest ions, which means it acts as the host for the addition compounds. The complex formed between the guest cation and the cryptand is labeled as a cryptate. These ligands give rise to complexes with numerous hard cations (ammonium ion included), lanthanoids, alkaline earth metals, and alkali metals. Unlike crown ethers, cryptands get attached to the guest ions via both oxygen, as well as, nitrogen sites. This kind of 3-dimensional encapsulation bonding tells us about size-selectivity and enables us to distinguish different alkali ions like K^+ or Na^+ cation.

Cryptands are quite expensive and very problematic to synthesize, nevertheless, very valuable because of better strength and selectivity than their crown ethers counterparts.

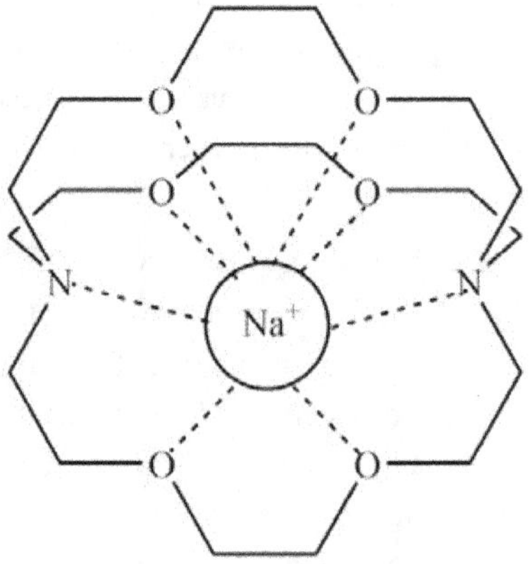

More stable Cryptate

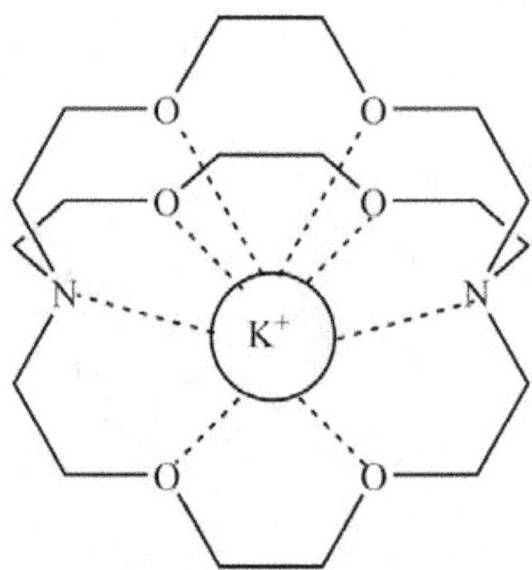

Less stable Cryptate

It is also worthy to note that cryptands are also capable of binding with insoluble salts into the organic phase, making them quite useful as phase transfer catalysts like crown ethers. These ligands enabled the preparation of the electrides and alkalides, and have also been employed to crystallize Zintl ions like Sn_9^{4-}.

> *Inclusion Compounds*

An inclusion compound in host-guest chemistry may simply be defined as a complex where one compound provides a void to accommodate the other.

The bonding between the guest and host molecules is purely van der Waals interactions. The idea of inclusion compounds is quite broad, including the channels created between molecules in a crystal lattice where the guest molecules are acceptable.

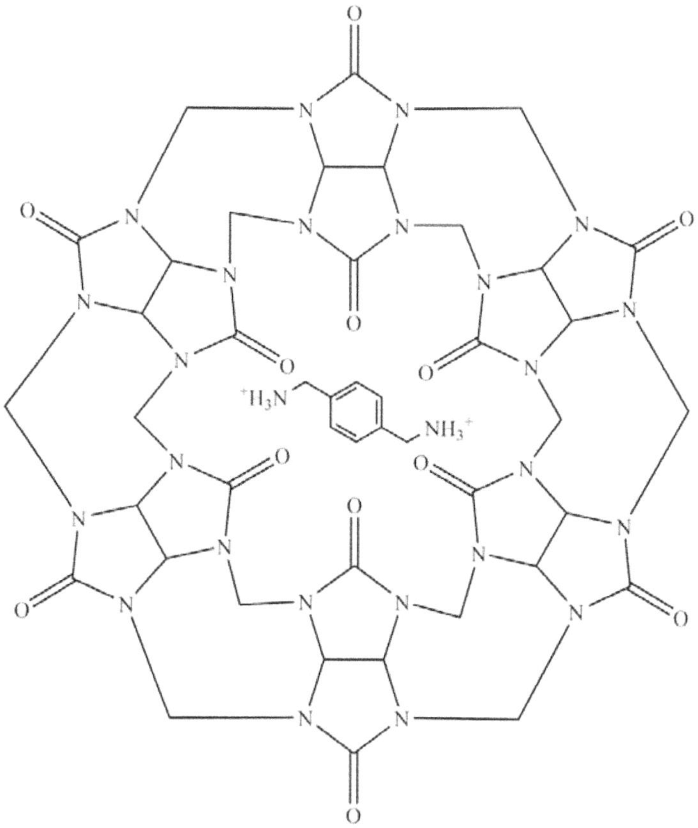

Inclusion compound consisting of a
p-xylylenediammonium bound
within a cucurbituril

Calixarenes and associated formaldehyde-arene condensates form an important class of host molecules that can give inclusion compounds. One of the most popular examples is the cyclobutadiene adduct, which would be is unstable otherwise. Cyclodextrins are quite proven hosts for the generation of inclusion compounds. For instance, the ferrocene can be inserted into the cyclodextrin at 100 °C if conditions are hydrothermal.

Cyclodextrin also gives inclusion compounds when treated with fragrances. Consequently, the fragrance molecules have a reduced vapor pressure and become more stable to the exposure to air and light.

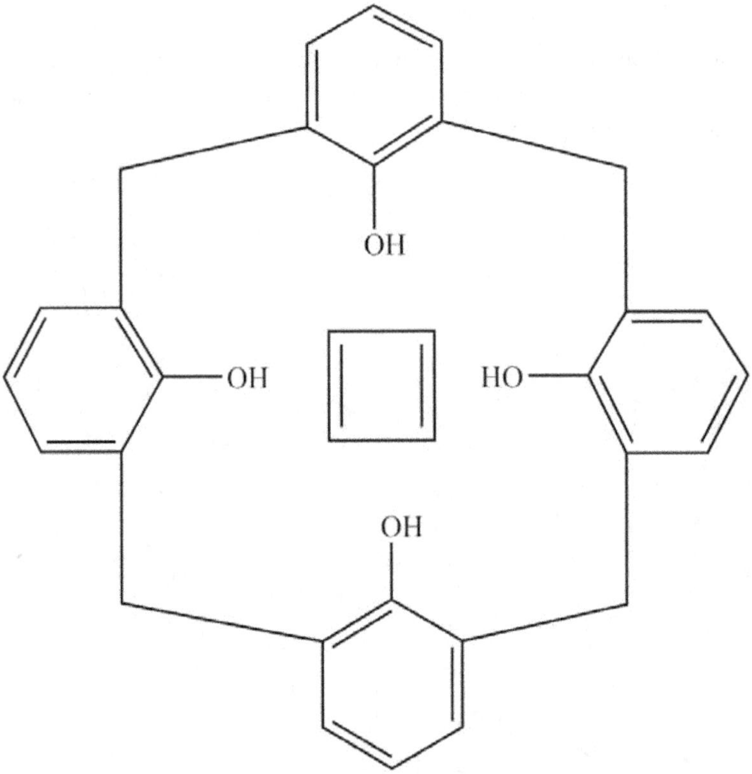

Cyclobutadiene adduct
with

Calix[4]arene-25,26,27,28-tetrol

Furthermore, if incorporated into textiles the fragrance lasts for a longer period because of the slow-release process. Also, crown ethers and cryptands generally don't give inclusion compounds because the guest is bound by attraction forces that are stronger than van der Waals interactions.

It is also worthy to note that if the guest is surrounded in such a way that it is 'trapped', the compound class is known as a clathrate, which is different from than inclusion complex. This encapsulation at the molecular level traps a guest molecule inside another molecule.

> ### Cyclodextrins

Cyclodextrins may simply be defined as a family of cyclic oligosaccharides which are consisted of a macrocyclic ring having glucose subunits united by α-1,4 glycosidic linkage. These compounds are generated by the enzymatic conversion from starch. Cyclodextrins are quite useful in pharmaceutical, food, chemical industries, and drug delivery, as well as environmental and agriculture engineering. These compounds are made up of 5 or more α-D-glucopyranoside units which are joined at 1->4, like in amylose (a starch's fragment). The biggest cyclodextrin molecule has 32 1,4-anhydroglucopyranoside units, whereas as a poorly characterized mixture, 150-membered cyclic oligosaccharides have also been proved. Archetypal cyclodextrin molecules have several glucose monomer units ranging from 6 to 8 parts in a cycle, giving rise a cone shape structures.

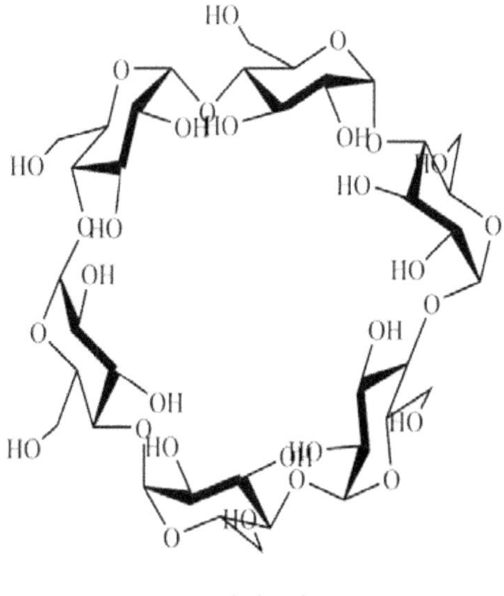

α-cyclodextrin

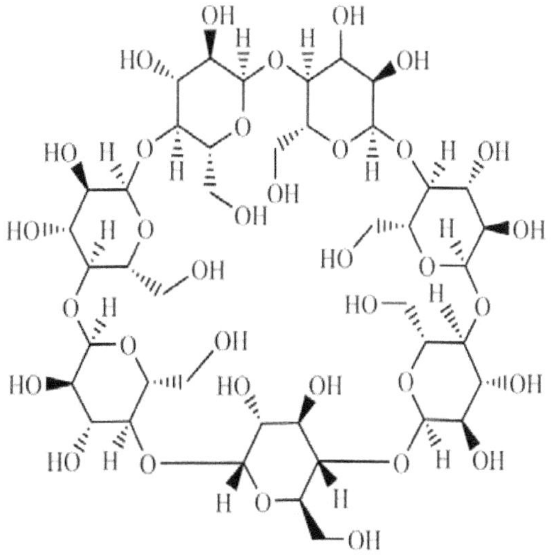

β-Cyclodextrin

With a hydrophilic exterior and hydrophobic interior, these molecules give rise to complexes with hydrophobic systems. The FDA-approved compounds are α-, β-, and γ-cyclodextrin that have been employed for delivery of different kinds of drugs, including prostaglandin, hydrocortisone, itraconazole, nitroglycerin, chloramphenicol. These compounds confer stability and solubility to these drugs. The inclusion complexes of cyclodextrins with hydrophobic fragments can penetrate tissues, and therefore, can be used to release biologically important compounds under explicit conditions. In many cases, the mechanistic behavior of controlled decay of such compounds is based upon the pH change of aqueous solutions, yielding the loss of ionic or hydrogen bonds between the guest and host systems. The other means for the disturbance of the compound take benefit of enzymatic or heating action which is capable to break the α-1,4 bond between glucose monomeric units. These compounds have also been shown to improve mucosal penetration of different kinds of drugs.

Cyclodextrin compounds are synthesized by the enzymatic reaction on starch. Normally cyclodextrin glycosyltransferase is used along with α-amylase. The starch is liquified first either via heating action or by employing α-amylase, then cyclodextrin glycosyltransferase is added for the enzymatic treatment. cyclodextrin glycosyltransferase gives rise to the mixtures of different kinds of cyclodextrins, and therefore, the three main types of cyclic compounds are produced, where the ratios are strongly dependent on the enzyme employed. Also, each cyclodextrin glycosyltransferase has its characteristic ratio of α:β:γ yield.

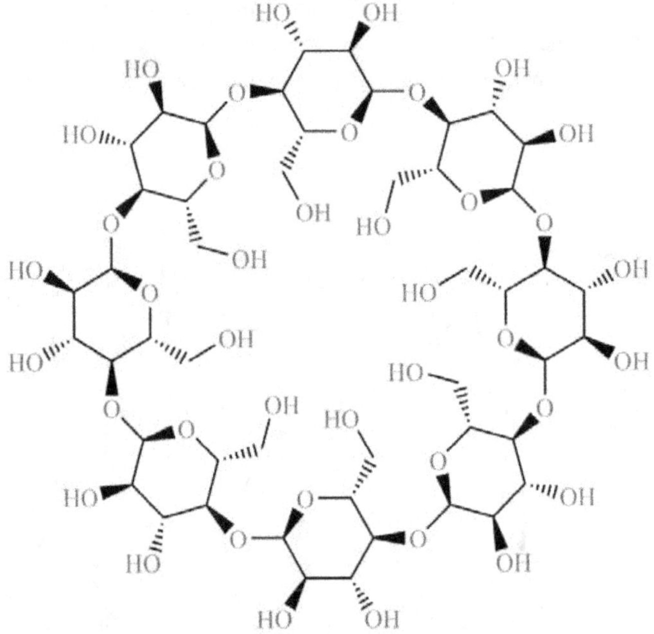

γ-cyclodextrin

The distillation of the 3 kinds of cyclodextrins takes the benefit of their dissimilar water solubility; β- cyclodextrins is very poorly soluble in water and can be easily obtained via crystallization whereas the more soluble α- and γ- cyclodextrins are typically obtained by using expensive and time-consuming techniques of chromatography. Alternatively, some complexing agents (like acetone toluene, or ethanol) can also be added in the course of the enzymatic treatment step to yield a complex with the desired cyclodextrin, which in turn, can easily be precipitated subsequently. The formation of the complex drives the transformation of starch into the generation of the precipitated cyclodextrin, and so enriching its amount in the final products. W. Chemie practices some dedicated enzymes, that can yield α-, β- or γ-cyclodextrin explicitly. This is quite important particularly in food engineering because the only α-, β- or γ-cyclodextrin can be consumed without any limit on the daily intake.

❖ Catenanes and Rotaxanes

Besides the addition compounds, some compounds contain two or more sovereign parts which are linked but without any valence forces. A more general discussion on these two types of organic compounds is given below.

➢ *Catenanes*

A catenane may simply be defined as a mechanically interlocked molecular construction that is consisted of two or more macrocycles, i.e. a molecular assembly having two or more intertwined cycles.

These interlocked cycles cannot be parted without the cleavage of covalent bonds of one of the macrocycles. These compounds are obtained their name from the Latin word 'catena' which literally means 'the chain'. Catenanes can be abstractly associated with other mechanically interlocked molecular constructions, like rotaxanes, molecular Borromean rings, or molecular knots.

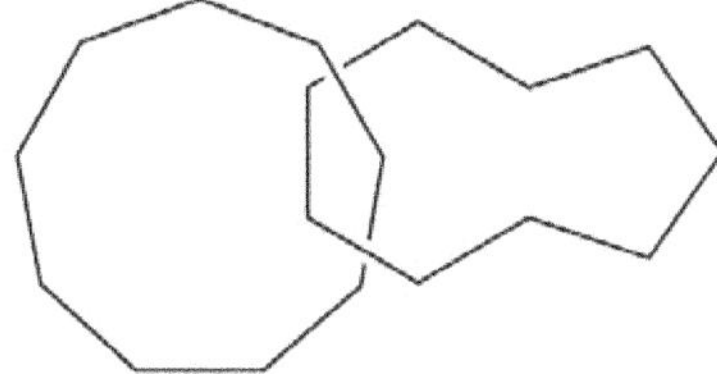

A [2]-Catenane

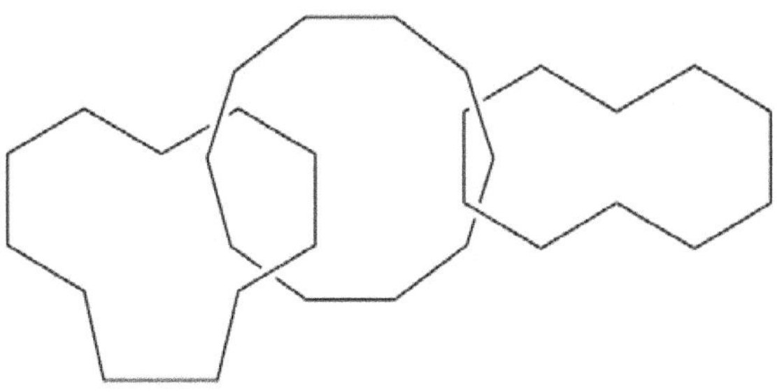

A [3]-Catenane

In recent years, the term "mechanical bond" has been devised that defines the correlation between the macrocycles of a single catenane.

Catenanes can be synthesized via two different routes; the first one is the statistical synthesis, whereas the second one is called as the template-directed method. The first approach is to simply carry out a ring-closing reaction and hoping that some of the reactnt will be generated around other cyclic system yielding the required catenane produce. This approach is very inefficient, needing high dilution of the closing ring and a huge excess of the pre-formed ring, and therefore, this methos is used in rare cases.

The 2nd approach depends upon the supramolecular preorganization of the macrocyclic reactants by employing properties like metal coordination, hydrogen bonding, coulombic interactions, or hydrophobic effect.

➢ *Rotaxanes*

A rotaxane may simply be defined as a mechanically interlocked molecular assembly that is consisted of a dumbbell-shaped molecule, which is threaded through a macrocycle. These compounds got their name from the Latin word wheel (rota) and axle (axis). Rotaxane's two components are kinetically trapped because the dumbbell's ends are bigger than the internal diameter of the cycle, and therefore, forbid the dissociation of the components as it would need a huge magnitude of distortion of the covalent interactions.

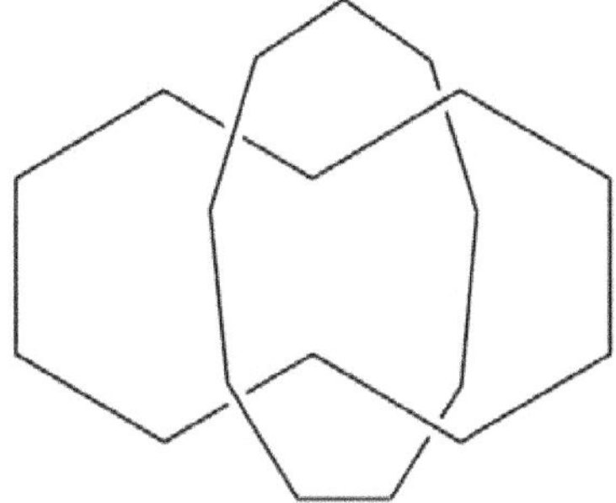

A [2]-rotaxane

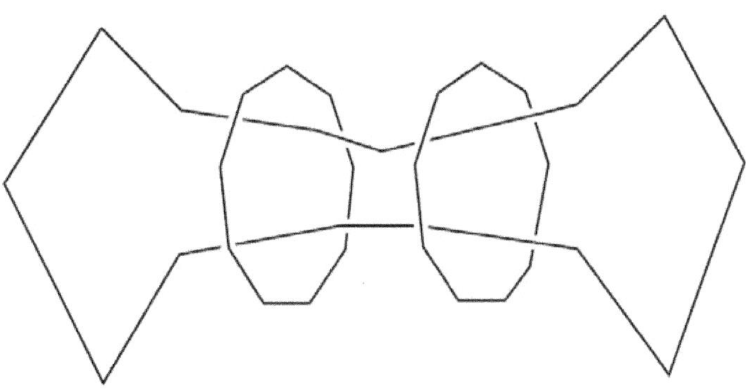

A [3]-rotaxane

Also, most of the research about rotaxanes has been focused on their efficient preparation or their use as artificial machines at the molecular level.

Nevertheless, many rotaxane examples have been found in naturally occurring substances like peptides (i.e. cyclotides or lasso-peptides and cystine knot peptides).

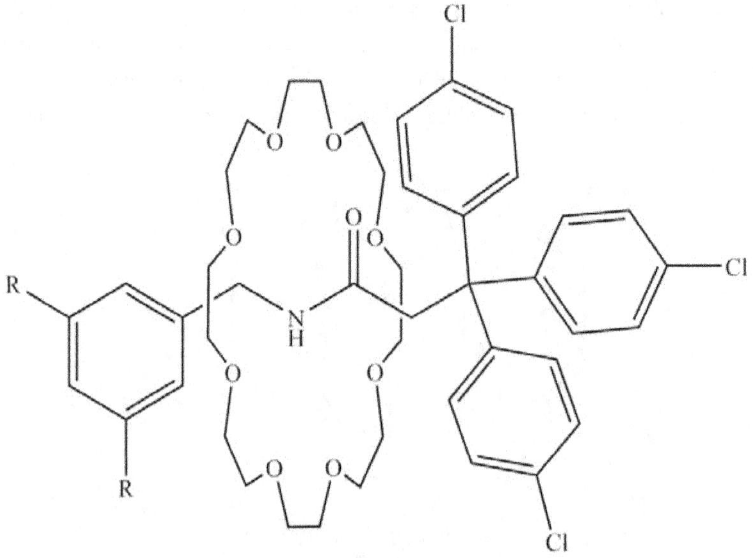

A [2]-rotaxane formed by
crown ether

❖ Problems

Q 1. Define the delocalized chemical bond in the valence bond framework with a suitable example.

Q 2. What is resonance? How is it different from tautomerism?

Q 3. What is hyperconjugation? How can it be used to rationalize the relative stability of carbocations?

Q 4. Write a short note on alternate and non-alternate hydrocarbons.

Q 5. Define aromaticity with special reference to the benzenoid compounds.

Q 6. What is the Huckel rule of aromaticity? Discuss its theoretical foundation.

Q 7. Comment on the aromatic character of different annulenes.

Q 8. What are homoaromatic compounds?

Q 9. Discuss the 'perturbed molecular orbital' approach for aromaticity.

Q 10. What are antiaromatic compounds? How they can be rationalized in terms of planarity.

Q 11. Discuss the bonds weaker than covalent with special reference to Crown ether complexes.

❖ Bibliography

1. M. B. Smith, *March's Advanced Organic Chemistry: Reactions, Mechanisms, and Structure*, John Wiley & Sons, Inc., New Jersey, USA, 2013.

2. D. Klein, *Organic Chemistry*, John Wiley & Sons, Inc., New Jersey, USA, 2015.

3. C. A. Coulson, B. O'Leary, R. B. Mallion, *Hückel Theory for Organic Chemists*, Academic Press, Massachusetts, USA, 1978.

4. H. Zimmerman, *Quantum Mechanics for Organic Chemists*, Academic Press, New York, USA, 1975.

5. J. Clayden, N. Greeves, S. Warren, *Organic Chemistry*, Oxford University Press, Oxford, UK, 2012.

6. R. L. Madan, *Organic Chemistry*, Tata McGraw Hill, New Delhi India, 2013.

7. L. H. Gade, *Encapsulated, Sandwiched, or Sticking Out: Closed-Shell Interactions of d10 Metal Centers with Thallium(I)*, Angewandte Chemie, 40 (2001) 3573-3575.

8. C. Tian, S. D. P. Fielden, G. F. Whitehead, I. J. Yrezabal and D. A. Leigh, *Weak Functional Group Interactions Revealed Through Metal-Free Active Template Rotaxane Synthesis*, Nature Communications, 11 (2020) 744.

CHAPTER 2

Stereochemistry

❖ Chirality

All the crystals or molecular geometries that cannot be superimposed on their mirror images have a special property to rotate the plane of polarized light when passed through them; such compounds are called as chiral compounds and the property is called as chirality or optical activity.

In 1811, a French physicist named François Jean Dominique Arago observed the rotation of the orientation of linearly polarized light when it is passed through quartz. A French physicist, Jean Baptiste Biot, also observed the rotation of the plane of polarized light in 1815 with some liquids and vapors of organic compounds like turpentine. After that, an English astronomer, Sir John F. W. Herschel, found in 1820 that all the different individual quartz crystals were actually the mirror images of each other, and rotated the plane of polarized light by equal magnitude but the directions of rotation were opposite. The crystal that rotates the plane of polarized light towards the right is called dextrorotatory, while the mirror image will rotate to left imparting a label of levorotatory. Now since all the crystals that were capable of rotating the plane of polarized light could not be superimposed on their mirror image crystals by any mean, the term 'chiral' (derived from Greek word, cheir = hand) become extremely popular to define such crystals as our hands also the non-superimposable mirror images of each other.

➤ *Experimental Measurement of Chirality or Optical Activity*

Optical activity or chirality is measured by an instrument called the 'polarimeter'. The basic experimental setup to measure the magnitude of optical activity is shown below.

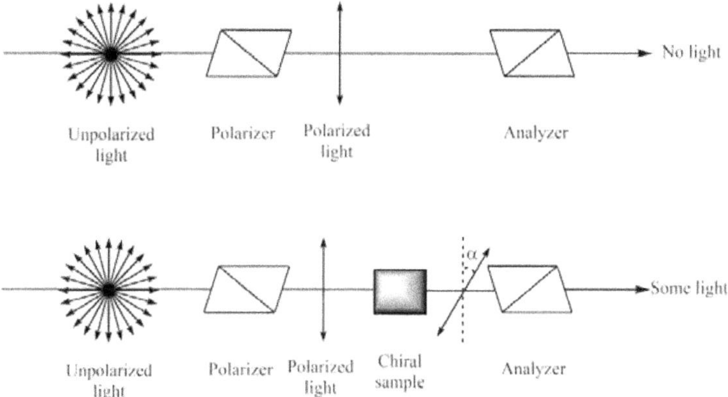

Figure 1. Schematic diagram of a polarimeter to measure chirality.

DALAL INSTITUTE

The angle through which the plane of polarized light gets rotated by the chiral molecules is symbolized by α and can be formulated as given below.

$$\alpha = [\alpha]_\lambda^T . l . c \tag{1}$$

Where l is the path length of the sample in cm whereas c represents the concentration of the sample in g/ml. The symbol represents the specific angle of rotation $[\alpha]_\lambda^T$ at temperature T and wavelength λ. It is also worthy to note that the observed angle of rotation also depends upon the nature of the solvent. Furthermore, the wavelength of the radiation used and the temperature of the system should be kept constant throughout the experiment. In order to determine the specific angle of rotation, use path length and concentration unity in the rearranged form of equation (1), i.e.,

$$[\alpha]_\lambda^T = \frac{\alpha}{l . c} \tag{2}$$

$$[\alpha]_\lambda^T = \frac{\alpha}{1 \, cm \times 1 \, g/ml} \tag{3}$$

or

$$[\alpha]_\lambda^T = \alpha \tag{4}$$

Therefore, the specific angle of rotation of any chiral molecule may simply be defined as the observed angle of rotation when the solution of the same compound with unit concentration and unit path length is placed in the path of polarized light.

> ➢ *Conditions for Chirality*

In order to be optically active, any one of the following conditions can be used to define the chirality of a molecule to somewhat less or more strictly.

Condition 1: Since All the crystals or molecular geometries that are capable of rotating the plane of polarized light could not be superimposed on their mirror images, the primary condition for a molecule to be optically active or chiral can be summarized as follows.

If a molecular geometry wants to be optically active, it must not be superimposable on its mirror image.

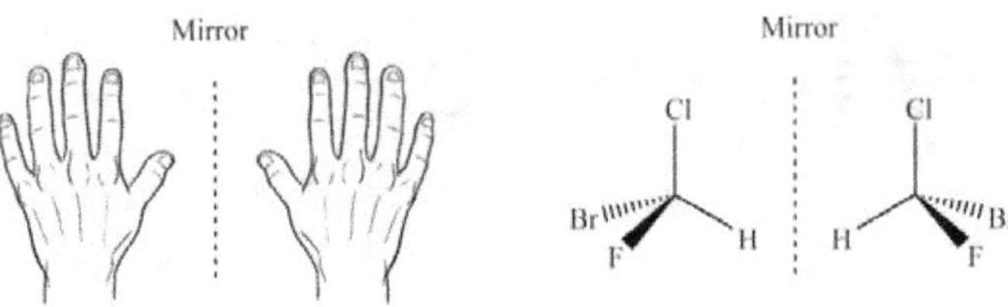

This condition defines chirality in an absolute sense and will be chiral if it is satisfied.

Condition 2: Now although the first condition is successful in defining chirality, its application is quite difficult. This is because it is very difficult to imagine the mirror image of the molecular geometry in three dimensions (especially complex molecules), and then the confirmation of their superimposition is even more difficult to tackle for human imagination. Therefore, an alternate route is necessary to check the first condition. This problem can be solved by the fact that any molecular geometry lacking plane of symmetry cannot be superimposed on its mirror image; and therefore, the problem of 'visualizing the mirror image and its subsequent superimposition' is reduced to finding the plane of symmetry only. If a plane of symmetry is present, the molecule would be superimposable of its mirror image, and hence, will be achiral. Conversely, if a plane of symmetry is absent, the molecule would not be superimposable of its mirror image, and hence, will be chiral. Hence, this condition of chirality can be summarized as follows;

If a crystal molecular geometry wants to be optically active (or chiral), there should be no plane of symmetry.

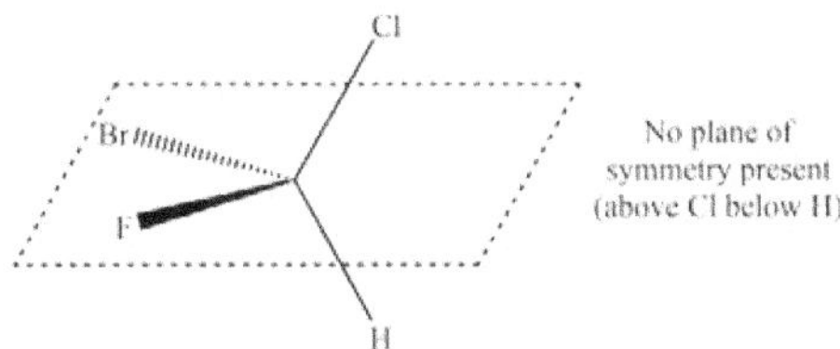

This condition defines chirality in part because there are molecules that don't have any plane of symmetry but still inactive.

Condition 3: Since there are molecules that don't have any plane of symmetry but still inactive, a more inclusive approach is needed to confirm the optical activity with ever treating mirror images. This problem can be solved by the fact that any molecular geometry lacking secondary symmetry elements (plane of symmetry, inversion center, and alternating axis of symmetry) can never be superimposed on its mirror image; and therefore, will be achiral. Hence, this condition of chirality can be summarized as follows;

If a crystal molecular geometry wants to be optically active (or chiral), there should secondary symmetry elements i.e., plane of symmetry (σ), center of symmetry (i), and alternating axis of symmetry (S_n).

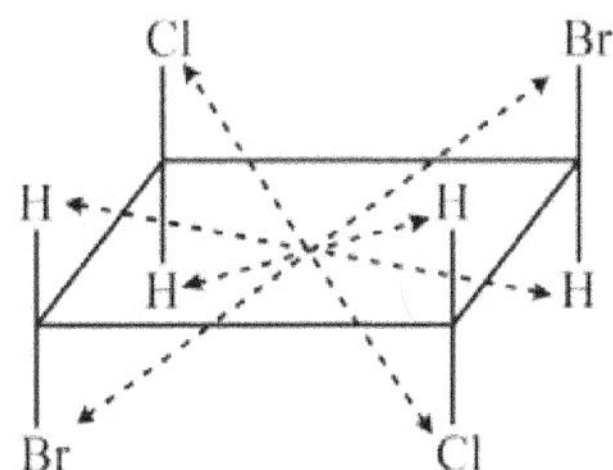

No plane of symmetry present
yet achiral due to inversion (i.e. *i* or S_2)

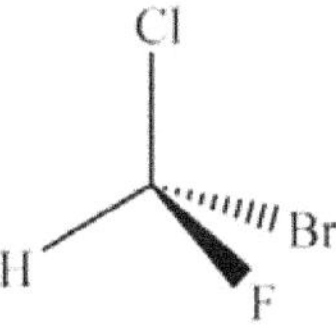

No secondary symmetry element (σ, *i*, S_n) present;
and therefore, chiral in nature.

This condition defines chirality in an absolute sense and will be chiral if it is satisfied.

> *Types of Chirality*

Depending upon the geometrical profile of molecular species, chiral compounds or chirality can primarily be divided into four categories as given below.

1. Chirality arising from a center (chiral center): This type of chirality arises when all the four groups around tetrahedrally coordinated carbon atom become different. In other words, an organic molecule can no longer be superimposed on its mirror image if it has a center with all different groups.

H₃C—Br, Cl, H   Et—OH, Me, H   Et—Br, Cl, H

1-bromo-1-chloroethane   butan-2-ol   1-bromo-1-chloropropane

Some of the most simple examples of organic molecules with this type of chirality are CH_3-CHClBr, butan-2-ol, and 1-bromo-1-chloropropane.

2. Chirality arising from an axis (chiral axis): This type of chirality arises when a tetrahedrally coordinated prochiral molecule becomes chiral by extending the center along an axis. In other words, a prochiral molecule can no longer be superimposed on its mirror image if its center has been extended to a line with same groups at different ends.

(R)-penta-2,3-diene

2,2'-dimethyl-1,1'-biphenyl

Some of the most simple examples of organic molecules with this type of chirality are penta-2,3-diene and 2,2'-dimethyl-1,1'-biphenyl (a biphenyl derivative).

3. Chirality arising from a plane (chiral plane): This type of chirality arises when relacing a group in a plane makes the molecule chiral. In other words, an organic molecule can no longer be superimposed on its mirror image if the replacement of a particular group induces chirality.

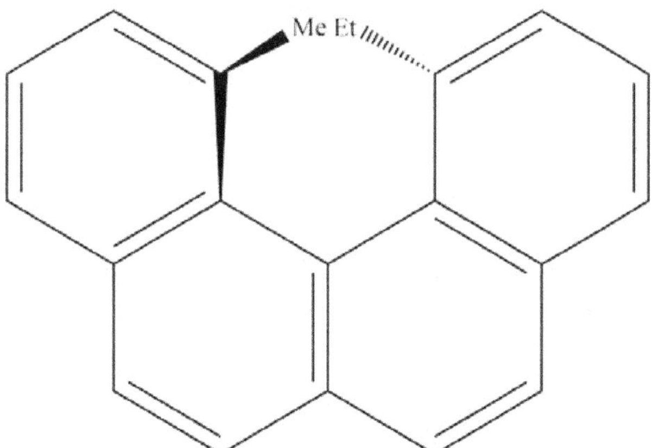

13-bromo-1,10-dioxa[8]paracyclophane

Some of the most simple examples of organic molecules with this type of chirality are ansa compounds like 13-bromo-1,10-dioxa[8]paracyclophane.

4. Chirality arising from a spiral (helical chirality): This type of chirality arises when the molecule has a helical structure. In other words, an organic molecule can no longer be superimposed on its mirror image if its geometry resembles a helix.

1-ethyl-12-methylbenzo[c]phenanthrene

Some of the most simple examples of organic molecules with this type of chirality are helical compounds like 1-ethyl-12-methylbenzo[c]phenanthrene.

❖ Elements of Symmetry

Elements of symmetry may simply be defined as the point, line or plane inside or passing through the molecular geometry about which some operations like rotation, inversion, or reflection generate indistinguishable images. These operations are generally labeled as symmetry operations. A general discussion on different symmetry elements is given below.

➤ *Axis of Rotation (C_n)*

The axis of rotation or simply the symmetry axis may simply be defined as the line passing through a molecular geometry about which the rotation through a certain angle generates indistinguishable images.

The axis of rotation is generally symbolized by C_n where n can have the value from 1, 2, 3, 4… and so on. The expression for *n* is

$$n = \frac{360°}{\theta} \tag{5}$$

Where θ is the minimum angle required to generate indistinguishable images. For instance, in the case of a regular trigone or BF_3 molecule, the geometry must be rotated through 120° minimum about the line perpendicular to the molecular plane to get indistinguishable images. Therefore, we can say that it is a C_3 (three-fold) axis of symmetry.

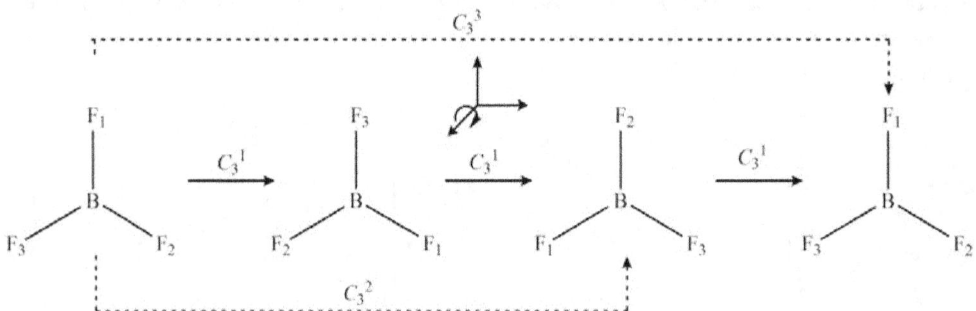

Where C_3^1, C_3^2 and C_3^3 are the symmetry operations via 120°, 240°, and 360°, respectively. Similarly, for the H_2O molecule, the minimum angle required to generate indistinguishable images is 180°, giving a C_2 axis of symmetry.

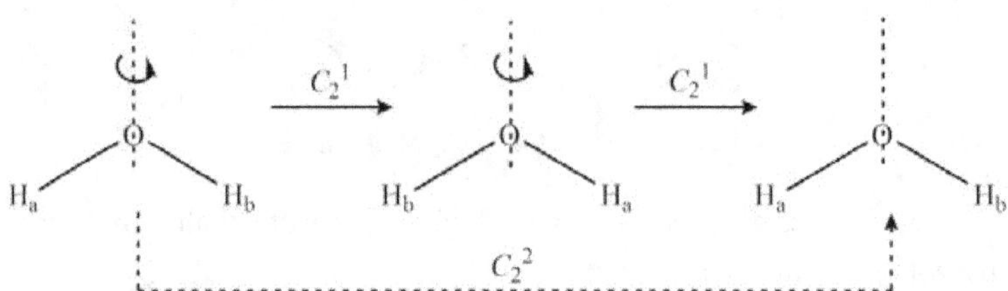

> ➢ *Plane of Symmetry (σ)*

The plane of symmetry or simply the symmetry plane may be defined as the plane bisecting the molecular geometry in such a way that one half is the mirror image of the other.

The axis of rotation is generally symbolized by σ_h or σ_v where the subscript h or v is to denote wheater the plane is parallel or perpendicular to the principal axis (symmetry axis of the highest order). There is also a third kind of plane of symmetry called the dihedral plane (σ_d): In other words, we can say that a dihedral plane bisects two σ_v planes. On a final note, a plane of symmetry can also be designated by the Cartesian orientation encompassing it, e.g., (*yz*-plane) or (*xz*-plane). For instance, there are two σ_v planes in water molecules as shown below.

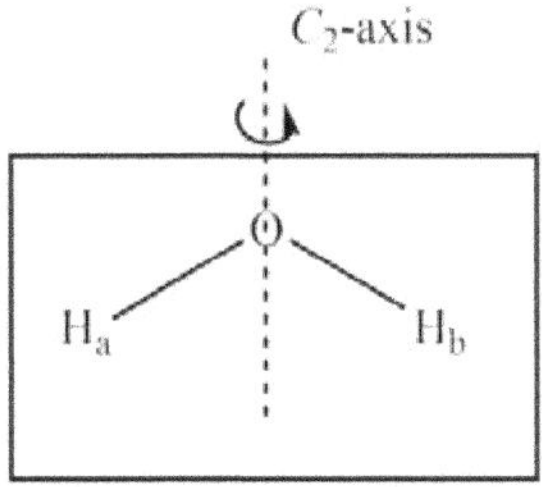
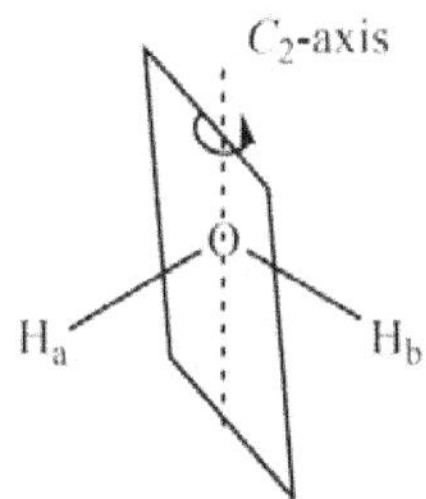

Similarly, there are a total of three vertical (σv) planes and one horizontal (σ_h) plane in the case of BF_3 molecules as shown below.

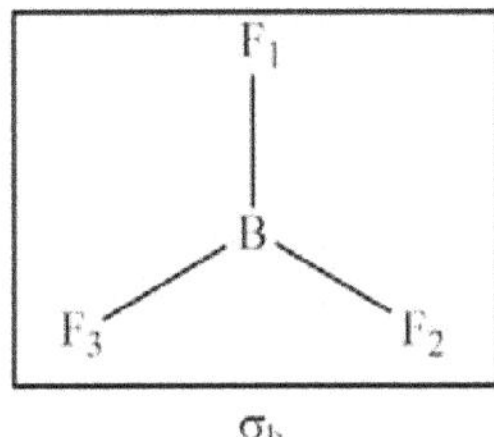
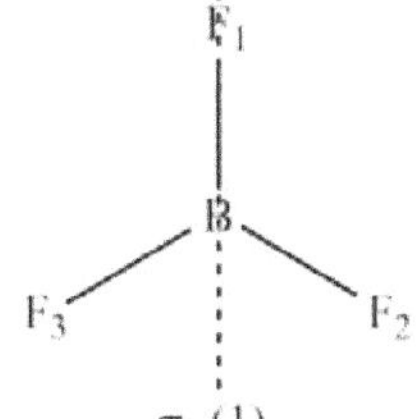
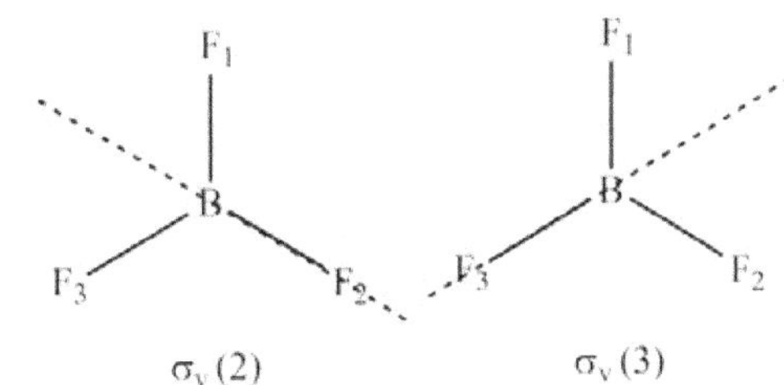

Similarly, there are a total of two vertical (σv) planes and one horizontal (σ_h) and two dihedral planes in the case of XeF_4 molecules as shown below.

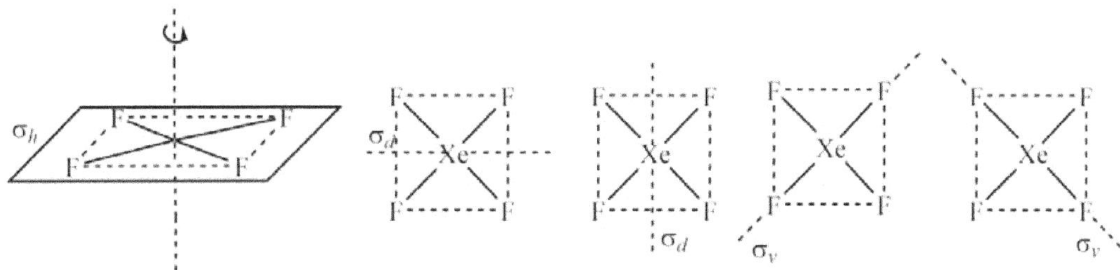

➤ *Center of Symmetry (i)*

The molecular geometry is said to possess the center of symmetry if a rotation through 180° followed by the perpendicular reflection generates an indistinguishable image.

The center of symmetry or simply the 'inversion center is denoted by the symbol '*i*', which is a point inside the geometry at such a position that if an object is inverted about this point, the position vector of any point in an object ⟨say x, y, z⟩ is also inverted ($-x, -y, -z$). For instance, consider the case of the SF_6 molecule.

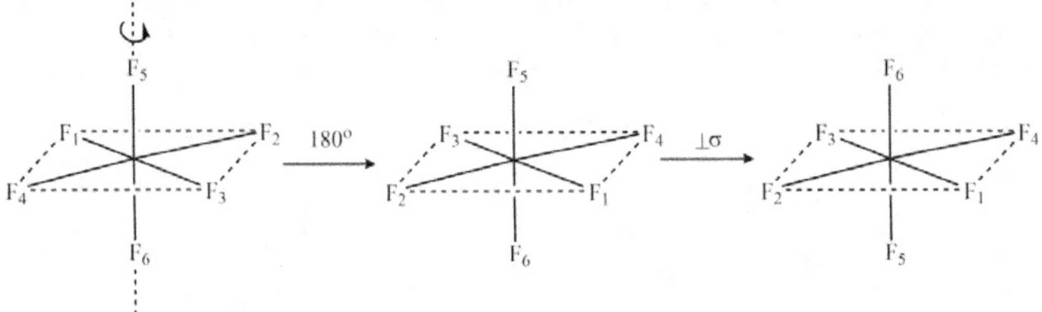

Similarly, the complete staggered form of CHClBr–CHClBr also possesses the center of symmetry as it produces indistinguishable images after inverting through the center. Furthermore, the center of symmetry in any molecular geometry can also found by drawing lines of equal length in the opposite direction from the center, provide that similar points are observed.

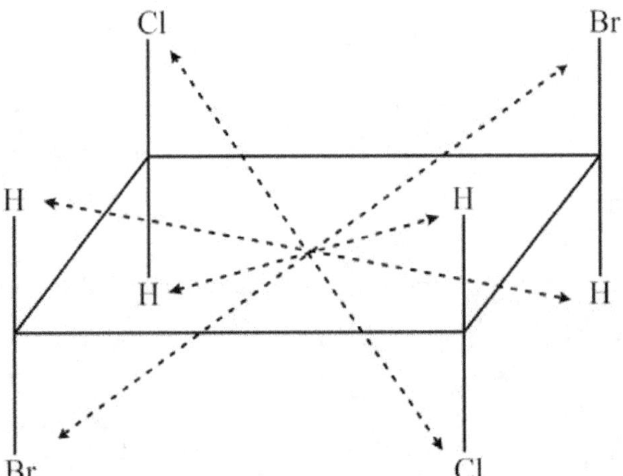

Similarly, other examples of molecules with the center of symmetry are acetylene, a staggered form of ethane and ethylene.

> *Alternating Axis of Symmetry (S_n)*

   The alternating axis of symmetry or improper axis of rotation may simply be defined as the line passing through a molecular geometry about which a rotation followed by a perpendicular reflection generates indistinguishable images.

   The improper axis of rotation is generally symbolized by S_n where n can have the value from 1, 2, 3, 4… and so on. The expression for n is

$$n = \frac{360°}{\theta} \tag{6}$$

Where θ is the minimum angle required before perpendicular reflection to generate indistinguishable images. For instance, in the case of a regular trigone or CH_4 molecule, the geometry must be rotated through 90° before the reflection in a perpendicular plane is carried out to get indistinguishable images. Therefore, we can say that it is an S_4 (four-fold) alternating axis of symmetry.

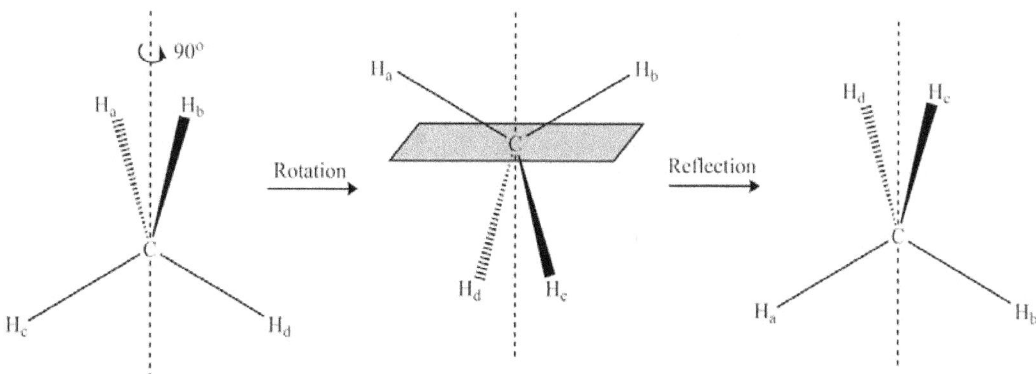

Similarly, for the BF_3 molecule, indistinguishable images can also be obtained by rotating the molecule through 120° about a line perpendicular to the molecular plane followed by the reflection. Therefore, we can say that it is an S_3 (three-fold) alternating axis of symmetry.

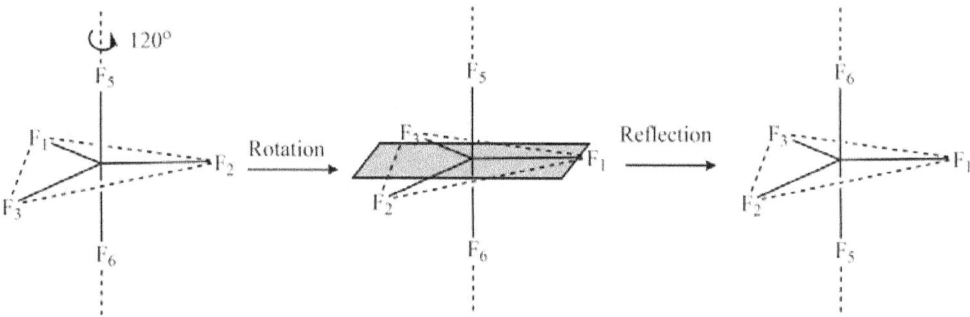

❖ Molecules with More Than One Chiral Centre: Diastereomerism

In most organic molecules, the optical activity or the chirality arises when all the four groups around an sp^3 hybridized carbon become different. Such carbon centers are typically called as chiral centers. Furthermore, it is also a quite well-known fact that the odd number of exchanges at a chiral center produce enantiomers. For instance, consider the case of CHClBrF.

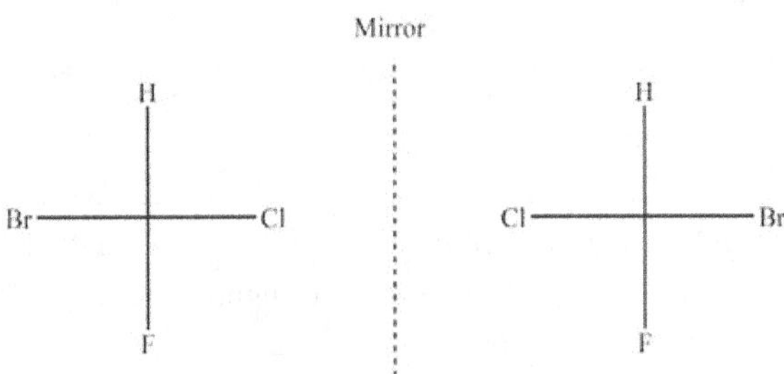

However, if the organic molecule contains more than one chiral center, different possibilities of odd and even exchanges at these stereocenters may generate identical, enantiomers, or diastereomers. Unlike enantiomers, which are non-superimposable mirror images; diastereomers are simply the non-mirror-image stereomers. For instance, consider the case of an organic molecule with two chiral centers unsymmetrical ends.

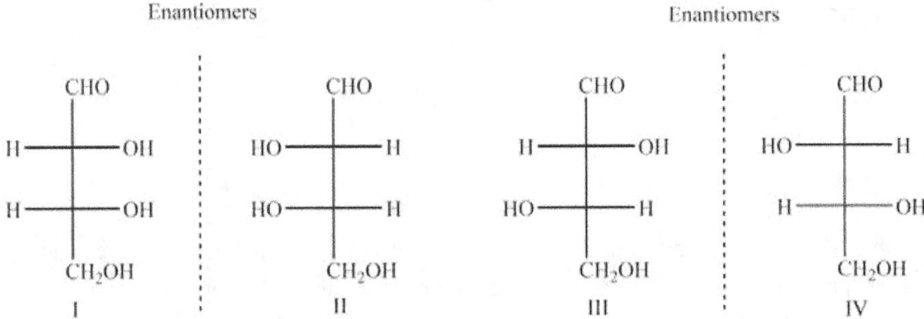

It is obvious that the odd number of exchanges at both chiral centers is creating non-superimposable mirror images (enantiomers); whereas the odd exchange at one chiral center and even exchange at other chiral centers is producing non-mirror-image isomers (diastereomers). It is also worthy to note that the even number of exchanges at both chiral centers will generate identical molecules just like in the case of molecules with a single chiral center. The calculation of the total number of optically active and meso forms for organic molecules with more than n chiral centers with unsymmetrical ends can be obtained by the following relations.

$$Number\ of\ optically\ active\ isomers = 2^n$$

$$Number\ of\ meso\ forms = 0$$

$$Total\ number\ of\ isomers = 2^n + 0$$

On the other hand, if we consider the case of an organic molecule with two chiral centers and symmetrical ends, the possibility of a meso compound also arises. The most popular example to prove the concept is tartaric acid.

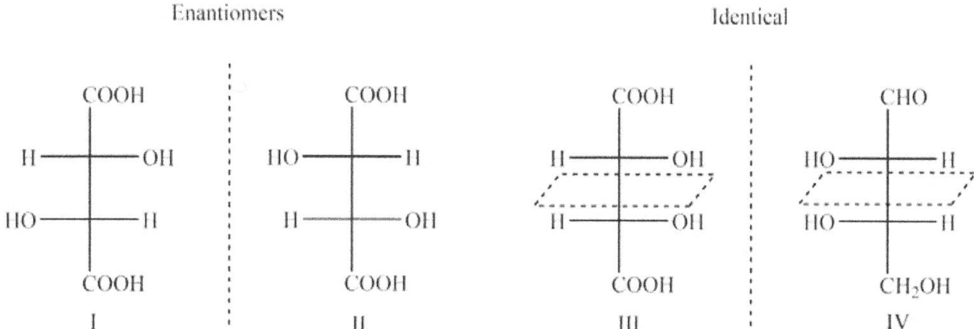

It is obvious that the odd number of exchanges at both chiral centers in L-tartaric acid is creating a non-superimposable mirror image (R-tartaric); whereas the odd exchange at one chiral center and even exchange at other chiral center is producing non-mirror-image isomers (diastereomers). However, structures III and IV are identical although they are produced by odd-odd exchanges at both chiral centers. The obviously because of the presence of the symmetry plane perpendicular to the carbon-carbon bond. Such compounds are called as meso form. It is also worthy to note that the even number of exchanges at both chiral centers will generate identical molecules just like in the case of molecules with a single chiral center.

The calculation of the total number of optically active and meso forms for organic molecules with more than n chiral centers with symmetrical ends can be obtained by using the following relations.

i) When the number of chiral centers is even:

$$Number\ of\ optically\ active\ isomers = 2^{(n-1)}$$

$$Number\ of\ meso\ forms = 2^{\left(\frac{n}{2}-1\right)}$$

$$Total\ number\ of\ isomers = 2^{(n-1)} + 2^{\left(\frac{n}{2}-1\right)}$$

ii) When the number of chiral centers is odd:

$$Number\ of\ optically\ active\ isomers = 2^{(n-1)} - 2^{\left(\frac{n}{2}-\frac{1}{2}\right)}$$

$$Number\ of\ meso\ forms = 2^{\left(\frac{n}{2}-\frac{1}{2}\right)}$$

$$Total\ number\ of\ isomers = 2^{(n-1)}$$

❖ Determination of Relative and Absolute Configuration (Octant Rule Excluded) with Special Reference to Lactic Acid, Alanine & Mandelic Acid

The three-dimensional character of organic molecules can be labeled absolutely or relative to some reference compound. In this section, we will study the relative and absolute configuration of various organic molecules with special reference to lactic acid, alanine & mandelic acid.

➢ *Relative Configuration*

The relative configuration of an organic stereoisomer may simply be defined as the correlation between two enantiomers even if the absolute configuration is unknown.

The elucidation of the absolute configuration of a chiral molecule was not possible before 1951 due to the absence of any such method. Therefore, the assignment of various groups in space was carried out relative to the groups of a standard reference compound.

D-(+)-glyceraldehyde   *L*-(–)-glyceraldehyde

One of the most popular reference compounds for the determination of stereochemical notation of chiral organic molecules is glyceraldehyde.

General route to assign D-L nomenclature: All the compounds that can be obtained or transformed to *D*-(+)-glyceraldehyde are said to belong to D-series, whereas all the compounds that can be obtained or transformed to *L*-(–)-glyceraldehyde are said to belong to L-series. These predictions find their base in the fact that if no bond breaking-formation occurs at the chiral center, the configuration is retained. For instance, consider the case of (–)-glyceric acid.

D-(+)-glyceraldehyde   *D*-(–)-glyceric acid

Since it can be obtained from D-(+)-glyceraldehyde, its name should be D-(–)-glyceric acid even though it is laevorotatory.

Determination of Relative Configuration Lactic Acid, Alanine & Mandelic Acid: The configuration of lactic acid, alanine & mandelic acid relative to glyceraldehyde (i.e., D-L configurations) can be obtained using the following chemical routes.

i) Determination of relative configuration of lactic acid:

As we know that lactic acid is optically active, and therefore, is bound to exist as enantiomeric pair. The configuration of one enantiomer relative to the glyceraldehyde can be obtained as given below.

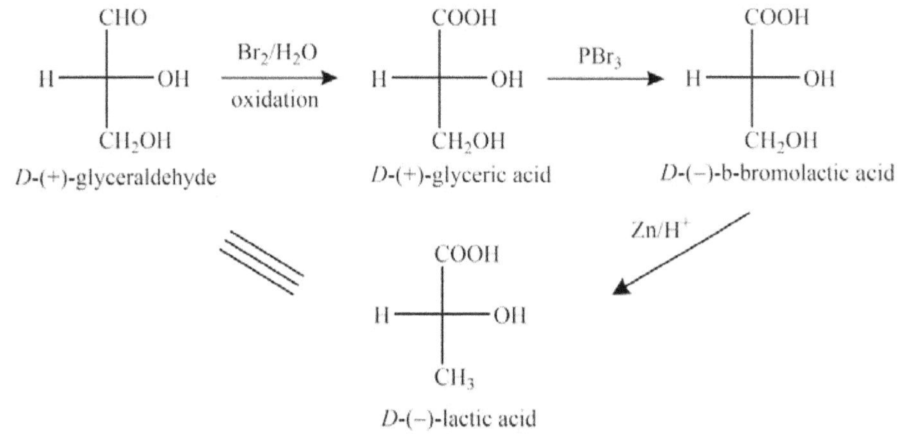

It is obvious that, unlike D-(+)-glyceraldehyde, the D configuration of lactic acid is found to be levorotatory.

Since lactic acid is optically active, the configuration of mirror image isomer relative to the glyceraldehyde can be obtained as given below.

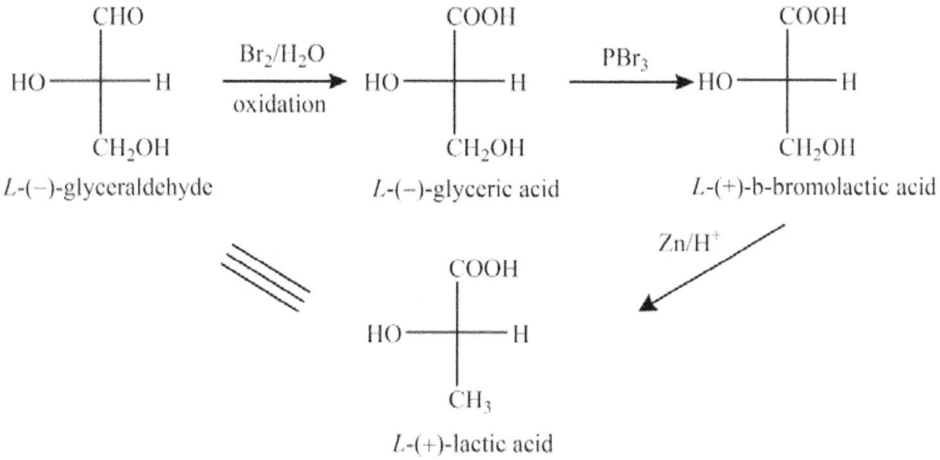

It is obvious that, unlike *L*-(−)-glyceraldehyde, the *L* configuration of lactic acid is found to be dextrorotatory.

ii) Determination of relative configuration of alanine:

As we know that alanine is optically active, and therefore, is bound to exist as enantiomeric pair. The configuration of one enantiomer relative to the glyceraldehyde can be obtained as given below.

$$
\begin{array}{ccccc}
\text{CHO} & \text{COOH} & \text{COOH} & & \text{COOH} \\
\text{HO}\!-\!\!\!-\!\text{H} \equiv & \text{HO}\!-\!\!\!-\!\text{H} \xleftarrow[\text{SN}_2]{\text{OH}^-} & \text{H}\!-\!\!\!-\!\text{Br} \xrightarrow[\text{SN}_2]{\text{N}_3^-} & \text{N}_3\!-\!\!\!-\!\text{H} \\
\text{CH}_2\text{OH} & \text{CH}_3 & \text{CH}_3 & & \text{CH}_3 \\
\textit{L-(–)-glyceraldehyde} & \textit{L-(+)-lactic acid} & \textit{L-(+)-2-bromo propanoic} & & \\
& & \text{acid} & &
\end{array}
$$

$$
\text{Pt/H}_2
$$

$$
\begin{array}{c}
\text{COOH} \\
\text{H}_2\text{N}\!-\!\!\!-\!\text{H} \\
\text{CH}_3 \\
\textit{L-(+)-Alanine}
\end{array}
$$

It is obvious that, unlike *L-(–)-glyceraldehyde*, the *L* configuration of alanine is found to be dextrorotatory.

Since alanine is optically active, the configuration of mirror image isomer relative to the glyceraldehyde can be obtained as given below.

$$
\begin{array}{ccccc}
\text{CHO} & \text{COOH} & \text{COOH} & & \text{COOH} \\
\text{H}\!-\!\!\!-\!\text{OH} \equiv & \text{H}\!-\!\!\!-\!\text{OH} \xleftarrow[\text{SN}_2]{\text{OH}^-} & \text{Br}\!-\!\!\!-\!\text{H} \xrightarrow[\text{SN}_2]{\text{N}_3^-} & \text{H}\!-\!\!\!-\!\text{N}_3 \\
\text{CH}_2\text{OH} & \text{CH}_3 & \text{CH}_3 & & \text{CH}_3 \\
\textit{D-(+)-glyceraldehyde} & \textit{D-(–)-lactic acid} & \textit{D-(–)-2-bromo propanoic} & & \\
& & \text{acid} & &
\end{array}
$$

$$
\text{Pt/H}_2
$$

$$
\begin{array}{c}
\text{COOH} \\
\text{H}\!-\!\!\!-\!\text{NH}_2 \\
\text{CH}_3 \\
\textit{D-(–)-Alanine}
\end{array}
$$

It is obvious that, unlike D-(+)-glyceraldehyde, the *D* configuration of alanine is found to be levorotatory.

iii) Determination of relative configuration of mandelic acid:

As we know that alanine is optically active, and therefore, is bound to exist as enantiomeric pair. The configuration of one enantiomer relative to the mandelic acid can be obtained as given below.

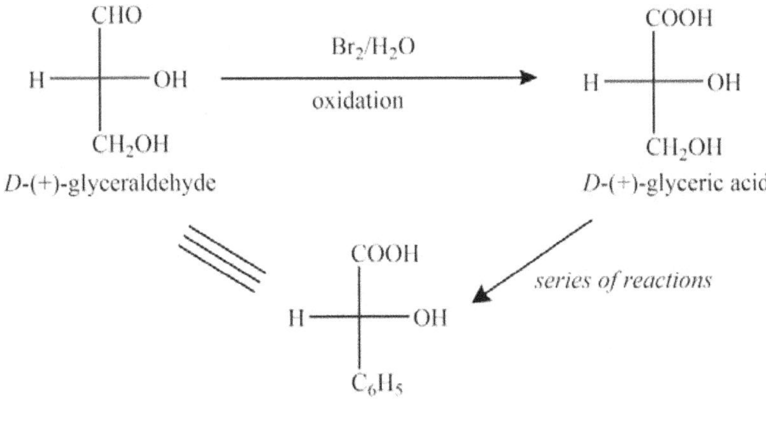

It is obvious that, unlike D-(+)-glyceraldehyde, the *D* configuration of mandelic acid is found to be levorotatory.

Since mandelic acid is optically active, the configuration of mirror image isomer relative to the glyceraldehyde can be obtained as given below.

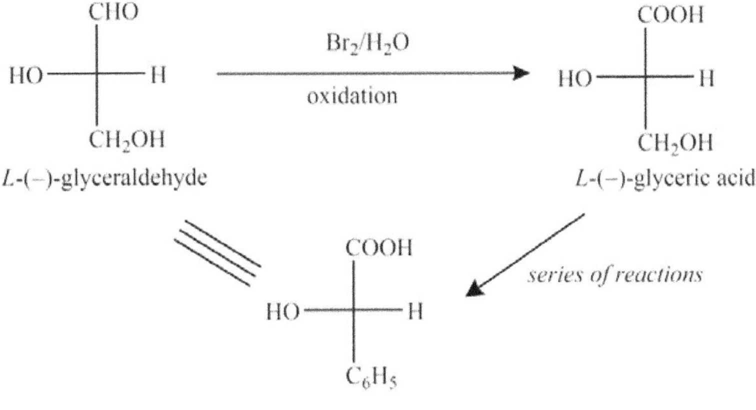

It is obvious that, unlike L-(−)-glyceraldehyde, the *L* configuration of mandelic acid is found to be dextrorotatory.

➤ *Absolute Configuration*

The absolute configuration of a particular stereoisomer of an organic molecule may simply be defined as the actual arrangement of atoms or groups in space.

Since the D-L system of stereoisomeric nomenclature was relative and difficult to apply, a more simplistic and robust approach was needed. The problem was solved by R. S. Chan, C. K. Ingold and V. Prelog by developing a nomenclature method that was free from limitations posed by the relative approach. This nomenclature technique is generally called as Chan-Ingold-Prelog or R-S system. The whole procedure includes two steps; the first is the priority assignment of different groups and the second step involves the assignment of absolute configuration.

Rules for priority assignment of different groups: A priority sequence of (1), (2), (3), and (4) is assigned to all the four groups attached to the chiral center using the following set of rules.

i) If the substituents attached with the chiral center are simply atoms, substituents with a higher atomic number of bonded atoms are given higher priority and vice-versa.

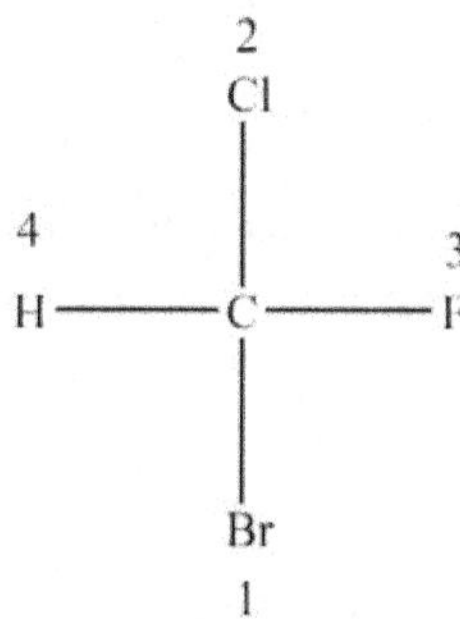

ii) If the bonded atoms of two groups are isotopes of the same element, a higher priority will be given to the group with a higher atomic mass.

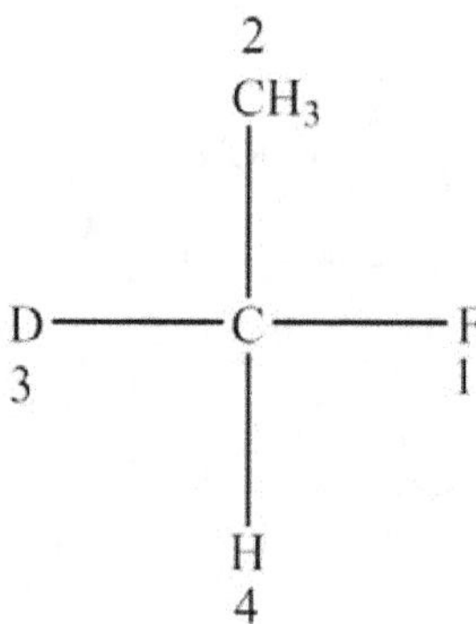

iii) Atoms next to the bonded atoms must be considered if the two above-mentioned rules are not able to distinguish the groups concerned.

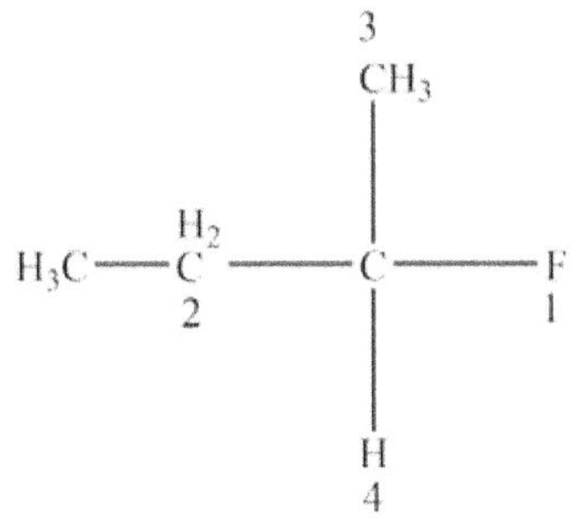

$$
\overset{\displaystyle \overset{3}{CH_3}}{\underset{\displaystyle \underset{4}{H}}{H_3C \overset{H_2}{\underset{2}{C}} - \underset{1}{C} - F}}
$$

iv) For priority assignment, multiple bonds must be considered as multiple numbers of single bonds, i.e., double and triple bonds must be treated as two single and three single bonds, respectively.

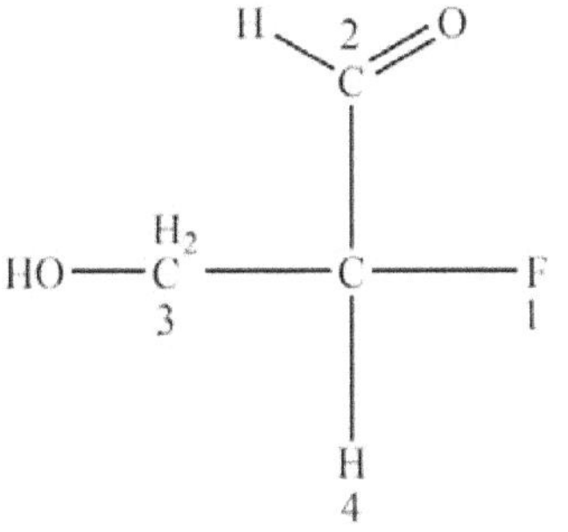

$$
\overset{\displaystyle \overset{2}{C}\underset{H}{\overset{O}{\diagup}}}{\underset{\displaystyle \underset{4}{H}}{HO - \overset{H_2}{\underset{3}{C}} - \underset{1}{C} - F}}
$$

Now using all these rules, let us try to determine the priority sequence of phenyl and acetylenic groups.

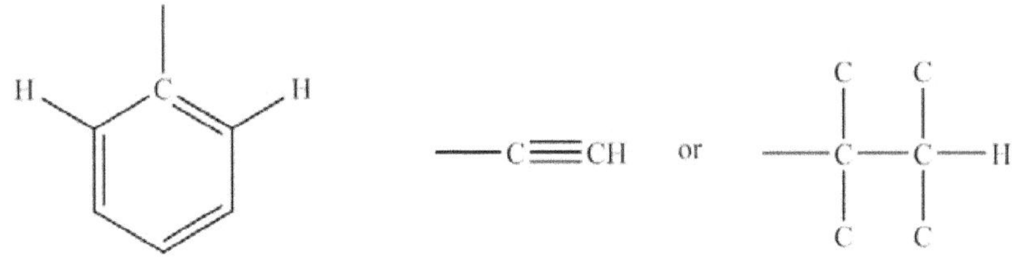

Phenyl group   Vinyl group

According to the sequence rule, phenyl and should be prioritized over acetylenic; which can be attributed to the fact that the carbon in the former pair is attached with three other carbons whereas in the latter pair it binds with only two carbons. However, only one carbon has successive bonds in acetylenic (2C and 1H); whereas in the phenyl group, two carbons have successive bonds (2C and 1H) allotting it a higher priority.

Assignment of absolute configuration: Since two of the most popular depiction of three-dimensional molecules on two-dimensional paper are Flying-Wedge and Fischer projection, we will study the determination of absolute configuration for both.

i) Assignment of absolute configuration in Flying-Wedge representation:

After assigning priorities to different groups the molecule is oriented in space such that the group of lowest priority goes away from the observer. Now if the tracking of decreasing priority of the remaining three groups comes gives rise to clockwise flight, the molecules should be labeled as R. However, if the tracking of decreasing priority of the remaining three groups comes gives rise to anticlockwise flight, the molecules should be labeled as S.

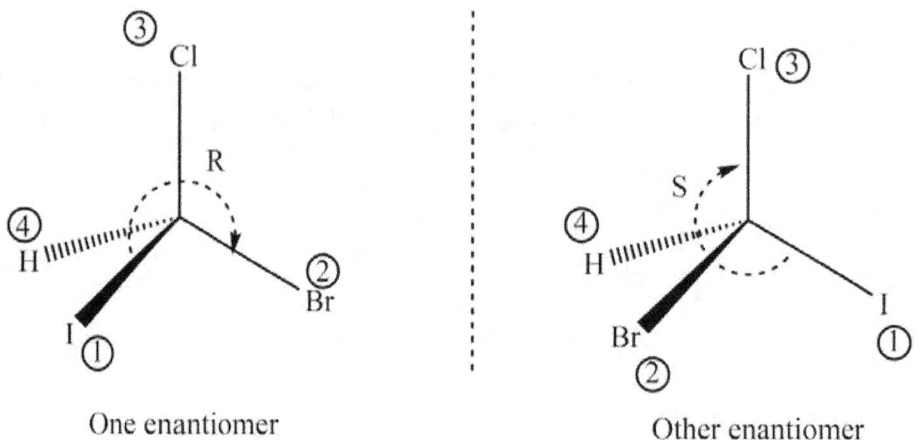

One enantiomer   Other enantiomer

Furthermore, if the group with the lowest priority is toward the observer or in the plane of the paper, carry out an even number of exchanges to through the lowest-priority-group away from the observer before the R-S labeling.

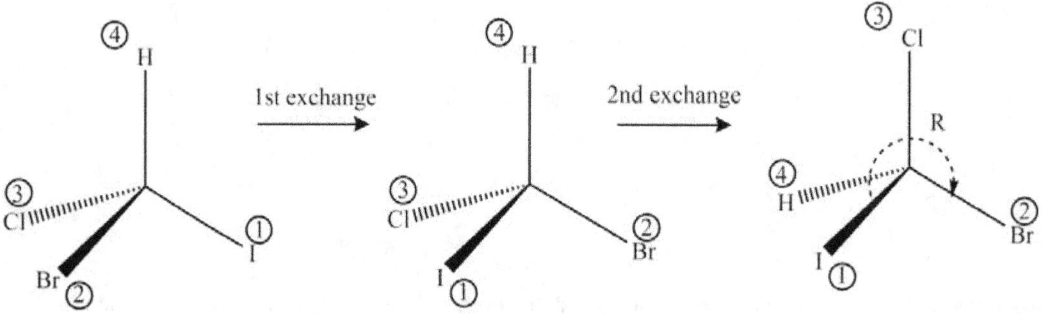

It is also worthy to note that the actual stereochemical notation can also be found by reverting the answer directly if the group with the lowest priority is toward the observer.

ii) Assignment of absolute configuration in Fischer representation:

After assigning priorities to different groups the Fischer projection of the molecule is transformed to an identical one by an even number of exchanges so that the group of lowest priority is at the vertical position. Now if the tracking of decreasing priority of the remaining three groups comes gives rise to clockwise flight, the molecules should be labeled as R. However, if the tracking of decreasing priority of the remaining three groups comes gives rise to anticlockwise flight, the molecules should be labeled as S.

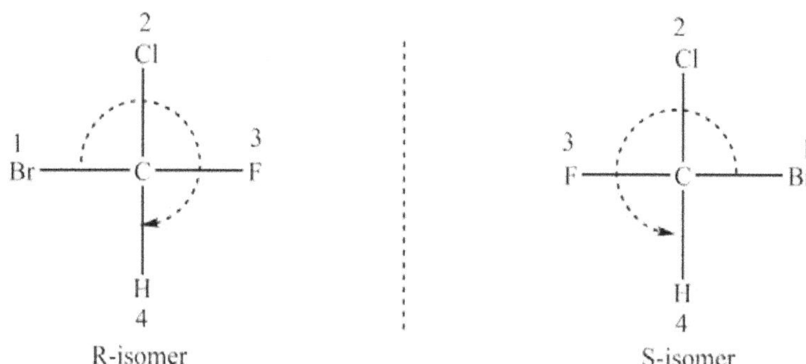

Furthermore, if the group with the lowest priority is already located at the vertical position, nothing is needed but priority tracking only.

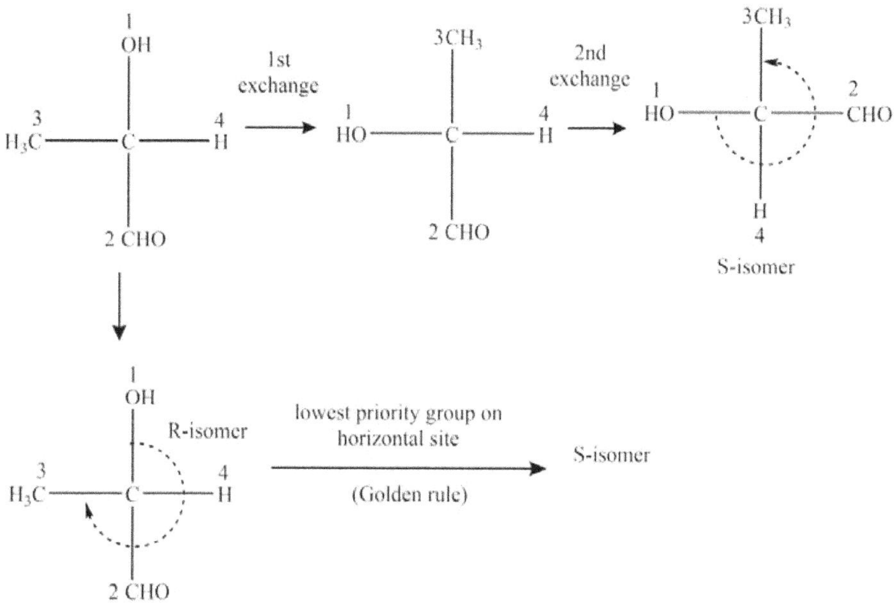

It is worthy to note that the actual stereochemical notation can also be found by reverting the answer directly if the group with the lowest priority is at the horizontal position, which is also called as the golden rule.

Determination of Absolute Configuration Lactic Acid, Alanine & Mandelic Acid: The absolute configuration of lactic acid, alanine & mandelic acid (i.e., R-S configurations) can be obtained using the following chemical routes.

i) Determination of absolute configuration of Lactic Acid:

As we know that lactic acid is optically active, and therefore, is bound to exist as enantiomeric pair. The absolute configuration of both enantiomers can be obtained as given below.

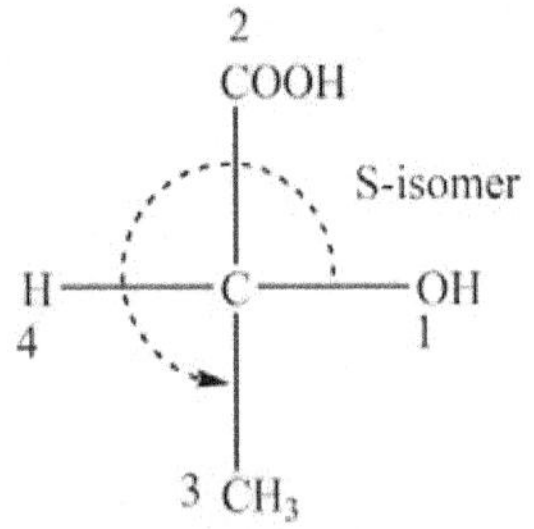 2 $COOH$ S-isomer H —— C —— OH 4 1 3 CH_3 (–)-Lactic acid Lowest priority group on horizontal side ↓ R-isomer	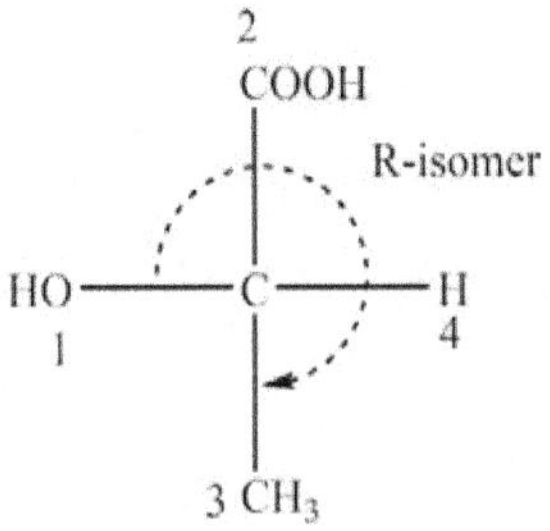 2 $COOH$ R-isomer HO —— C —— H 1 4 3 CH_3 (+)-Lactic acid Lowest priority group on horizontal side ↓ S-isomer

ii) Determination of absolute configuration of alanine:

As we know that alanine is optically active, and therefore, is bound to exist as enantiomeric pair. The absolute configuration of both enantiomers can be obtained as given below.

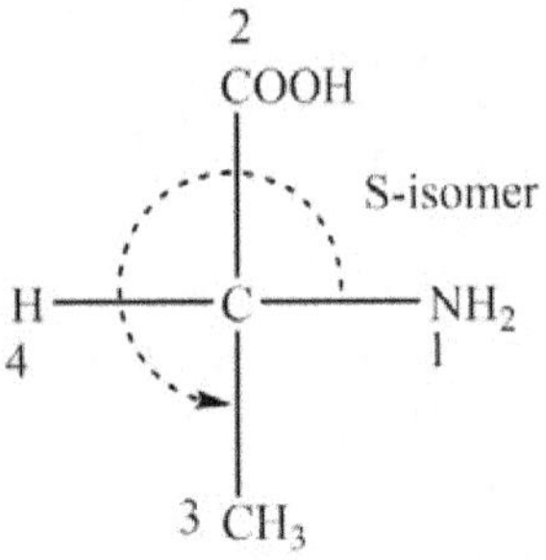 2 $COOH$ S-isomer H —— C —— NH_2 4 1 3 CH_3 (–)-Alanine Lowest priority group on horizontal side ↓ R-isomer	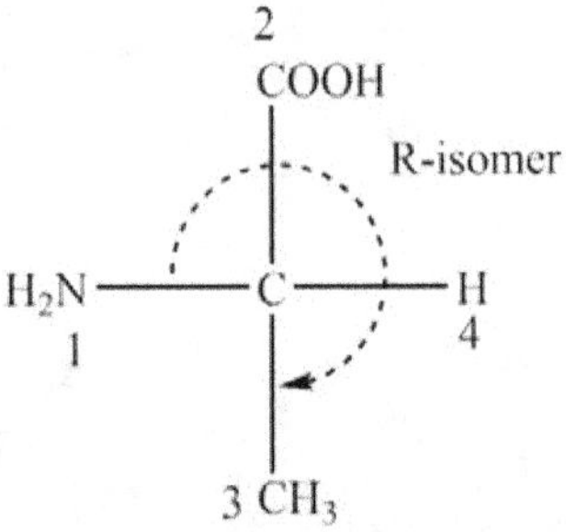 2 $COOH$ R-isomer H_2N —— C —— H 1 4 3 CH_3 (+)-Alanine Lowest priority group on horizontal side ↓ S-isomer

ii) Determination of absolute configuration of Mandelic acid:

As we know that mandelic acid is optically active, and therefore, is bound to exist as enantiomeric pair. The absolute configuration of both enantiomers can be obtained as given below.

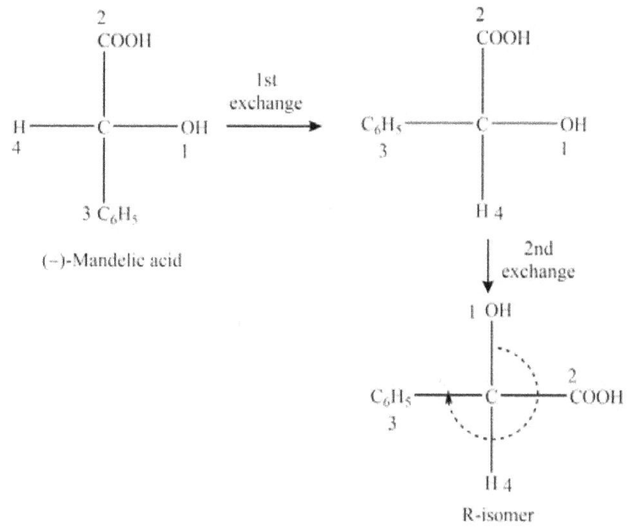

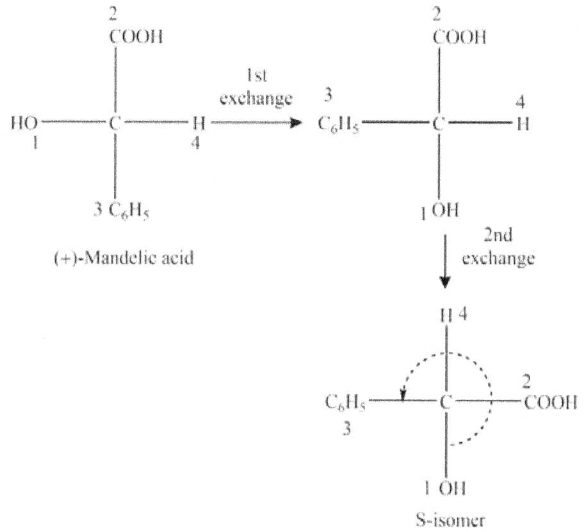

(−)-Mandelic acid → 1st exchange → 2nd exchange → R-isomer

(+)-Mandelic acid → 1st exchange → 2nd exchange → S-isomer

It is also worthy to note that *R*- and *S*-configuration are not bound to be dextro- or levorotatory in particular; in other words, *R*-configuration can be dextro-, as well as levorotatory. If R is dextro-, *S*-configuration will be levorotatory, and if *R* is levorotatory, *S*-configuration will be dextrorotatory.

❖ Methods of Resolution

If the amount of laevorotatory and dextrorotatory enantiomers of a chiral molecule are equal in a solution, it will be called as a racemic mixture or racemate. One of the first racemic mixtures known was racemic acid, which is a mixture of the two enantiomeric forms of tartaric acid. A solution with only one enantiomer is called an enantiomerically pure or simply the enantiopure compound.

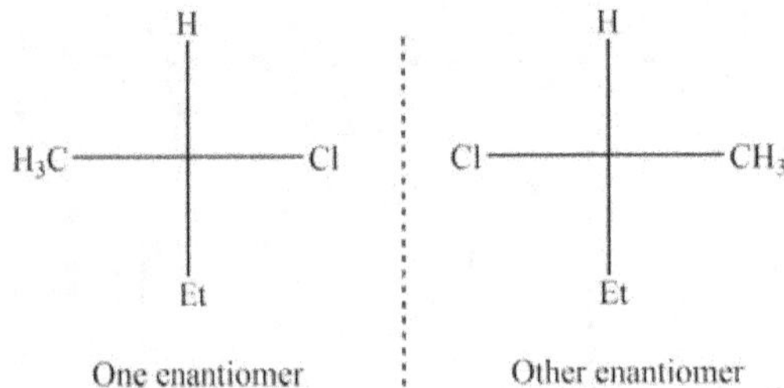

➢ *The Phenomenon of Racemization*

The phenomenon of racemization in most of the organic compounds can take place via the mechanisms given below.

1. By the rotation about carbon-carbon single bond: When an enantiomer of optically active biphenyl derivative is heated, some of it can easily be converted into another enantiomer just by the rotation about carbon-carbon single bond.

2. By the phenomenon of enolization: When an enantiomer of optically active biphenyl derivative is heated, some of it can easily convert into another enantiomer just by the rotation about carbon-carbon single bond.

3. By SN₁ mechanism: The racemic mixture can also be obtained by subjecting an enantiomer in a typical SN₁ attack with the same nucleophile.

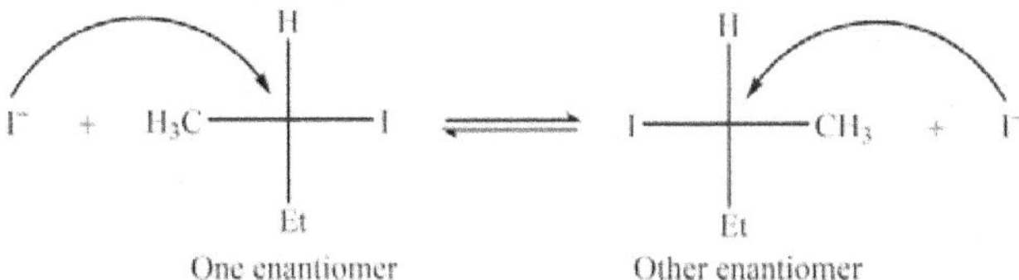

4. By SN₂ mechanism: The racemic mixture can also be obtained by subjecting an enantiomer in a typical SN₂ attack with the same nucleophile.

> ➢ *The Resolution of Racemic Mixture*

The process of getting individual enantiomers from a typical racemic mixture by any physical or chemical route is called as the resolution of racemic mixtures. Some methods of resolution are given below.

1. Mechanical separation: This is the most popular and easy method to resolve an enantiomeric mixture into it components. This method involves the hand-picking of single crystals of R and S enantiomers using tweezers and a magnifying glass.

2. Chemical method: We know that enantiomers have the same physical properties but different chemical properties towards chiral reagents and differences in physical properties must be used for the act of separation. Therefore, is very much favorable to convert the enantiomers into corresponding diastereomers, which in turn, can be converted into corresponding enantiomers by using the difference between the physical properties.

3. Biochemical extraction: It is quite a well-known fact that many enantiomers are quite consumable by certain types of bacteria; and therefore, only one enantiomer of the racemic mixture will be left behind if the same is subjected to such conditions. For instance, penicillium glaucum eats (+)-tartaric acid leaving behind (+)-tartaric acid only. Now although the method is quite easy to follow, it suffers from the drawback of the destruction of almost half of the compound.

4. Chromatographic separation: The individual enantiomers of a racemic mixture can also be separated by employing the route of "column chromatography' when the adsorbent-taken is an optically active compound. Now since only one enantiomer will get attached strongly to the adsorbent, the elution of the column will result in the earlier extraction of weakly bound enantiomer.

❖ Optical Purity

Optical purity or the enantiomeric excess (ee) may simply be defined as the purity measurement used for chiral compounds and reflects the extent to which a sample contains one enantiomer in greater amounts than another enantiomer.

The enantiomeric excess of the racemic mixture 0%, whereas a single completely pure enantiomer has an enantiomeric excess of 100%. A sample having 60% of one enantiomer and 40% of the other has an ee of 20% (60% − 40%). The general expression for optical purity may be given by the following relation.

$$Optical\ purity = \%\ ee = \frac{\alpha_{obs}}{\alpha_{max}} = \frac{[R] - [S]}{[R] + [S]} \times 100 \tag{7}$$

Where α_{obs} and α_{max} are the observed angle of rotation of plane-polarized light by the racemic mixture under consideration and maximum angle of rotation that it could rotate when R is replaced by S enantiomer and vice-versa, respectively. The symbol $[R]$ and $[S]$ are simply the percentage of R and S enantiomer, respectively.

Furthermore, the percentage of major and minor enantiomers can be obtained if the enantiomeric excess is known as given below.

$$\%\ of\ major\ enantiomer = \frac{100 + \%\ ee}{2} \tag{8}$$

Similarly

$$\%\ of\ minor\ enantiomer = \frac{100 - \%\ ee}{2} \tag{9}$$

In an ideal situation, each component's contribution to the total magnitude of optical rotation is directly proportional to the corresponding mole fraction, and therefore, the optical purity should be identical to the enantiomeric excess. This gives rise to the informal usage of the two terms as interchangeable, especially due to the fact that optical purity was the conventional route of measuring enantiomeric excess. Nevertheless, other methods like NMR spectroscopy and chiral column chromatography are now quite popular for measuring the amount of each enantiomer separately.

The success of asymmetric synthesis is also quantified using enantiomeric excess. In the case of diastereomers mixtures, analogous definitions (diastereomeric excess) are employed for measurement. The term 'enantiomeric excess' was presented by Morrison and Mosher in 1971 in their paper entitled "Asymmetric Organic Reactions", indicating its historic ties with optical rotation. It has been proposed that the concept of 'enantiomeric excess' should be changed by that of 'enantiomeric ratio' which means S:R or S/R because the determination of optical purity has already been replaced by other experimental techniques which directly measure R and S for simplistic mathematical treatments. The same can be said for replacing the 'diastereomeric excess by 'diastereomeric ratio; though it seems very far from practice.

❖ Prochirality

The prochirality in stereochemistry may simply be defined as the property of a molecule by which can be converted from achiral to a chiral entity in a single step, and such molecules are called as prochiral molecules.

This can be understood by taking the example of propanoic acid where two identical substituents are attached to an sp^3-hybridized carbon atom, and the pro-R and pro-S descriptors are used to differentiate between the two.

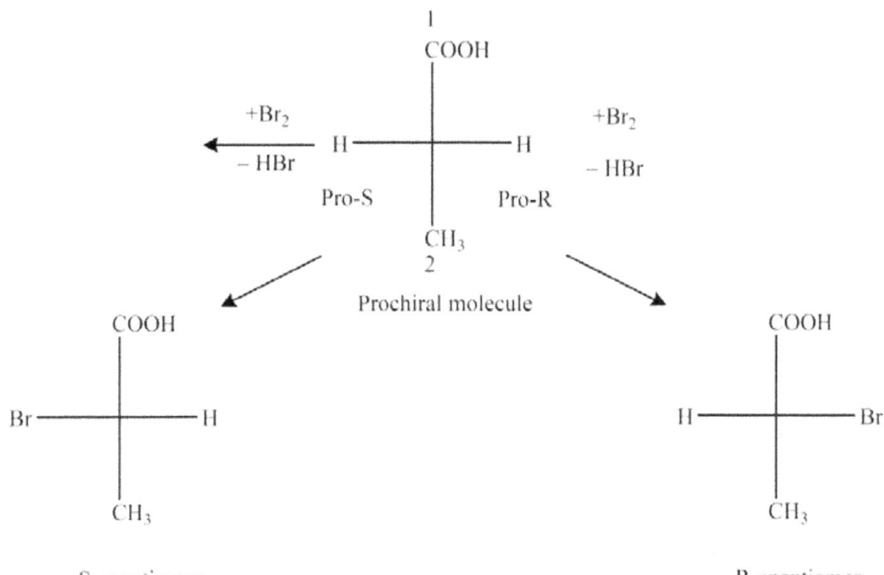

In other words, if we promote the pro-R substituent to a higher priority than the other identical substituent, we will get an R chirality center at the sp^3-hybridized carbon, and vice-versa is also true.

An *sp²*-hybridized carbon atom with trigonal planar coordination can also be converted to a chiral center if a group is attached to the '*re*' or '*si*' face of the organic molecule under consideration. For instance, imagine the case of benzaldehyde where the attack from the front and rear sides results in an enantiomeric pair.

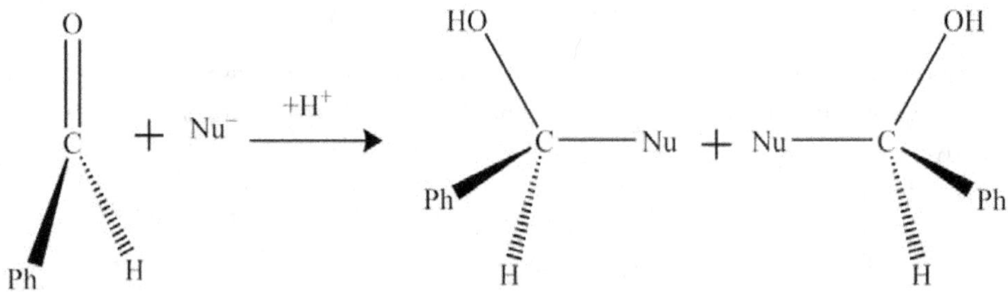

Enantiomeric Pair

The face will be labeled '*re*' if the substituents priority decreases in clockwise order at the trigonal atom when looking at that face; the face will be labeled '*si*' if the substituents priority decreases in anticlockwise order at the trigonal atom when looking at that face. Also, the designation of the resulting optically active carbon as S or R is a function of the priority of the incoming substituent.

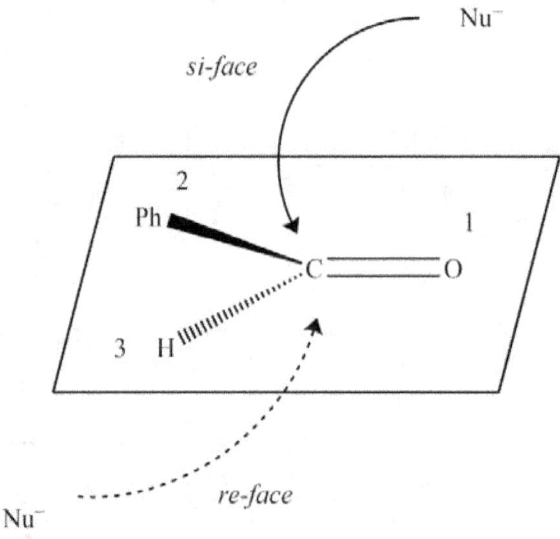

Furthermore, if an achiral species can be converted to a chiral one in two steps, it will be called a proprochiral. converted to a chiral one in two steps, it will be called a proprochiral.

❖ Enantiotopic and Diastereotopic Atoms, Groups and Faces

The topicity in stereochemistry may simply be defined as the stereochemical relationship between substituents and the structure with which they are bonded. Based on such relationships, groups can be classified as homotopic, enantiotopic, or diastereotopic. A general discussion on these three types is given below.

➢ *Homotopic Groups and Faces*

Homotopic groups and faces in an organic molecule can be found either based on chemical replacement or by using the symmetry criteria.

1. Homotopic groups: Homotopic groups in an organic molecule are equivalent groups. Two groups A and B are said to be homotopic if the resulting molecule remains the same (counting stereochemical notation also) when the groups are replaced with some other atom or group (such as bromine) whilst the remaining portion of the molecule is kept intact. Homotopic atoms or groups are always identical whether the environment is chiral or not. Also, NMR-active homotopic groups have the same chemical shift in NMR spectra. For instance, the four H of CH_4 methane are homotopic, like two H or the two Cl groups in CH_2Cl_2).

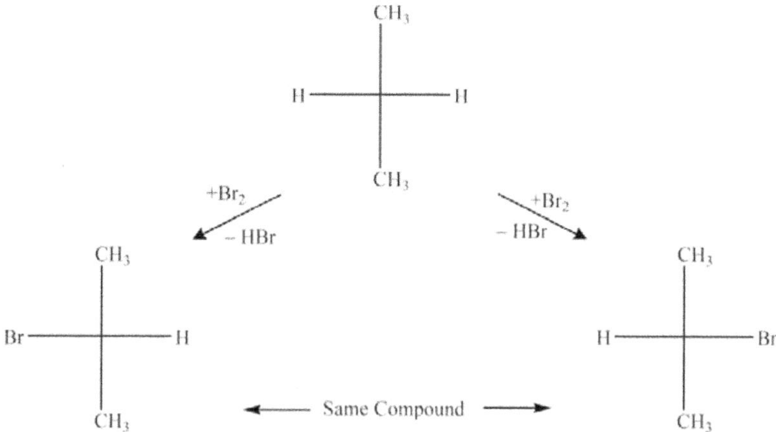

On the basis of molecular symmetry, two groups are said to be homotopic in nature if they are exchangeable by a primary symmetry element i.e. proper axis of rotation (C_n).

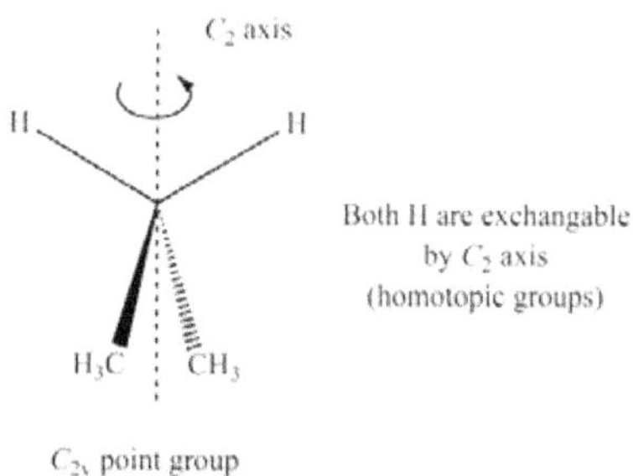

2. Homotopic faces: Homotopic faces in an organic molecule are equivalent faces, i.e., two faces A and B are homotopic if the molecule remains the same (including stereochemically) when the faces are attacked with some reagent (such as Cl^-) while the remaining parts of the molecule stay fixed. Homotopic faces are always identical, in any environment. For instance, two faces of methyl carbocation are homotopic, as the attack on two faces by an incoming nucleophile generates the same products.

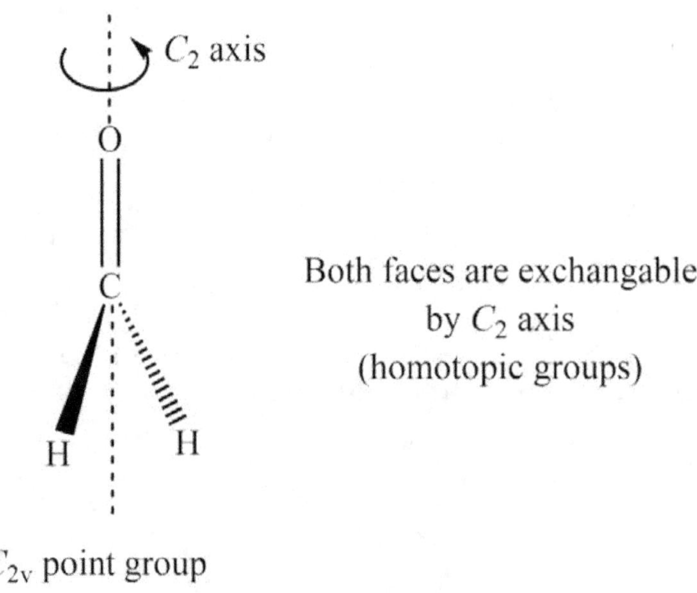

Same compound

On the basis of molecular symmetry, two faces are said to be homotopic in nature if they are exchangeable by a two-fold axis symmetry (C_2).

> *Enantiotopic Groups and Faces*

Homotopic groups and faces in an organic molecule can be found either on the basis of chemical replacement or by using the symmetry criteria.

1. Enantiotopic groups: Enantiotopic groups in a chemical compound are non-equivalent groups. Two groups A and B are enantiotopic if the enantiomers are obtained when the groups are interchanged with some other atom (such as bromine) while the remaining parts of the molecule stay fixed. Enantiotopic atoms are always identical in any achiral media while different in chiral media. Enantiotopic NMR-active nuclei have the same chemical shift in an NMR spectrum in achiral media and different chemical shifts in achiral media. For example, the two hydrogen atoms in CH_2ClBr are enantiotopic with one another, as the replacement by a third group gives rise to the enantiomeric pair.

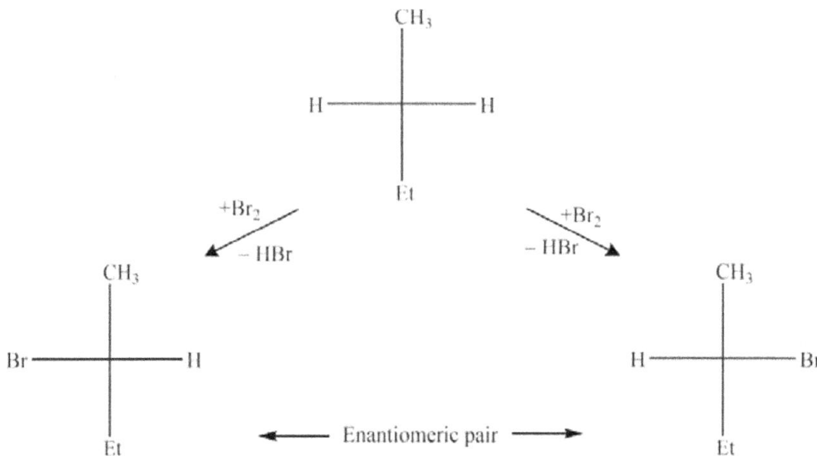

On the basis of molecular symmetry, two groups are said to be enantiotopic in nature if they are exchangeable by a secondary symmetry element i.e. plane of symmetry (σ), the center of symmetry (i), or alternating axis of symmetry (S_n).

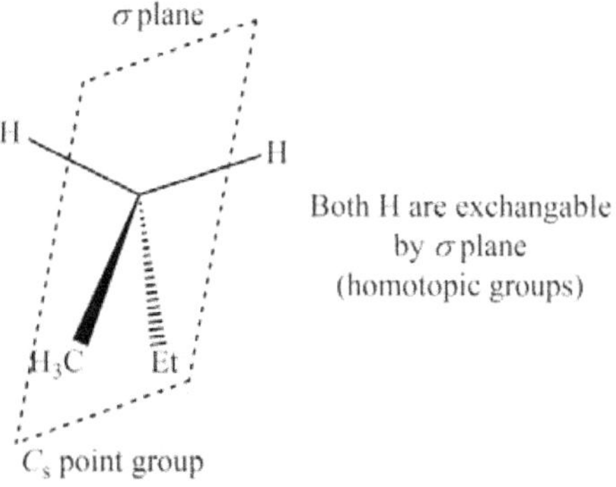

2. Enantiotopic faces: Enantiotopic faces in a chemical compound are non-equivalent faces, i.e., two faces A and B are enantiotopic if the molecule gives rise to enantiomeric pair when the faces are attacked with some reagent (such as Cl^-) while the remaining parts of the molecule stay fixed. Enantiotopic faces are always identical in the achiral environment and different in chiral media. For instance, two faces of primary carbocation are homotopic, as the attack on two faces by an incoming nucleophile generates enantiomeric pair.

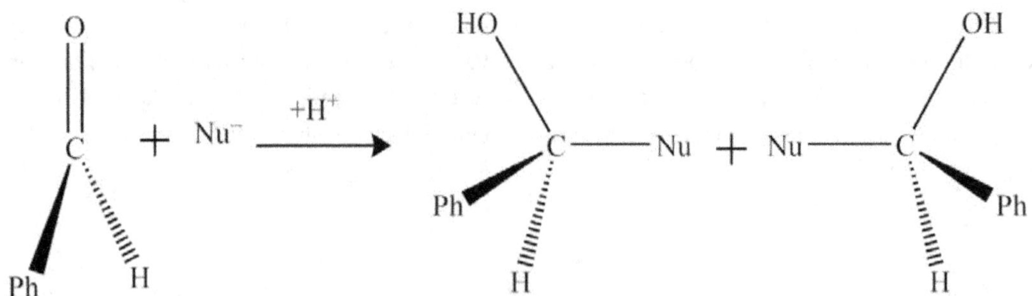

Enantiomeric pair

On the basis of molecular symmetry, two faces are said to be enantiomeric in nature if they are exchangeable by a secondary symmetry element i.e. plane of symmetry (σ), the center of symmetry (i), or alternating axis of symmetry (S_n).

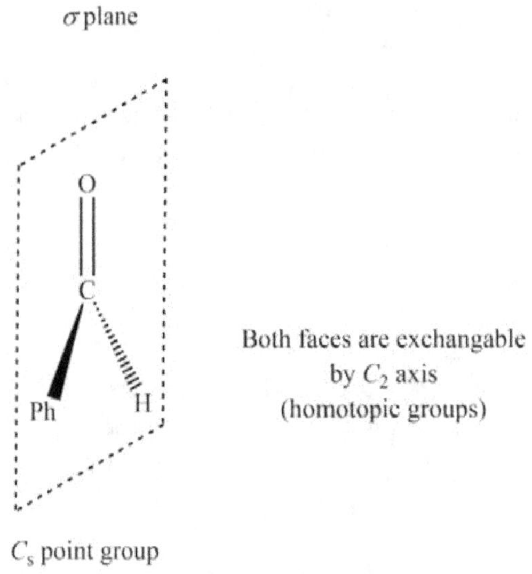

> ### *Diastereotopic Groups and Faces*

Diastereotopic groups and faces in an organic molecule can be found either on the basis of chemical replacement or by using the symmetry criteria.

1. Diastereotopic groups: Diastereotopic groups in a chemical compound are non-equivalent groups. Two groups A and B are diastereotopic if the diastereomers are obtained when the groups are interchanged with some other atom (such as bromine) while the remaining parts of the molecule stay fixed. Diastereotopic atoms are always different in any type of media whether it is chiral or achiral. Diastereotopic NMR-active nuclei have different chemical shifts in an NMR spectrum in any medium and different chemical shifts in achiral media. For example, the two hydrogen atoms in $(Et)(OH)(Me)C–CH_2ClBr$ are $CH2ClBr$ with one another, as the replacement by a third group gives rise to the diastereomeric pair.

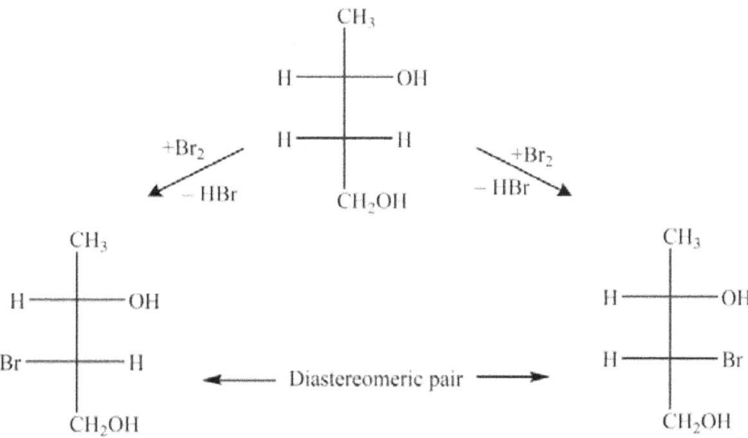

Based on molecular symmetry, two groups are said to be diastereotopic if they are not exchangeable by any symmetry element whether it is primary or secondary.

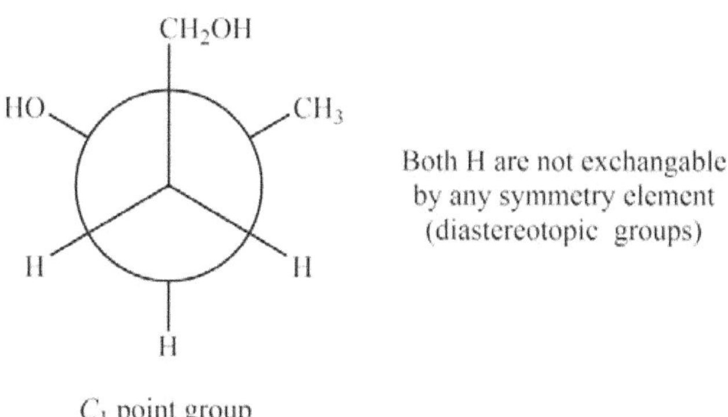

C_1 point group

2. Diasterotopic faces: Diastereotopic faces in a chemical compound are non-equivalent faces, i.e., two faces A and B are diastereotopic if the molecule gives rise to diastereotopic pair when the faces are attacked with some reagent (such as Cl^-) while the remaining parts of the molecule stay fixed. Diastereotopic faces are always non-identical in any type of environment whether it is chiral or achiral. For instance, two faces of 2-methylbutanal are diastereotopic, as the attack on two faces by an incoming nucleophile generates diastereomeric pair.

2-methylbutanal

$+ \ Nu^- \xrightarrow{+H^+}$

Diastereomeric pair

On the basis of molecular symmetry, two faces are said to be diastereotopic in nature if they are not exchangeable by any symmetry element, i.e. neither by primary nor by secondary symmetry element.

Both faces are not exchangable
by any symmetry element
(diastereotopic faces

C_1 point group

❖ Asymmetric Synthesis: Cram's Rule and Its Modifications, Prelog's Rule

Asymmetric synthesis or the stereoselective synthesis may simply be defined as the chemical synthesis in which one stereoisomer (enantiomer or diastereomer) is formed more predominantly than the other one, i.e. unequal amounts.

The phenomenon of stereoselectivity can primarily be classified into two categories; one as the enantioselective reactions and the other as the diastereoselective reactions. In enantioselective reactions, one enantiomer is formed in more amount than the other; whereas, in the case of diastereoselective reactions, the formation of one diastereomer dominates the other.

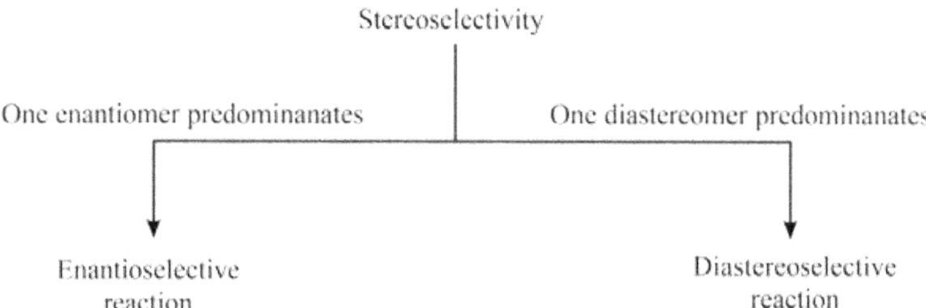

Now although the concept is quite wide and important, here we will only discuss the fundamental aspects of both types of stereoselective routes.

➢ *Diastereoselective Synthesis*

The diastereoselective synthesis may simply be defined as the chemical synthesis in which one diastereomer is formed more predominantly than the other one.

The diastereomer that will form in a higher amount can be predicted by Cram's rule in chiral ketones and by Prelog's rule in the case of chiral β-keto-esters.

1. Cram's Rule: The Crams rule states that when the attacking group approaches the trigonal face of the double bond in a chiral ketone, it will prefer the side of the plane with a smaller group on the asymmetric carbon.

Furthermore, it is also worthy to note that the reactive conformation will have the carbonyl group oriented between the medium and large group.

The modeling of the transition state of diastereoselective synthesis was also proposed by H. Felkin and N. Prudent and was different than the transition state given by the Cram as shown below.

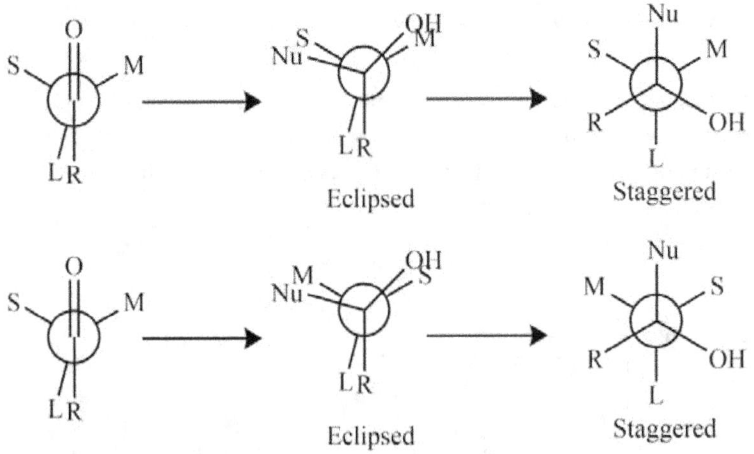

2. Prelog's Rule: The extension of Cram's rule to rationalize the formation of unequal amounts of two diastereomers in the case of chiral α-keto-esters is called as Prelog's rule. This rule states that when the attacking group approaches the trigonal face of the double bond in a chiral α-keto-ester, it will prefer the side of the plane with a smaller group on the asymmetric carbon.

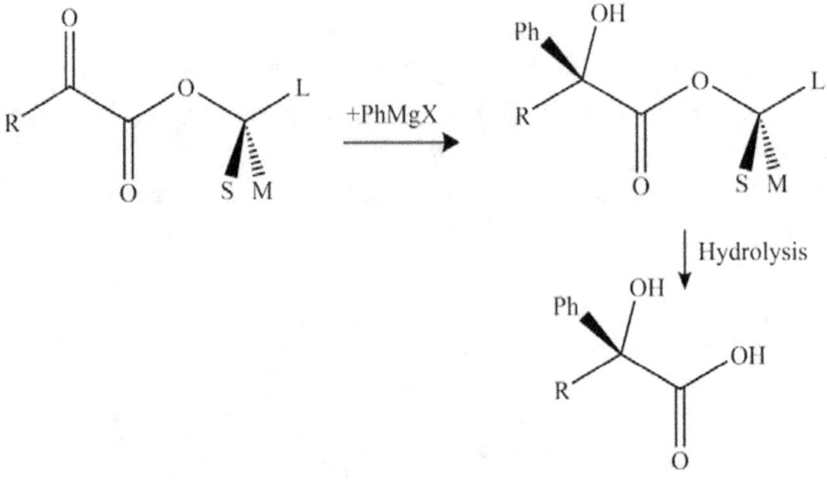

Furthermore, it is also worthy to note that the reactive conformation will have the carbonyl group oriented between the medium and large group.

> ### Enantioselective Synthesis

The enantioselective synthesis may simply be defined as the chemical synthesis in which one enantiomer is formed more predominantly than the other one.

The enantiomer that will form in a higher amount can be predicted by the attacking group itself. One of the most common examples to demonstrate the concept of enantioselective reactions is Sharpless asymmetric epoxidation in which allyl alcohols are converted into epoxy alcohols via a complex cycle.

The typical catalytic cycle of the Sharpless asymmetric epoxidation is shown below.

❖ Conformational Analysis of Cycloalkanes (Upto Six Membered Rings)

Most of the cyclic compounds can be categorized on the basis of the number of atoms participating in the cyclic skeleton and their special properties. Rings with 3 or 4 atoms are quite rigid and extremely strained because of their large deviation from the normal tetrahedral angle (109°28') which gives rise to a very high angle strain and the presence of a very large magnitude of torsional strain as well.

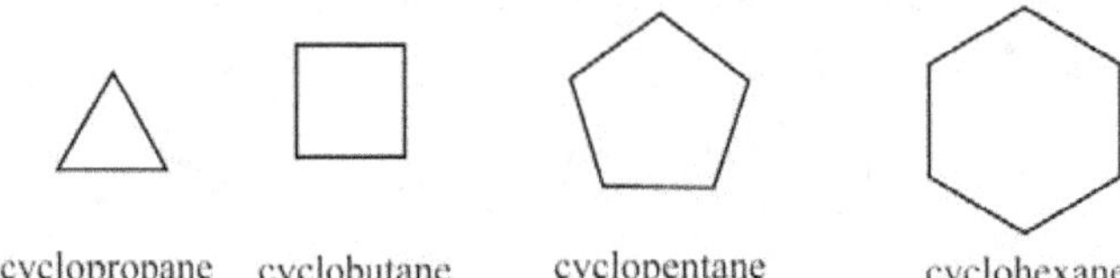

cyclopropane   cyclobutane   cyclopentane   cyclohexane

Most common rings have 5–7 members and can be characterized by the normal tetrahedral angles and groups oriented outward from the cycle. Rings with 8–11 members are labeled as medium rings and can be characterized by transannular interactions which is a strain produced by groups pointing inward from the cycle. Furthermore, if the number of atoms is equals to or greater than 12 (large rings), very low strain exists and such rings are pretty much comparable with the corresponding acyclic hydrocarbons. In this section, we will discuss the conformational analysis of cycloalkanes up to six-membered rings.

➤ *Conformations of Cyclopropane*

Cyclopropane is a kind of cycloalkanes with C_3H_6 as the molecular formula and is consisted of three carbon atoms that are connected with each other to give a cyclic structure. Each carbon atom is also bound with two hydrogens symmetrically so that the point group becomes D_{3h}. There is a huge ring strain in the cyclopropane due to its smaller size.

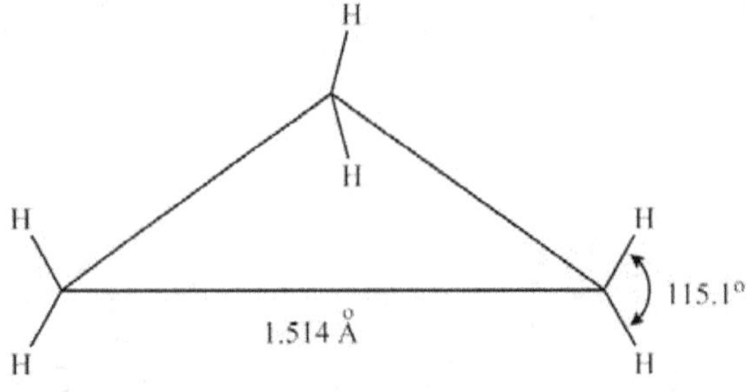

cyclopropane

The trigonal structure of cyclopropane needs the carbon-carbon bond angles to be at 60°; however, it is much smaller than the ideal bond angle of 109.5° (most stable thermodynamically and is generated for bonds with sp^3 hybridized orbitals) and gives rise to the large magnitude of ring strain.

The cyclopropane molecule also possesses a very high torsional strain arising from the eclipsed conformation of its H atoms. Per se, the bonds between the C atoms are significantly weaker than in an archetypal alkane, giving rise to much higher reactivity in this case. Also, the nature of bonding between the carbon atoms is usually described in terms of bent or banana bonds, where the carbon-carbon bonds are bent outwards so that the inter-orbital angle becomes equal to 104°.

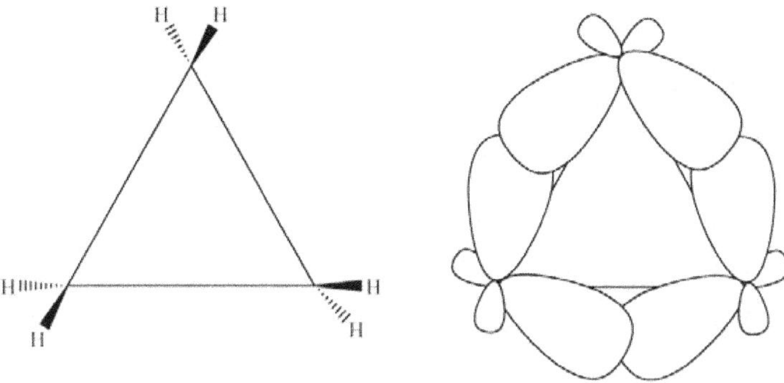

Banana bonds in cyclopropane

All this decreases the magnitude of bond strain and is obtained by distorting the sp^3 hybridized orbitals of carbon atoms to theoretically an sp^5 hybridized orbitals (i.e. 1/6 contribution of s orbital and 5/6 contribution of p orbital) so as to make C-C bonds have more π-character than usual (and giving s-character same time the carbon-to-hydrogen at the more bonds gain).

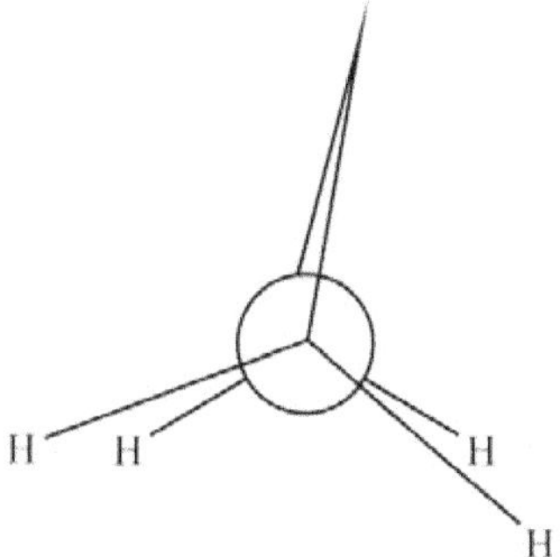

Newman projections of cyclopropane

One uncommon consequence of bending of bonds is that while the C–C bonds in cyclopropane are weaker than usual, the C atoms are also closer together than in a normal alkane bond: 151 pm versus 153 pm, when the average alkene bond length is about 146 pm.

➢ *Conformations of Cyclobutane*

The angles between carbon-carbon bonds in cyclobutane are quite strained, and therefore, have lower bond dissociation energies than corresponding unstrained or linear hydrocarbons like cyclohexane or butane. Furthermore, the cyclobutane molecule is very unstable above 500 °C. As far as the structure is concerned, four carbons in cyclobutane are not in a single plane but adopts a puckered or somewhat "folded" conformation.

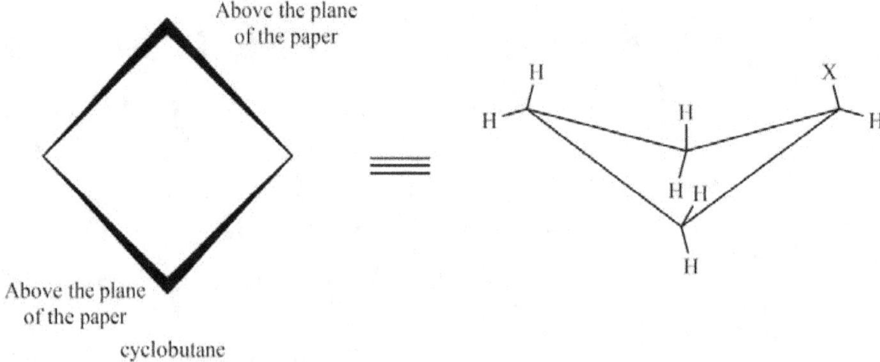

Any single carbon also has an angle of 25° with the plane of the remaining three carbon atoms. Like this, some of the interactions due to eclipsing are also diminished.

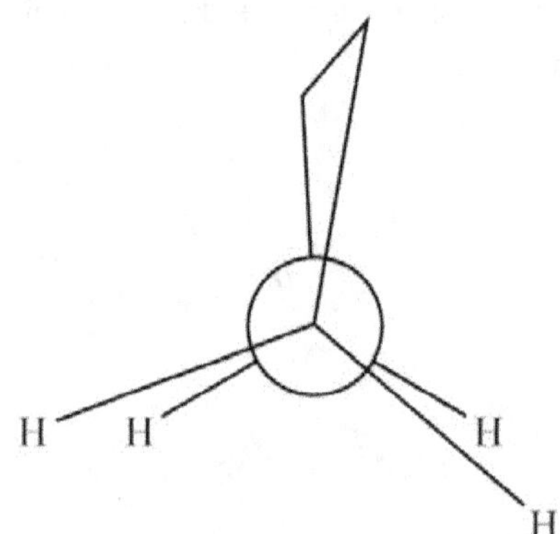

Newman projections of cyclobutane

The conformation of cyclobutene is also known as the "butterfly" structure. Furthermore, it is also worthy to note that equivalent puckered conformations of cyclobutene interconvert within each other at room temperature due to an appropriate supply of thermal energy.

> *Conformations of Cyclopentane*

The cyclopentane is the second-most common cycloalkane (after cyclohexanes) and stabler than cyclobutanes. Its planar conformation has almost zero angle strain but a huge magnitude of torsional strain resulting in a non-planar conformation, which in turn, actually increases angle strain slightly but still favorable due to stability from reduced torsional strain. The cyclopentane undergoes a rapid bond rotation process At room temperature where every carbon has its turns of being at the endo site eventually.

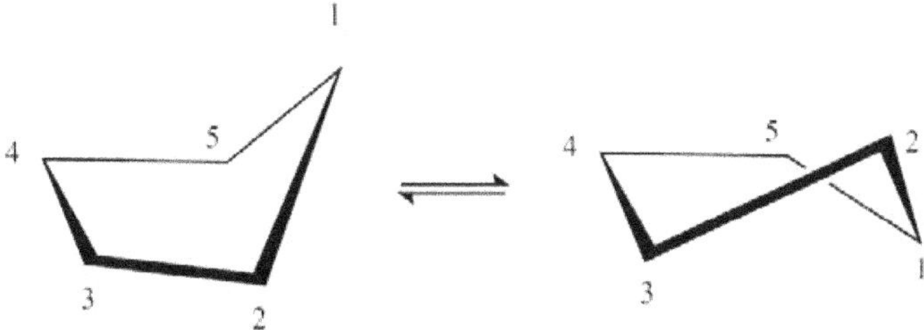

Conformational equilibrium of cyclopentane

The 'envelope' conformation (left) of cyclopentane is of the lowest energy where the out-of-plane carbon (i.e., 1 in the left side structure) is said to be in the 'endo or inside' site. The enveloping eliminates the torsional strain along the sides and the envelope's flapping happens. Nevertheless, the neighboring C atoms are eclipsed at the envelope's plane and are situated away from the fold.

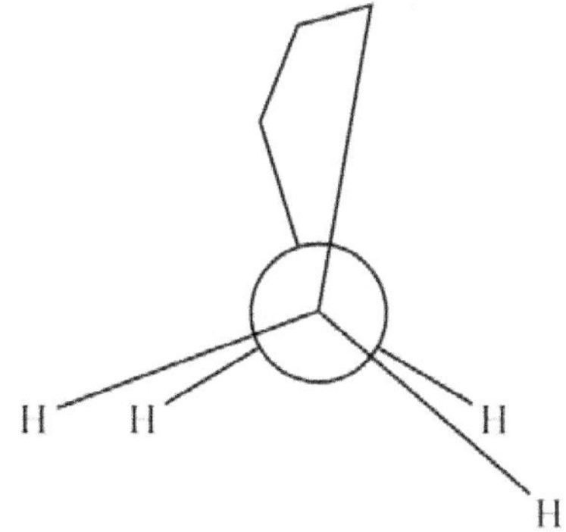

Newman projections of cyclopentane

Furthermore, if we view Newman's projection of cyclopentane molecule signed down one of the carbon-carbon bonds reveals the staggering nature of carbon-hydrogen bonds.

> ### *Conformations of Cyclohexane*

Many students confuse the polygon formula of cyclohexane molecule with its actual structure and assume that it must be planar. Nevertheless, its structure is far from a perfect hexagon because the conformation of a flat 2-dimensional planar hexagon would have a huge magnitude of angle strain. After all, its carbon-carbon bonds would not be at the ideal tetrahedral angle (i.e., 109.5°). Furthermore, the torsional strain would also be of significant magnitude in hexagonal structure because all of the bonds would be of eclipsed type. So, to decrease torsional strain, the cyclohexane adopts a 3-dimensional geometry (chair form), which interconvert at room temperature very rapidly via the chair flipping mechanism during which 3 other intermediate conformations are met: the first one is the half-chair (most unstable), a more stable boat conformation, and the third one is the twist-boat (more stable than the boat but less stable than the chair form).

Structure of cyclohexane

In the potential energy diagram, the half-chair and the boat are at transition states and signify energy maxima whereas the chair form and twist-boat form are at energy minima inferring their conformer nature. The chair conformation was first suggested by Hermann Sachse in 1890 which gained acceptance at a widespread level in the later period.

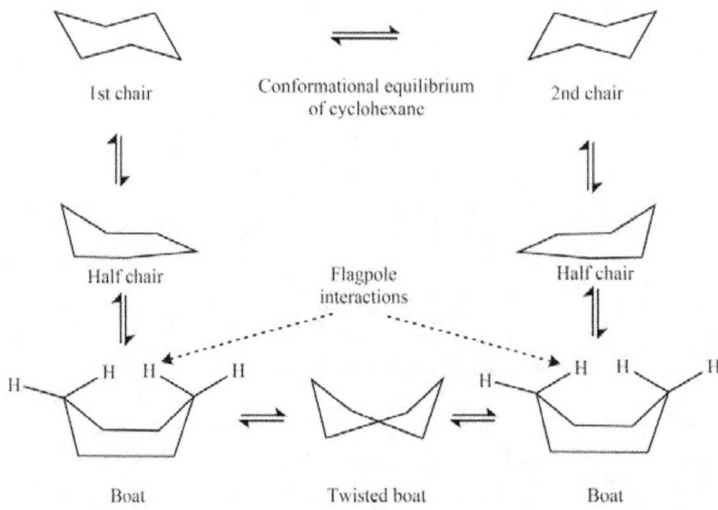

In the chair form, the carbon-carbon bond angles become 109.5°, and therefore, the angle strain was completely eliminated. Also, six out of twelve hydrogens were parallel to the principal axis (C_3-axis), and are called axial hydrogens; whereas the remaining hydrogens are at higher angles from the principal axis imparting the label of equatorial hydrogens.

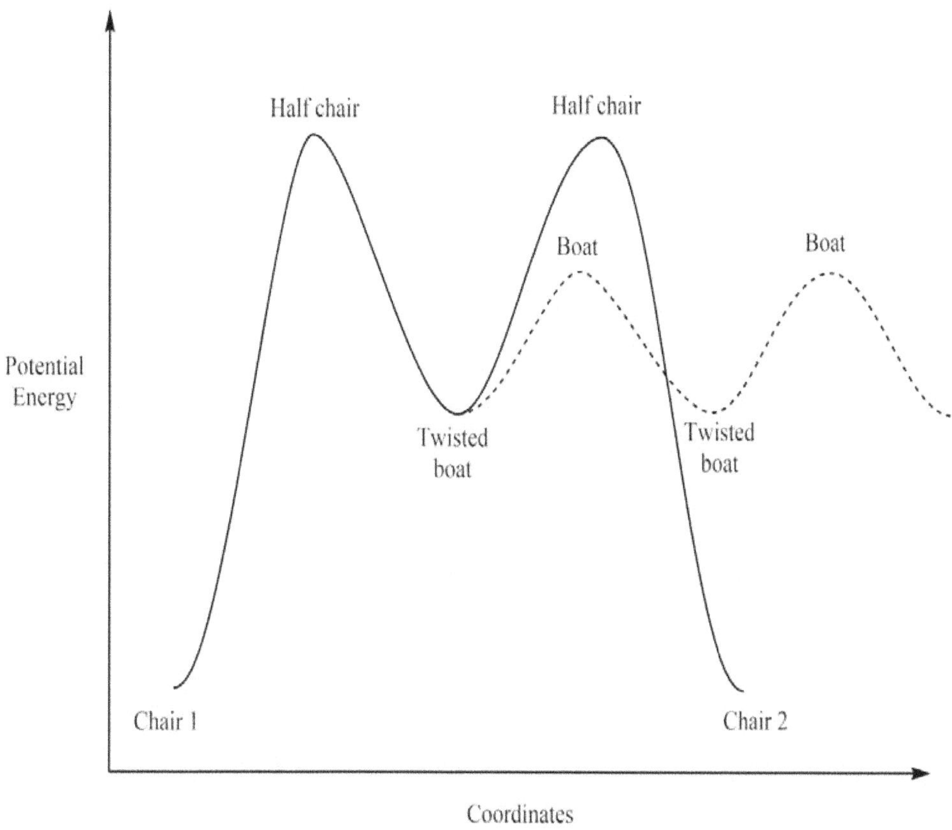

Figure 1. Potential energy diagram for different conformations of cyclohexane molecule.

Another important confirmation of cyclohexane molecule is known as the boat form which also gets interconverted to the more stable chair form. Furthermore, if the mono-substitution occurs at cyclohexane, it will most likely happen at the equatorial site because of the less torsional strain. Finally, the cyclohexane molecule has the lowest magnitude of torsional and angle strain of all cycloalkanes available making the cyclohexane a strain-free system.

❖ Decalins

If two cyclic compounds are fused together, we will get a bicyclic system. These bicyclic compounds can be either bridged or fused systems. Two cycles share two adjacent C atoms in a fused system, whereas one or more carbon atoms act as a bridge between two non-adjacent carbon atoms in bridged systems. Here we will study one special kind of fused ring system called decalin or bicyclo-[4,4,0]-decane.

Fused bicyclo compound   Bridged bicyclo compound

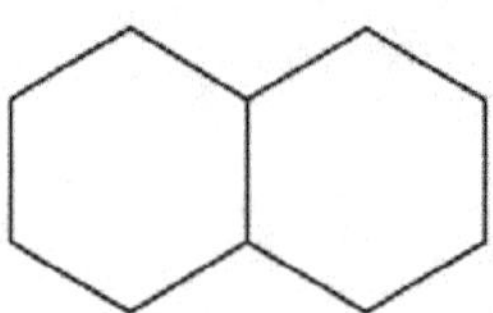

Bicyclo-[4,4,0]-decane   Bicyclo-[2,2,1]-heptane
(Decalin)   (Norbornane)

Also, as far as the IUPAC name of decalin is concerned i.e. bicyclo-[4,4,0]-decane, the numbers 4, 4, 0 show the number of C atoms in each cycle (excluding the bridgehead carbon atoms).

➢ *Synthesis of Decalin*

One of the most popular names of Decalin is decahydronaphthalene because it is a saturated counterpart of naphthalene molecule, and can also be obtained from the same by simple hydrogenation in a fused state.

The preparation gives two stereoisomers of decalin; where the first have both the hydrogens at the bridgehead carbons at 'cis' arrangement and trans hydrogens in the second one.

> *Geometrical Isomerism in Decalin*

Since there are two cyclohexane rings fused in decalin, and those cyclohexenes are most stable in their chair form; it is reasonable to think that most stable should also have them in chair form. Using the same rationale, Sachse and Mohr proposed the decalin should be a puckered structure with no strain which exists as two isomers that cannot be interconverted without any bond-breaking. And therefore, they must be treated as diastereomers or configurational isomers. Both the puckered structures of decalin (cis and trans isomers) are shown below.

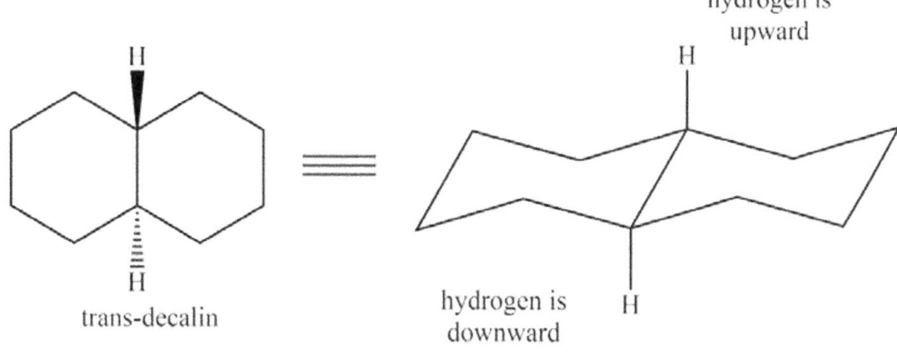

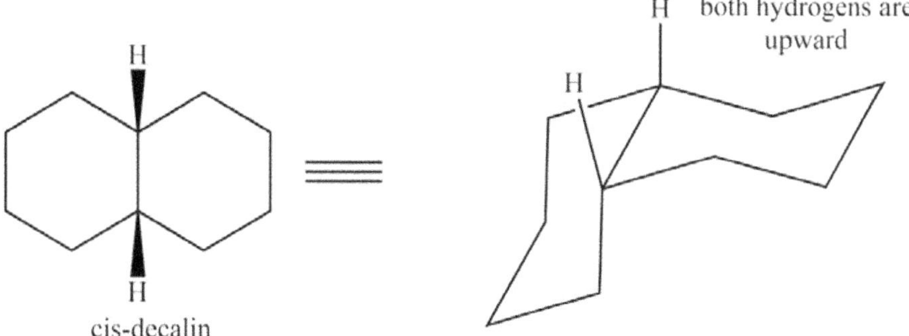

Now because both the stereoisomers have cyclohexane chairs, there should be no torsional or angle strain in either decalin molecule. It is also obvious that the two bridgehead hydrogens in trans decalin are directed in opposite directions, and therefore, both are axial hydrogens. Conversely, the two bridgehead hydrogens in cis decalin are pointing in the same direction with one axial and other of equatorial nature.

> *Conformational Analysis of Decalin*

After studying the geometrical isomerism in decalin, we need to discuss the conformational behavior of the same in a more comprehensive and sophisticated way. To do so, we will first discuss the ring flipping in trans and cis-decalin one by one.

1. Ring flipping in trans-decalin: The two cyclohexane rings in trans decalin are connected via equatorial positions. Since we know that the flipping of the ring in cyclohexane converts all the axial bonds to equatorial bonds and the vice-versa is also true, the flipping of the ring in trans-decalin would give a conformation where the two cyclohexane units would be connected via axial bonds. Nevertheless, this ring flipping is strongly forbidden because it is not conceivable to build a 6-membered ring with two diagonally opposite bonds. We are bound to fail if we attempt to build a molecular model in which the cyclohexane units are connected via axial bonds.

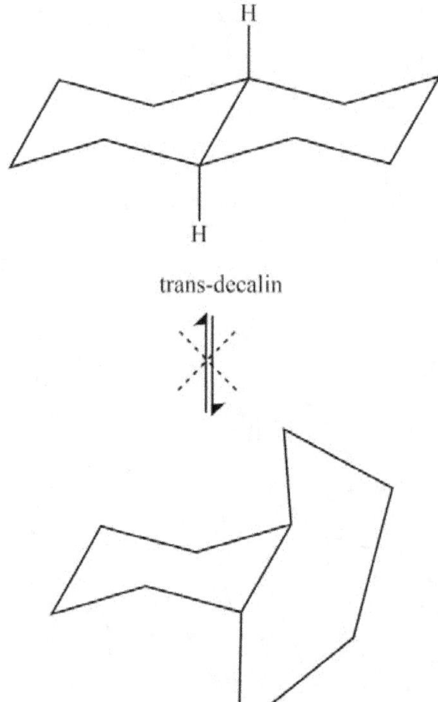

Hence, we may conclude that the trans-decalin is actually 'conformationally locked' making ring flipping impossible. So, it keeps its equatorial confirmation for both the rings. Also, e,e-trans-decalin has an inversion center, and therefore, it is an optically inactive or achiral compound. In other words, trans-decalin is superimposable on its mirror image. There are also two C_2 symmetry axes, the first one through the equator and the other one passes through the axis.

2. Ring flipping in cis-decalin: Unlike trans-decalin, which is relatively flat, the cis-decalin resembles a tent-like geometry with less hindered (convex)and more hindered (concave) sides. Though in both cases the cyclohexanes are stable chair-like conformations, the cis decalin has them with equatorial and axial bond joining. The flipping of the ring is allowed in this situation which changes one cis form into another one and translates the equatorial bonds to axial ones.

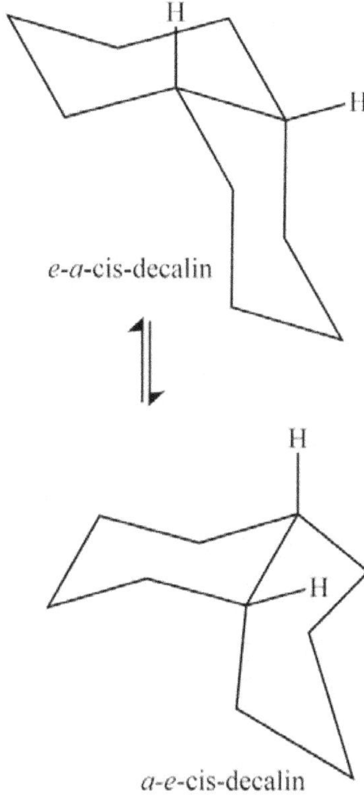

The conformational study of cis-decalin also proved that it is an optically active or chiral molecule even if it has no chiral center. In other words, the cis-decalin cannot be superimposed on its mirror image. Nevertheless, a rapid flipping of the ring system cancels its optical activity and changes it into its mirror image. Consequently, we can not resolve the two chiral forms due to this rapid ring inversion, and it exists as a racemic mixture. The flipping of the ring in cis-decalin has also been supported by the NMR studies proving that only one peak H-NMR spectra; whereas, two proton peaks are observed in trans decalin. Two H-NMR peaks in the spectra of trans-decalin can be credited to its inflexibility. The equatorial and axial hydrogens are present in different chemical, as well as magnetic environments, and therefore, show the different magnitude of chemical shifts. On the other hand, the cis-decalin is able of very fast interconversion resulting in a single chemical shift for all the hydrogens.

❖ Conformations of Sugars

It is quite a well-known fact that carbohydrates can primarily be classified into three categories; monosaccharides, oligosaccharides, and polysaccharides. The monosaccharides are the simplest carbohydrates that cannot be further hydrolyzed to simpler molecules. The general formula of monosaccharides is $(CH_2O)_n$ where n = 3–8. The oligosaccharides are the carbohydrate molecules that can produce 2–10 molecules of monosaccharides. Polysaccharides are carbohydrate molecules that can produce a very large number of monosaccharides' molecules upon hydrolysis.

Furthermore, in addition to the number of hydrolysis produce, the carbohydrates can also be classified on the basis of their taste. It has been found that all the monosaccharides and oligosaccharides (dia-, tri-, tetra-saccharides etc.) are crystalline compounds, soluble in water and sweet in taste; and typically labeled as sugars. On the other hand, polysaccharides are amorphous compounds, insoluble in water, and don't have any taste; and therefore, these carbohydrates are typically called as non-sugars. In this section, we will discuss the conformation of different types of sugars i.e. monosaccharides and oligosaccharides. The term "conformation" here refers to the overall three-dimensional structure adopted by a sugar (saccharide) molecule as a result of the through-bond and through-space physical forces it experiences arising from its molecular structure. The physical forces that dictate the three-dimensional shapes of all sugar molecules are sometimes summarily captured by such terms as "steric interactions" and "stereoelectronic effects".

➤ *Conformation of Monosaccharides*

In 1883, Tollan proposed that the glucose molecule does not have a free aldehydic group but a cyclic structure via a hemiacetal carbon. The thought that the terminal aldehydic carbon may participate in hemiacetal formation by using the hydroxyl group of 4th carbon in open-chain glucose molecule, giving rise to a five-membered furan-like ring structure. Later on, in 1926, Haworth proposed that the formation of hemiacetal carbon takes place via 5th carbon, giving rise to a six-membered pyran-like ring structure.

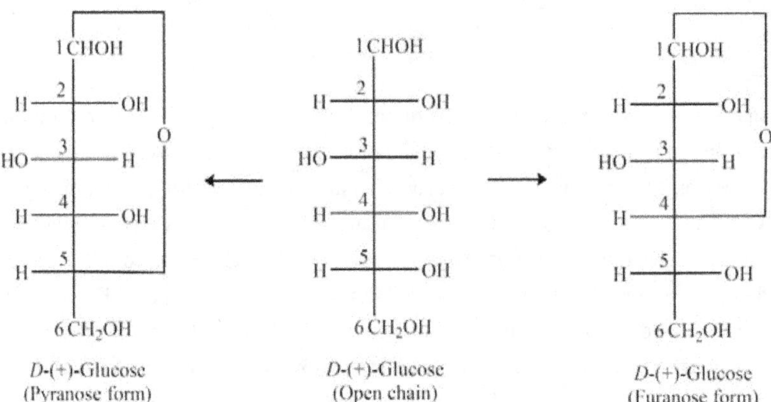

D-(+)-Glucose
(Pyranose form)

D-(+)-Glucose
(Open chain)

D-(+)-Glucose
(Furanose form)

The general discussion on the conformational analysis of some typical monosaccharide sugar molecules is given below.

1. Glucose: The terminal aldehydic carbon in open-chain glucose molecule may participate in hemiacetal formation by using the hydroxyl group of 4th and 5th carbon in open-chain glucose molecule, giving rise to a five-membered furan-like and six-membered pyran-like ring structure, respectively. In solutions, the open-chain form of glucose (either "D-" or "L-") exists in equilibrium with several cyclic isomers, each containing a ring of carbons closed by one oxygen atom. In an aqueous solution, however, more than 99% of glucose molecules, at any given time, exist as pyranose forms. The open-chain form is limited to about 0.25% and furanose forms exist in negligible amounts.

i) Pyranose form: The terminal aldehydic carbon participate in hemiacetal formation by using the hydroxyl group of 5th carbon in open-chain glucose molecule to give pyranose form.

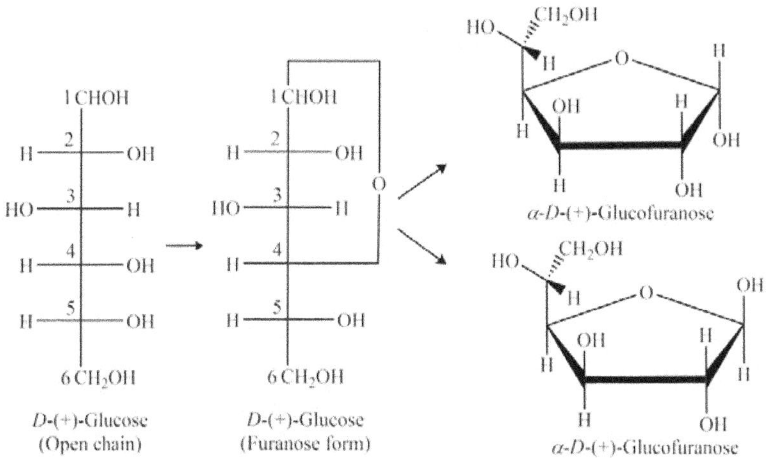

ii) Furanose form: The terminal aldehydic carbon participate in hemiacetal formation by using the hydroxyl group of 4th carbon in the open chain glucose molecule to give furanose form.

2. Fructose: The ketonic carbon in open-chain fructose molecule may participate in hemiketal formation by using the hydroxyl group of 5th and 6th carbon in open-chain glucose molecule, giving rise to a five-membered furan-like and six-membered pyran-like ring structure, respectively.

i) Pyranose form: The ketonic carbon participate in hemiketal formation by using the hydroxyl group of 6th carbon in open-chain glucose molecule to give pyranose form.

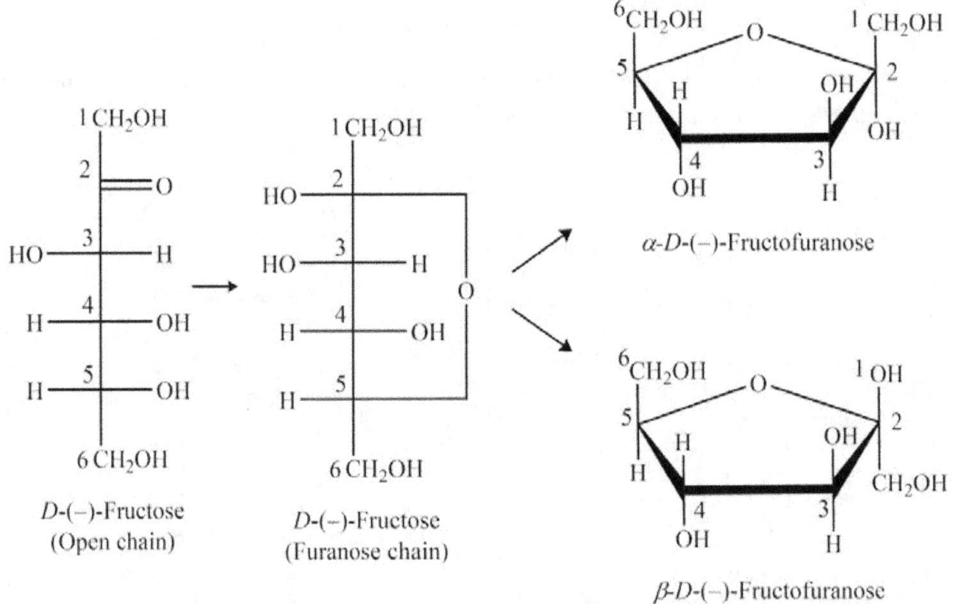

ii) Furanose form: The ketonic carbon participate in hemiketal formation by using the hydroxyl group of 6th carbon in open-chain glucose molecule to give furanose form.

3. Ribose: The aldehydic carbon in open-chain ribose molecule may participate in hemiacetal formation by using the hydroxyl group of 4th and 5th carbon, giving rise to a five-membered furan-like and six-membered pyran-like ring structure, respectively.

i) Pyranose form: The aldehydic carbon participate in hemiacetal formation by using the hydroxyl group of 5th carbon in open-chain ribose molecule to give pyranose form.

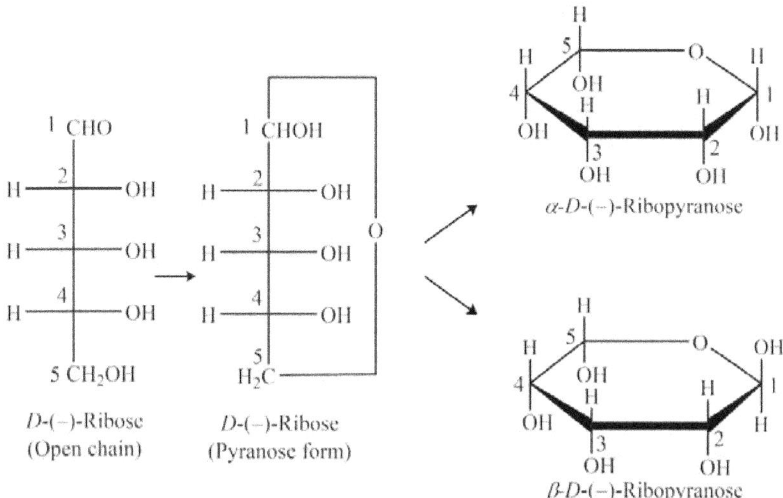

i) Furanose form: The aldehydic carbon participate in hemiacetal formation by using the hydroxyl group of 4th carbon in open chain ribose molecule to give furanose form.

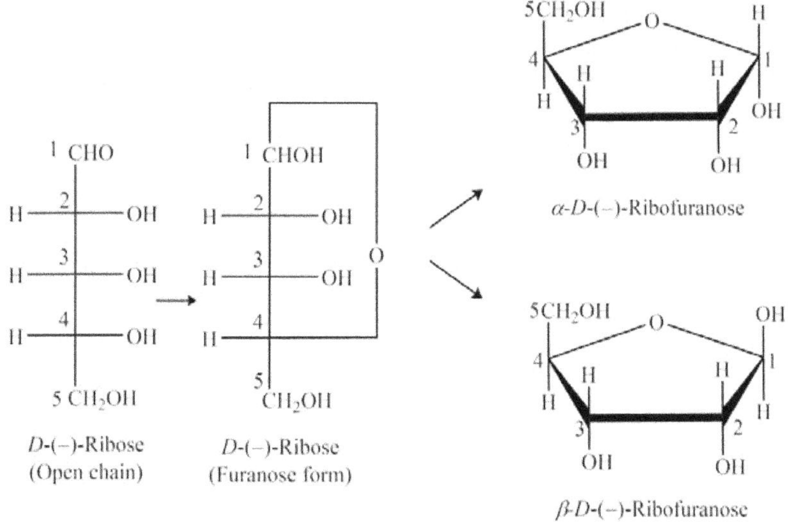

> *Conformation of Oligosaccharides*

In addition to the factors affecting monosaccharide residues, conformational analysis of oligosaccharides requires the consideration of some additional factors. One such major factor is the exo-anomeric effect, which is similar to the endo-anomeric effect. The difference is that the lone pair being donated is coming from the substituent at C-1. However, since the substituent can be either axial or equatorial there are two types of exo-anomeric effects, one from axial glycosides and one from equatorial glycosides as long as the donating orbital is anti-periplanar to the accepting orbital. The other one is Glycosidic torsion angles which angles are described by φ, ψ, and ω (in the case of glycosidic linkages via O-6). Steric considerations and anomeric effects need to be taken into consideration when looking at preferred angles.

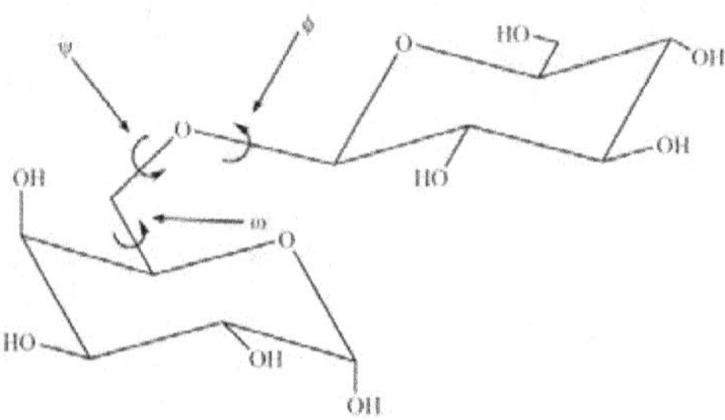

The general discussion on the conformational analysis of some typical monosaccharide sugar molecules is given below.

1. Sucrose: In sucrose, the components glucose and fructose are linked via an ether bond between C1 on the glucosyl subunit and C2 on the fructosyl unit. The bond is called a glycosidic linkage. Glucose exists predominantly as two isomeric "pyranoses" (α and β), but only one of these forms links to the fructose. Fructose itself exists as a mixture of "furanoses", each of which having α and β isomers, but only one particular isomer link to the glucosyl unit. What is notable about sucrose is that, unlike most disaccharides, the glycosidic bond is formed between the reducing ends of both glucose and fructose, and not between the reducing end of one and the nonreducing end of the other. This linkage inhibits further bonding to other saccharide units. Since it contains no anomeric hydroxyl groups, it is classified as a non-reducing sugar.

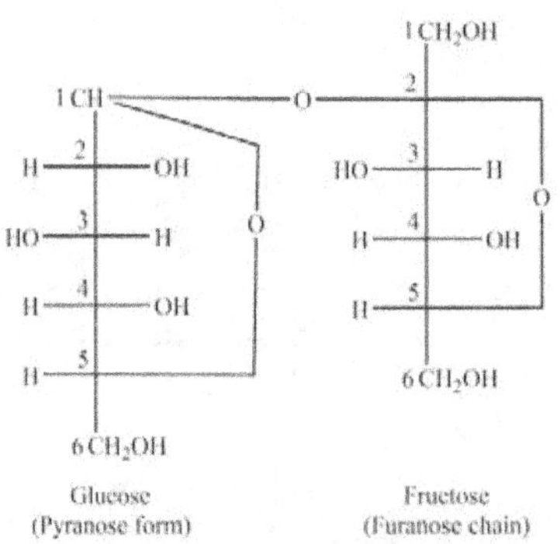

Glucose
(Pyranose form)

Fructose
(Furanose chain)

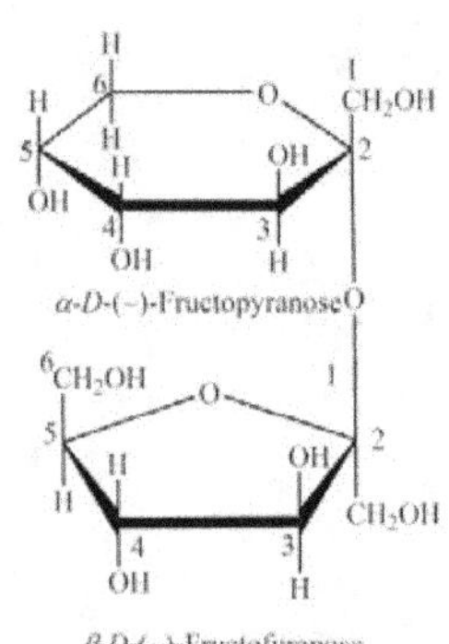

α-D-(–)-Fructopyranose

β-D-(–)-Fructofuranose

2. Maltose: Maltose is a disaccharide: the carbohydrates are generally divided into monosaccharides, oligosaccharides, and polysaccharides depending on the number of sugar subunits. Maltose, with two sugar units, is an oligosaccharide, specifically a disaccharide, because it consists of two glucose molecules. Glucose is a hexose: a monosaccharide containing six carbon atoms. The two glucose units are in the pyranose form and are joined by an O-glycosidic bond, with the first carbon (C1) of the first glucose linked to the fourth carbon (C4) of the second glucose, indicated as (1→4). The link is characterized as α because the glycosidic bond to the anomeric carbon (C1) is in the opposite plane from the CH2OH substituent in the same ring (C_6 of the first glucose). If the glycosidic bond to the anomeric carbon (C1) were in the same plane as the CH_2OH substituent, it would be classified as a β(1→4) bond, and the resulting molecule would be cellobiose. The anomeric carbon (C1) of the second glucose molecule, which is not involved in a glycosidic bond, could be either an α- or β-anomer depending on the bond direction of the attached hydroxyl group relative to the CH_2OH substituent of the same ring, resulting in either α-maltose or β-maltose. An isomer of maltose is isomaltose. This is similar to maltose but instead of a bond in the α(1→4) position, it is in the α(1→6) position, the same bond that is found at the branch points of glycogen and amylopectin.

3. Lactose: Lactose is a disaccharide derived from the condensation of galactose and glucose, which form a β-1→4 glycosidic linkage. Its systematic name is β-D-galactopyranosyl-(1→4)-D-glucose. The glucose can be in either the α-pyranose form or the β-pyranose form, whereas the galactose can only have the β-pyranose form: hence α-lactose and β-lactose refer to the anomeric form of the glucopyranose ring alone. Detection reactions for lactose are the Woehlk-[6] and Fearon's test.[7] Both can be easily used in school experiments to visualize the different lactose content of different dairy products such as whole milk, lactose-free milk, yogurt, buttermilk, coffee creamer, sour creme, kefir etc. Lactose is hydrolyzed to glucose and galactose, isomerized in alkaline solution to lactulose, and catalytically hydrogenated to the corresponding polyhydric alcohol, lactitol.[9] Lactulose is a commercial product, used for the treatment of constipation.

❖ Optical Activity in Absence of Chiral Carbon (Biphenyls, Allenes and Spiranes)

As we discussed earlier in this chapter, optically active compounds can primarily be divided into four categories on the basis of their geometrical profile; molecules with the chiral center, chiral axis, chiral plane, and helical chirality.

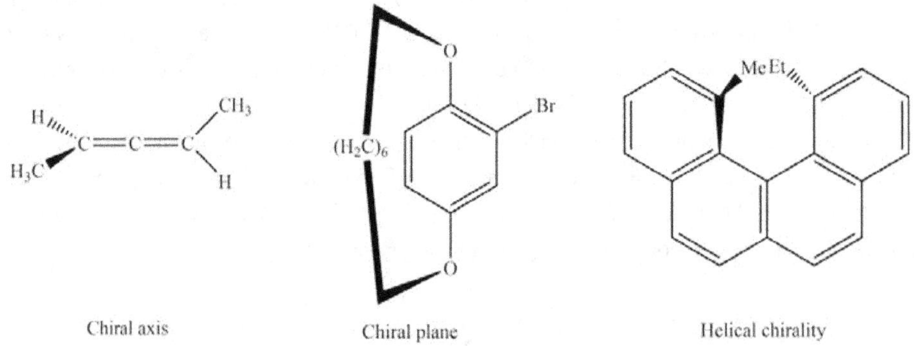

Chiral axis   Chiral plane   Helical chirality

In this section, we will study two kinds of optically active molecules without chiral carbon, compounds with chiral axis, and chiral plane.

> *Optically Active Compounds with Chiral Axis*

This type of chirality arises when a tetrahedrally coordinated prochiral molecule becomes chiral by extending the center along an axis. In other words, a prochiral molecule can no longer be superimposed on its mirror image if its center has been extended to a line with the same groups at different ends.

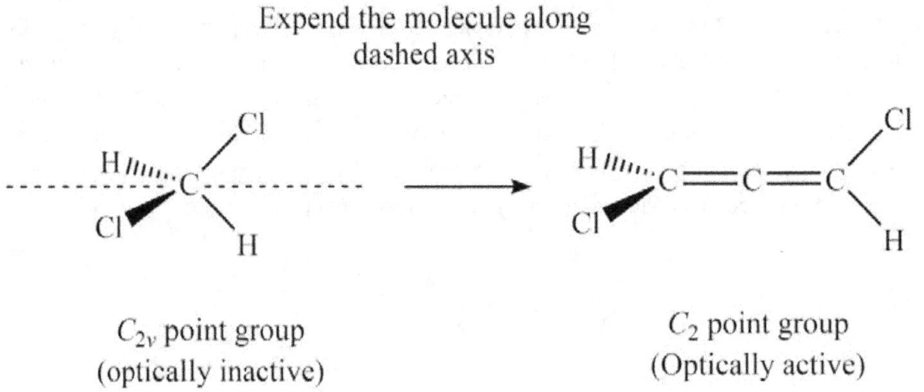

Expend the molecule along
dashed axis

C_{2v} point group
(optically inactive)

C_2 point group
(Optically active)

Figure 3. Conversion of an optically inactive molecule to optically active via chiral axis.

1. R-S nomenclature of Optically active compounds with chiral axis:

The whole procedure includes two steps; the first is the priority assignment of different groups at both ends using Chan-Ingold-Prelog and the second step involves the assignment of absolute configuration. It is worthy to note that the highest and lowest priorities (1, 4) should be assigned to the "out-of-plane" unit and intermediatory priorities (2, 3) must be assigned to the "in-plane" unit.

After assigning priorities to different groups, if the tracking of decreasing priority of the remaining three groups comes gives rise to clockwise flight, the molecules should be labeled as R and vice-versa. However, if the group of lowest priority is towards observer, revert the result from R to S and vice-versa.

2. Examples of Optically active compounds with chiral axis:

Some of the most common examples of organic molecules with this type of chirality are biphenyls, allenes, and spiranes derivatives.

i) Optically active biphenyls:

It is clear that biphenyl derivative will only be optically active groups on both ends are not the same.

ii) Optically active allenes:

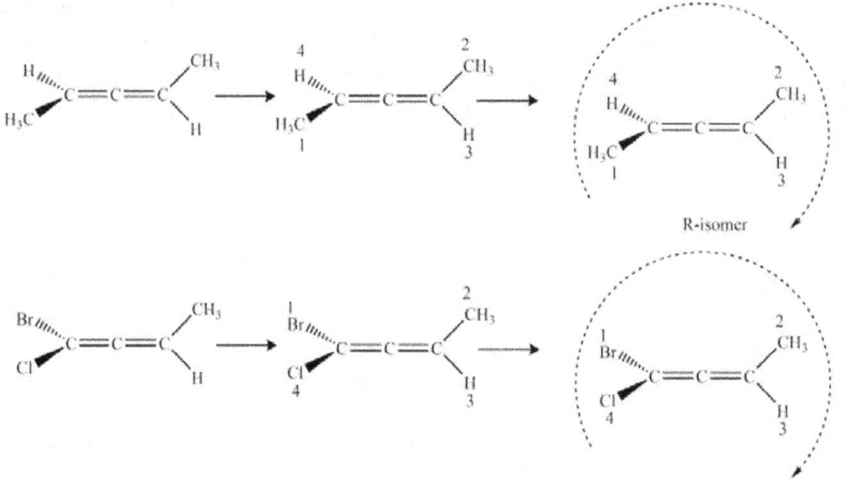

It is also obvious that the allene derivative compounds will only be optically active groups on both ends are not the same.

iii) Optically active Spiranes:

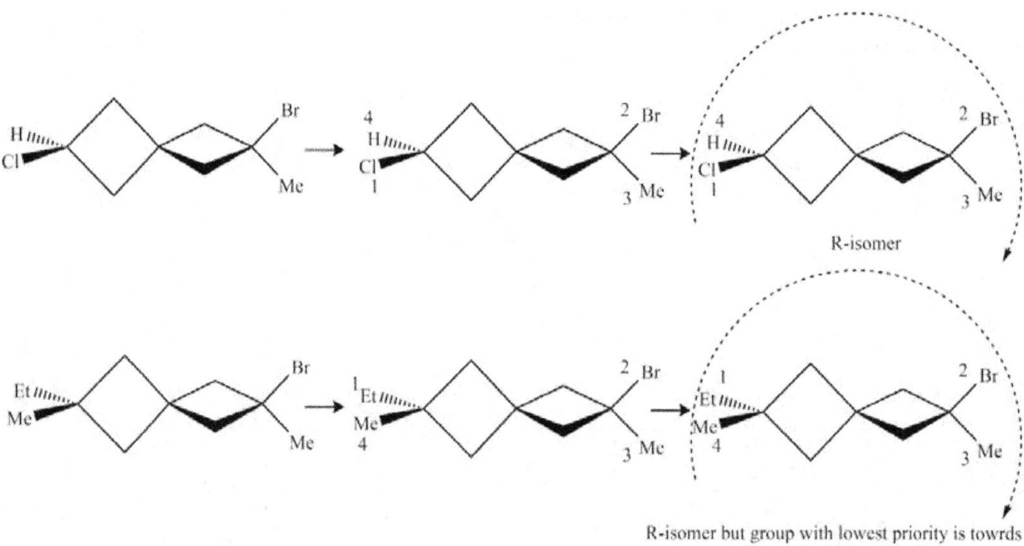

It is also obvious that the spirane derivative compounds will only be optically active groups on both ends are not the same.

> *Optically Active Compounds with Chiral Plane*

This type of chirality arises when relacing a group in a plane makes the molecule chiral. In other words, an organic molecule can no longer be superimposed on its mirror image if the replacement of a particular group induces chirality.

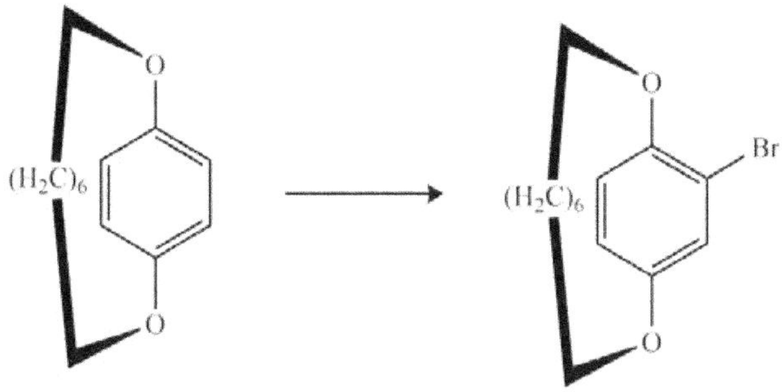

C_{2v} point group
(optically inactive)

C_1 point group
(optically active)

Figure 4. Conversion of an optically inactive molecule to optically active via chiral plane.

1. R-S nomenclature of Optically active compounds with chiral plane:

The whole procedure includes two steps; the first is the priority assignment of different groups next to the "pilot atom" (the directly bonded atom above the chiral plane). The atom next to the pilot atom is always assigned 1st priority, followed by the 2nd priority to next atom, and then deciding the priority of the next group using Chan-Ingold-Prelog rules.

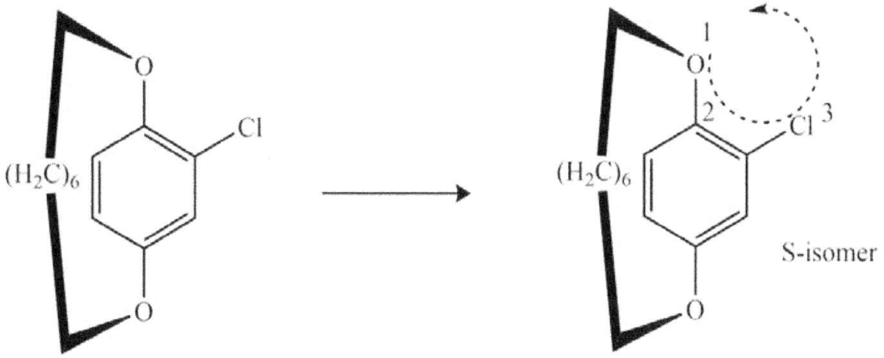

S-isomer

After assigning priorities to different groups, if the tracking of decreasing priority of the remaining three groups comes gives rise to clockwise flight, the molecules should be labeled as R. However, if the tracking of decreasing priority of the remaining three groups comes gives rise to anticlockwise flight, the molecules should be labeled as S.

2. Examples of Optically active compounds with chiral plane:

Some of the most common examples of organic molecules with this type of chirality are ansa compounds' derivatives.

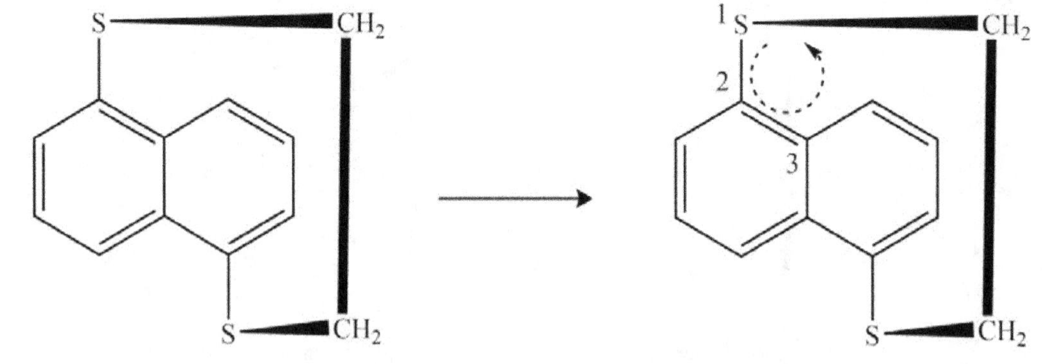

S-isomer

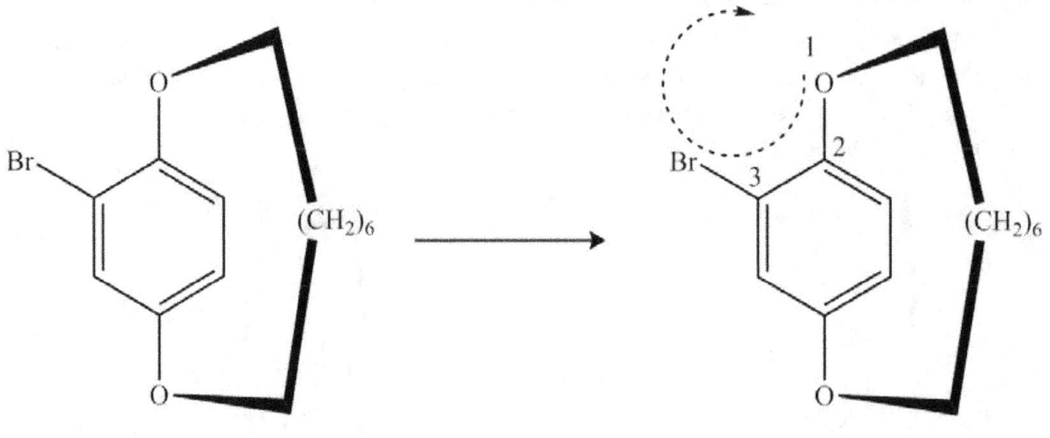

R-isomer

The enantiomer of any optically active ansa compound can simply be obtained by flipping the chain below the chiral plane.

❖ Chirality Due to Helical Shape

This type of chirality arises when the molecule has a helical structure. In other words, an organic molecule can no longer be superimposed on its mirror image if its geometry resembles a helix.

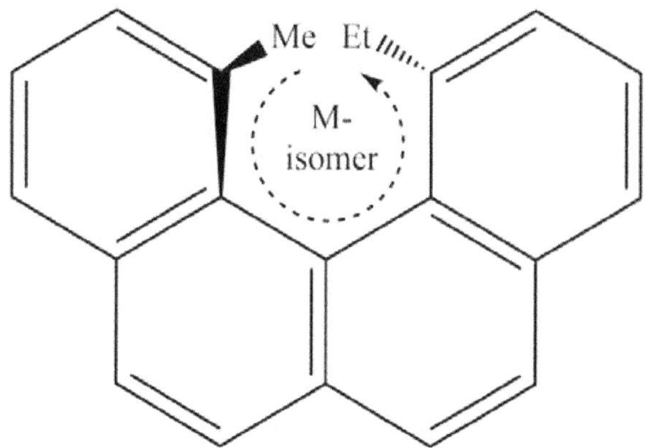

Figure 5. Conversion of an achiral molecule to optically active via the induction of helical shape.

➢ *R-S Nomenclature of Optically Active Compounds with Helical Chirality*

Many molecules (such as a helix) lack a chiral center, chiral axis, or chiral plane but still are optically active. Since we can view a helix along the axis, we need to check the behavior of the near and far end of the same. If the deboarding from the near end of the helix to the far end gives rise to clockwise flight, the molecules should be labeled as P. Conversely, if the deboarding from the near end to the far end gives rise to anticlockwise flight, the molecules should be labeled as M.

Anticlockwise deboarding

Furthermore, some optically active molecules; like allenes, biphenyls, and spiranes; can have axial as well as helical chirality. The route to do so is the same except for the fact that such molecules must be viewed along the chiral axis first, and priorities are assigned separately at both ends. Now if the deboarding from near highest priority group to the far highest priority group gives rise to clockwise flight, the molecules should be labeled as P. Conversely, if the deboarding from near highest priority group to the far highest priority group gives rise to anticlockwise flight, the molecules should be labeled as M.

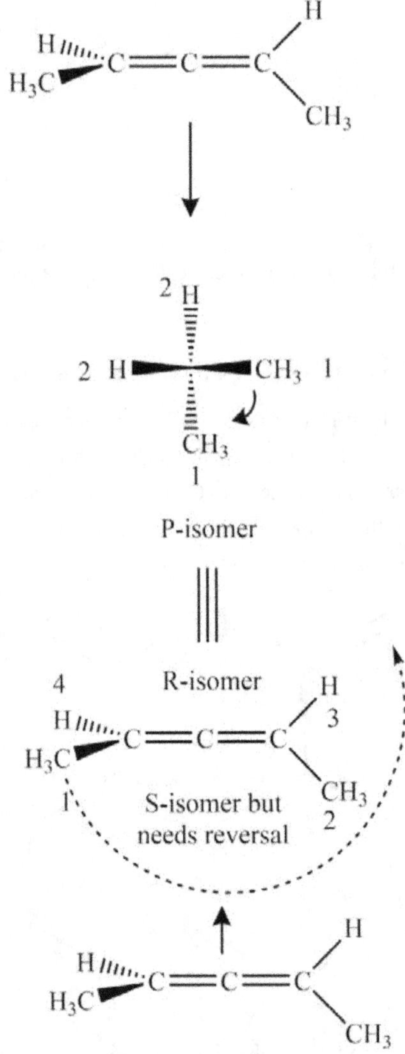

It is also very important to note the fact that the R and S labels for molecules with chiral axis translate P and M; respectively.

➤ ***Examples of Optically Active Compounds with Helical Chirality***

Some of the most common examples of organic molecules with this type of chirality (due to helical shape) are given below.

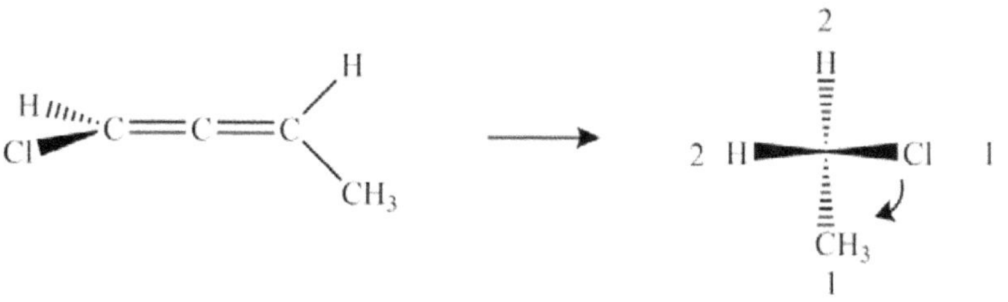

P-isomer

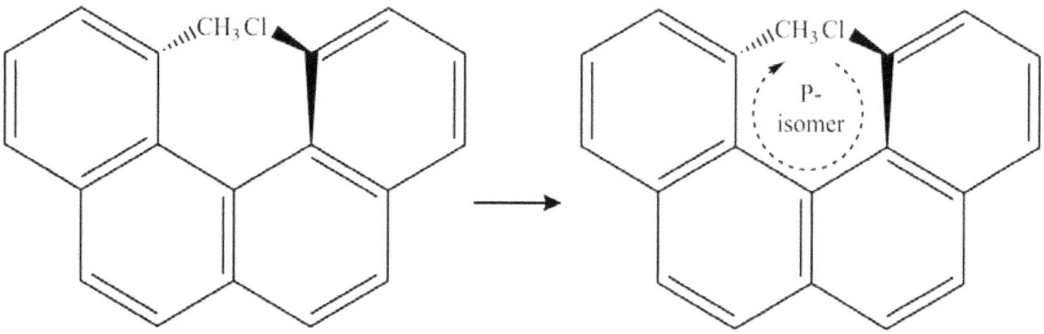

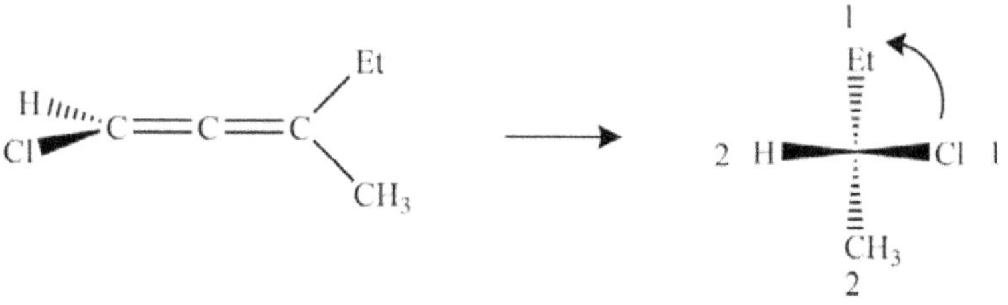

M-isomer

The enantiomer of any compound with helical chirality can simply be obtained by twisting the helix in the opposite direction.

❖ Geometrical Isomerism in Alkenes and Oximes

Before we study the geometrical isomerism in alkenes and oximes, we need to recall the general flow chart for different kinds of isomerisms first.

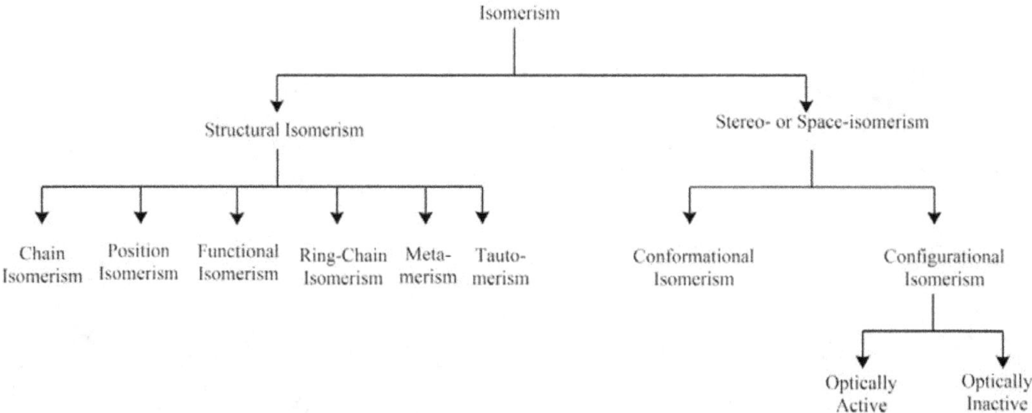

Since stereoisomerism can either be conformational or configurational, the latter possibility is of more importance in the case of oximes and alkenes as the rotation about the double bond is restricted. In this section, we will discuss the geometrical or configurational isomers of alkenes and oximes one by one.

> ### Geometrical Isomerism due to Double Bond

The carbon atom in alkenes is an sp^2-hybridized one, and therefore, it has a half-filled atomic orbital perpendicular to the molecular plane. After using all its three hybrid orbitals for σ-bonding, the half-filled p_z orbital can be used for side-wise overlap to form a π-bond. Nevertheless, if we rotate one of the half-filled p_z by an angle of 90°, it will not be able to do so anymore.

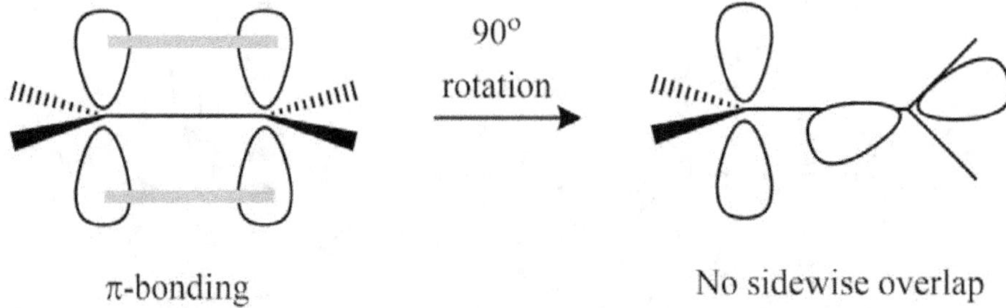

Hence, we can conclude that groups attached to sp^2-hybridized carbon cannot be exchanged simply by rotating about the double bond as it is restricted by the orbital picture. This eliminates the possibility of conformational isomerism, and therefore, we are only left with the case of the configurational one.

> *Condition for Geometrical Isomerism arising from Double Bond*

The presence of a double bond does not ensure the existence of geometrical isomers but some other conditions must also be satisfied. The primary condition is that two carbons of the double bond must have different kinds of substituents not only to carbons but to each other also.

cis-isomer   trans-isomer

No geomtrical isomerism possible

In other words, we can also say that no geometrical isomerism will be observed if one or both carbons carry the same kind of substituents.

> *Geometrical Isomerism in Alkenes*

The geometrical isomers of alkenes are primarily labeled as cis-trans or Z-E types, depending upon the nature of the groups on each side of the double bond.

1. Geometrical isomerism in disubstituted alkenes (cis-trans nomenclature): If the alkene under consideration is a disubstituted one, we can label it as cis or trans isomer, depending upon their mutual orientation.

cis-isomer   trans-isomer

2. Geometrical isomerism tri- or tetra-substituted alkenes (Z-E nomenclature): If the alkene under consideration is a tri- or tetra-substituted one, we cannot label it as cis or trans isomer, and therefore, we need to follow a special system for such compounds, called Z-E nomenclature. This system of nomenclature is also based upon the Chan-Ingold-Prelog system of priority assignment. The main postulates of the Z-E system of nomenclature are given below.

i) Priorities are assigned to different groups individually at both ends as per the sequence rule from the Chan-Ingold-Prelog system.

$$
\begin{array}{cc}
1 & 2 \\
H_3C & Cl \\
& \diagdown \diagup \\
& C = C \\
& \diagup \diagdown \\
H & Br \\
2 & 1
\end{array}
\qquad
\begin{array}{cc}
2 & 2 \\
H_3C & OH \\
& \diagdown \diagup \\
& C = C \\
& \diagup \diagdown \\
C_2H_5 & F \\
1 & 1
\end{array}
$$

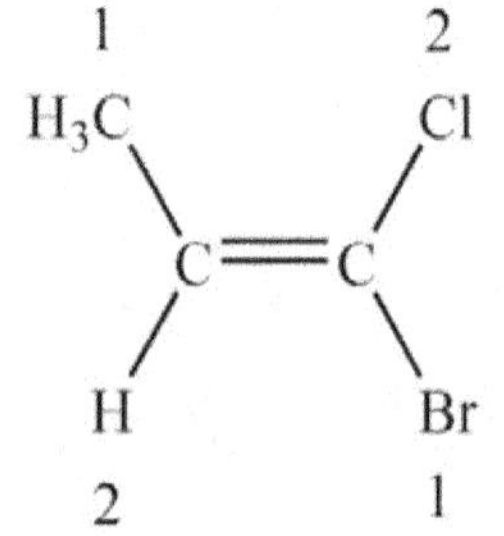

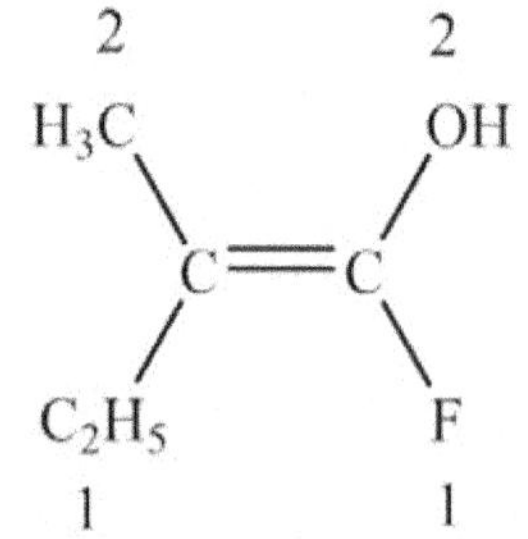

ii) Once the priorities are assigned, check if groups with higher priorities are on the same or opposite side of the double bond. If they are on the same side, the system is 'Z'; and if they are on the opposite side, the compound should be labeled as 'E' isomer.

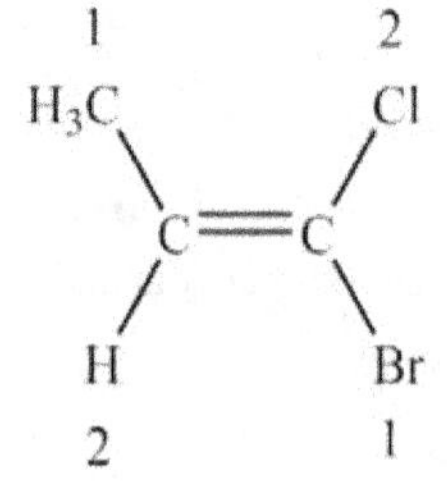

$$
\begin{array}{cc}
1 & 2 \\
H_3C & Cl \\
& \diagdown \diagup \\
& C = C \\
& \diagup \diagdown \\
H & Br \\
2 & 1
\end{array}
\longrightarrow
$$

Highest priorities are on
the opposite sides
= E isomer

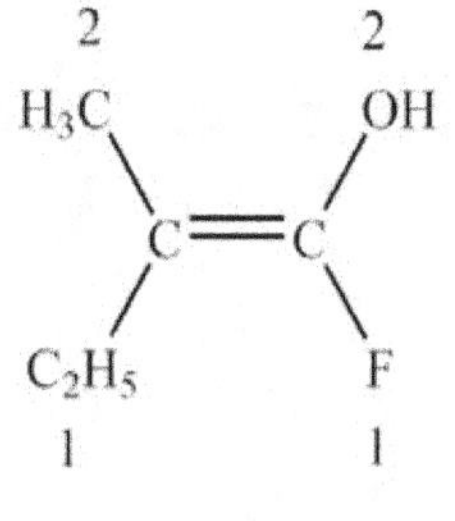

$$
\begin{array}{cc}
2 & 2 \\
H_3C & OH \\
& \diagdown \diagup \\
& C = C \\
& \diagup \diagdown \\
C_2H_5 & F \\
1 & 1
\end{array}
\longrightarrow
$$

Highest priorities are on
the same sides
= Z isomer

3. Geometrical isomerism in compounds with two or more double bonds: If the alkene under consideration has two or more double bonds, the number of geometrical isomers depends not only upon the number of double bonds only but also upon whether the ends are symmetrical or not. This can be classified into two categories as discussed below.

i) When ends are unsymmetrical: If the ends of the alkenes are unsymmetrical, the total number of geometrical isomers (whether the number is odd or even) is given by the following equation.

$$N_{isomer} = 2^n \tag{7}$$

Where n represents the number of double bonds.

ii) When ends are symmetrical: If the ends of the alkenes are symmetrical, the total number of geometrical isomers (whether the number is odd or even) is given by the following equation.

$$N_{isomer} = 2^{n-1} + 2^{(n/2-1)} \qquad if\ n = even \tag{8}$$

$$N_{isomer} = 2^{n-1} + 2^{(n/2-1/2)} \qquad if\ n = odd \tag{9}$$

Where n represents the number of double bonds.

(2*E*,4*E*)-3-bromohexa-2,4-diene

(2*Z*,4*E*)-4-bromohexa-2,4-dien-2-ol

(2*Z*,4*E*)-3-bromohexa-2,4-dien-2-ol

(1*E*,3*E*)-2-bromopenta-1,3-dien-1-ol

It is also obvious from the structures given above that besides tri- and tetra-substituted alkenes, the E-Z system of nomenclature also finds its application in compounds with many double bonds.

> ➢ *Geometrical Isomerism in Oximes*

Oximes are compounds that have a carbon-nitrogen double bond, and can easily be prepared by treating hydroxylamine with ketones or aldehydes in somewhat acidic solutions.

$$H_3C \diagdown C{=}O + H_2NHO \xrightarrow[pH\,=\,3.5-4.0]{H^+} H_3C \diagdown C{=}NOH + H_2O$$

acetaldehyde   acetaldehyde oxime

$$Ph \diagdown C{=}O + H_2NHO \xrightarrow[pH\,=\,3.5-4.0]{H^+} Ph \diagdown C{=}NOH + H_2O$$

benzaldehyde   benzaldehyde oxime

The carbon atom in oxime is an sp^2-hybridized one, and therefore, it has a half-filled atomic orbital perpendicular to the molecular plane. After using all its three hybrid orbitals for σ-bonding, the half-filled p_z orbital can be used for side-wise overlap to form a π-bond. Nevertheless, if we rotate one of the half-filled p_z by an angle of 90°, it will not be able to do so anymore because the nitrogen atom is also sp^2-hybridized with lone pair residing in one of the hybrid orbitals.

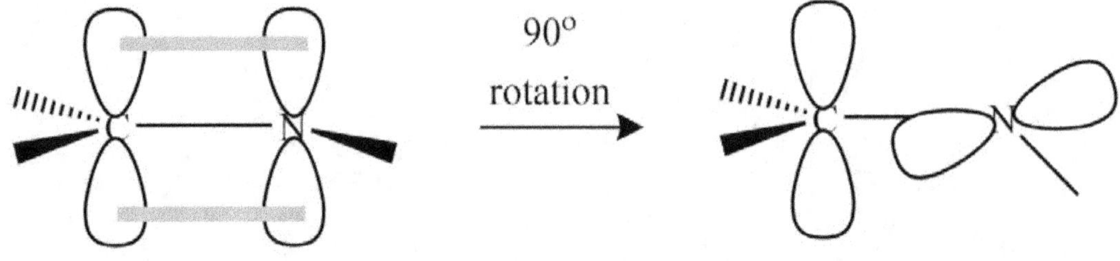

π-bonding   No sidewise overlap

Now, if the H and OH are on the same side of the double bond, the compound will be called as *syn*; whereas if H and OH are on the opposite side of the double bond, the compound should be labeled as *anti*.

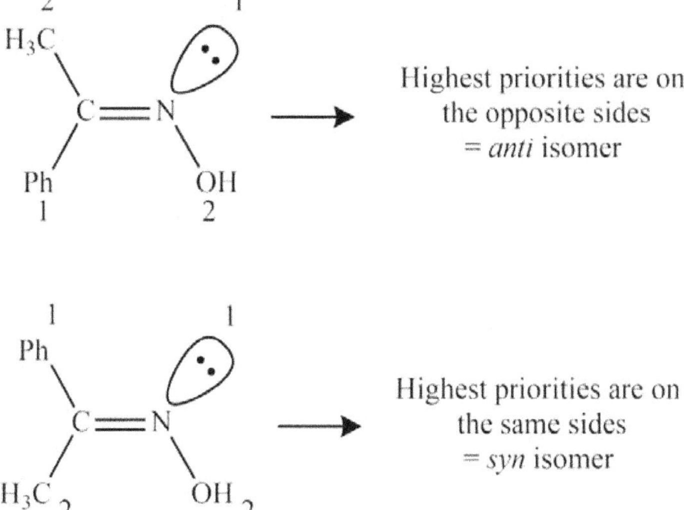

The nomenclature rule using H and OH is only applicable to aldoximes and cannot be applied to ketoximes because there is no H present. In such cases, we need to use priorities assignment using the Chan-Ingold-Prelog system with lone pair having the highest priority.

❖ Methods of Determining the Configuration

After studying the geometrical isomerism in alkenes, it's time to discuss the route by which know their configurations. The methods of determining the configuration of geometrical isomers are based upon either physical or chemical properties.

➢ *Determination of Configuration Using Physical Properties*

The geometrical isomers of alkenes can be identified with various kinds of physical properties like melting point, solubility, dipole moment, or boiling point, etc. Some of the important physical methods for determining the configuration of geometrical isomers are given below.

1. Melting point: The melting points of trans-isomers are generally higher than that of cis-isomers, which makes it a suitable tool for configuration prediction. The reason for such behavior is the fact that trans-isomers are symmetrical, and consequently, are well packed to give stronger intermolecular forces than cis-isomers.

Maleic acid
melting point = 403 K

Fumaric acid
melting point = 575 K

Allocinnamic acid
melting point = 341 K

Cinnamic acid
melting point = 406 K

2. Solubility: The solubility of cis-isomers are generally higher than that of trans-isomers, which makes it a suitable tool for configuration prediction. The reason for such behavior is the fact that cis-isomers are unsymmetrical, and consequently, are not well packed, giving weaker intermolecular forces than trans-isomers.

Maleic acid
Solubility = 790g per liter water

Fumaric acid
Solubility = 7g per liter water

3. Dipole moment: The dipole moments of cis-isomers are generally higher than that of trans-isomers, which makes it a suitable tool for configuration prediction. The reason for such behavior is the fact that trans-isomers have individual dipoles in the exact opposite direction which may or may not get canceled completely.

cis-dichloroethene
Dipole moment = 1.85D

trans-dichloroethene
Dipole moment = 0D

(Z)-pent-2-ene

More polar
Higher dipole moment

(E)-pent-2-ene

Less polar
Lower dipole moment

4. Boiling Point: The boiling points of cis-isomers are generally higher than that of trans-isomers, which makes it a suitable tool for configuration prediction. The reason for such behavior is the fact that the boiling points are typically the function of dipole moments, which are higher for cis-isomers.

(Z)-but-2-ene

Boiling point = 277K
Dipole moment = 0.4 D

(E)-but-2-ene

Boiling point = 274K
Dipole moment = 0.0 D

> ➢ *Determination of Configuration Using Chemical Properties*

The geometrical isomers of alkenes can be identified with various kinds of chemical properties like the formation of cyclic compounds, or the optical properties of the stereoisomers formed during bromination and hydroxylation reactions, etc. Some of the important chemical methods for determining the configuration of geometrical isomers are given below.

1. Generation of cyclic systems: Since the reactive groups of the substrate are on the same side in cis-isomer, the formation of cyclic systems will be an easy task. Conversely, if the reactive groups of the substrate are on the opposite side (in trans-isomer), the formation of cyclic systems will be much more difficult to carry out. This can be illustrated by the case of maleic acid and fumaric acid where the generation of maleic anhydride requires much more temperature for the later substrate.

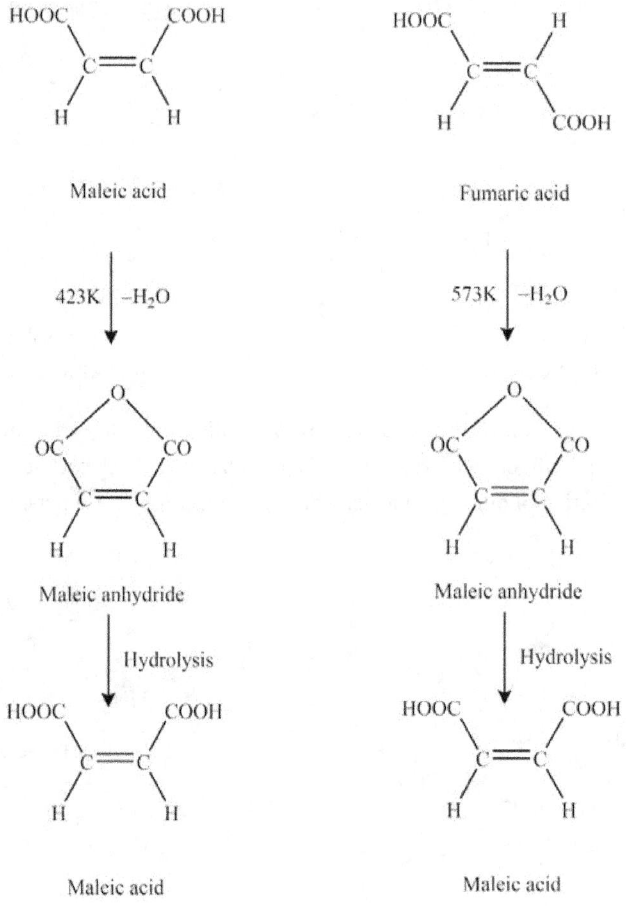

Maleic acid

Fumaric acid

$423K \quad -H_2O$

$573K \quad -H_2O$

Maleic anhydride

Maleic anhydride

Hydrolysis

Hydrolysis

Maleic acid

Maleic acid

Furthermore, it is also worthy to note that the hydrolysis of maleic anhydride gives back only maleic acid and no fumaric acid.

2. Optical properties of stereoisomers formed: The geometrical isomers of alkenes can be identified with the optical properties of the stereoisomers formed during bromination and hydroxylation reactions. Since the alkene's hydroxylation with dilute $KMnO_4$ or OsO_4 is cis-addition, we should get a meso-product with cis-isomer and racemic mixture for trans-substrate.

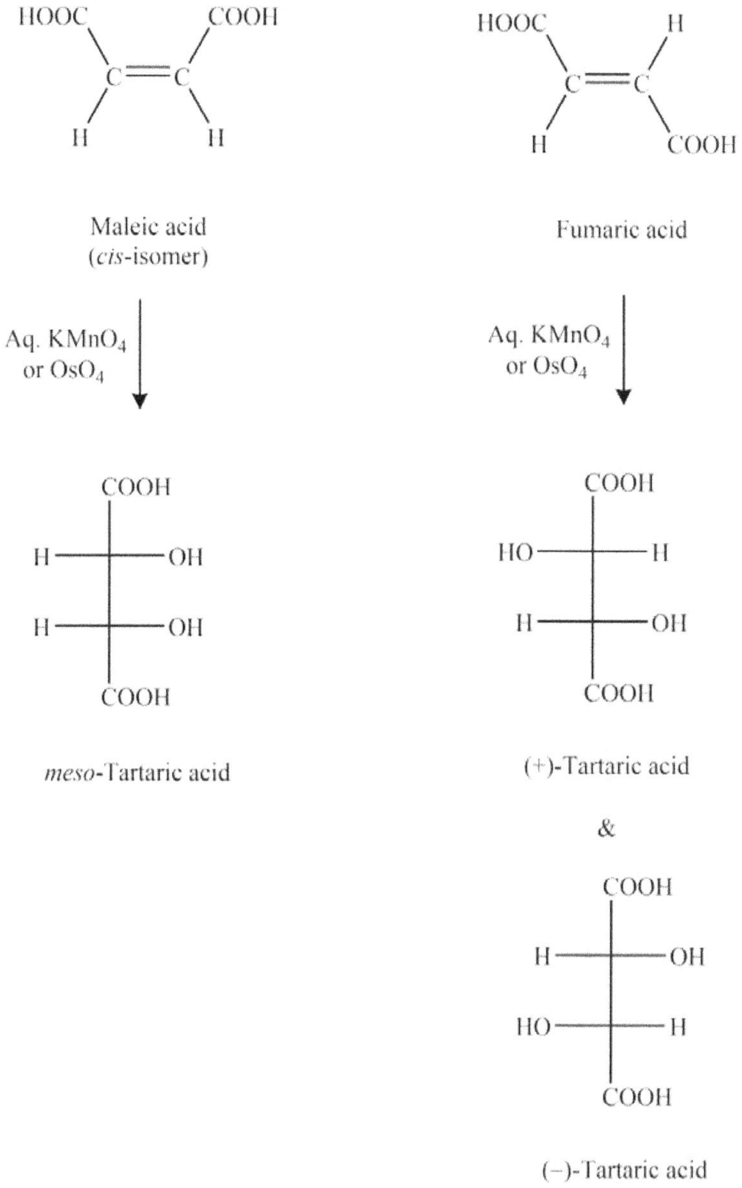

Maleic acid
(*cis*-isomer)

Fumaric acid

Aq. $KMnO_4$
or OsO_4

Aq. $KMnO_4$
or OsO_4

meso-Tartaric acid

(+)-Tartaric acid

&

(−)-Tartaric acid

On the other hand, the alkene's bromination with CCl_4 is trans-addition, we should get a meso-product with trans-isomer and racemic mixture for cis-substrate.

(Z)-but-2-ene $\xrightarrow[\textit{trans addition}]{Br_2/CCl_4}$ (±)-2, 3-dibromobutane

(E)-but-2-ene $\xrightarrow[\textit{trans addition}]{Br_2/CCl_4}$ meso-dibromobutane

❖ Problems

Q 1. What is the meaning of chirality in chemistry? Explain with necessary historical background.

Q 2. Define and discuss various symmetry elements.

Q 3. What is a secondary symmetry element? How is it related to the concept of optical activity?

Q 4. Discuss and differentiate the phenomenon of diastereomerism from enantiomerism.

Q 5. Discuss the absolute nomenclature of optically active compounds with suitable examples.

Q 6. Give various methods of resolution of racemic mixtures.

Q 7. Define optical purity. How is it calculated?

Q 8. What do you mean by the term 'prochirality'?

Q 9. What are enantiotopic groups and faces? How are they different from diastereotopic faces and groups?

Q 10. State and explain asymmetric synthesis.

Q 11. Draw and discuss various conformations of cyclohexane.

Q 12. What is the significance of the twisted boat form of cyclohexane?

Q 13. Define and discuss decalins.

Q 14. Discuss the conformational analysis of sugars.

Q 15. What is the chiral axis? How is it different from the chiral plane?

Q 16. Discuss the chirality arising from helical structures in detail.

Q 17. Discuss the geometrical isomerism in alkenes and oximes.

❖ Bibliography

1. K. Mislow, *The Relative Configuration of (levo)-Mandelic Acid*, J. Am. Chem. Soc., 73 (1951) 3954–3956.

2. M. B. Smith, *March's Advanced Organic Chemistry: Reactions, Mechanisms, and Structure*, John Wiley & Sons, Inc., New Jersey, USA, 2013.

3. D. Klein, *Organic Chemistry*, John Wiley & Sons, Inc., New Jersey, USA, 2015.

4. C. A. Coulson, B. O'Leary, R. B. Mallion, *Hückel Theory for Organic Chemists*, Academic Press, Massachusetts, USA, 1978.

5. R. L. Madan, *Organic Chemistry*, Tata McGraw Hill, New Delhi India, 2013.

6. H. Zimmerman, *Quantum Mechanics for Organic Chemists*, Academic Press, New York, USA, 1975.

7. J. Clayden, N. Greeves, S. Warren, *Organic Chemistry*, Oxford University Press, Oxford, UK, 2012.

CHAPTER 3

Reaction Mechanism: Structure and Reactivity

❖ Types of Mechanisms

Although the number of mechanisms by which different organic reactions proceed is very large, certain patterns can still be used to profile them for more systematic and simplistic analysis. Now before we classify different types of organic reaction mechanisms in brief, it is better to recall the basic terminology involved first.

1. Substrate and reagent: A typical organic reaction is thought to proceed via the breaking or making of one or more covalent bonds, and it is very suitable to call one reactant "substrate" and the other as "reagent". Generally, more reactive species is labeled as the reagent whereas the less reactive species is considered as the substrate.

2. Molecularity: The molecularity of a chemical reaction may simply be defined as the number of colliding molecular species involved in a single step. The most common types are unimolecular and bimolecular reactions that involve one and two molecular entities, respectively.

3. Electrophiles and nucleophiles: An electrophile is an electron-loving species or electron pair acceptor. Generally, the electrophiles are positively charged or neutral entities with vacant orbitals that are attracted to an electron-rich center. The electrophiles participate in the chemical reactions by accepting an electron pair to form a bond with the nucleophilic reactant. Now since the electrophiles accept electrons, they are seen as Lewis acids. A nucleophile is a nucleus (i.e. positive center) loving species or electron-pair donor. Normally, the nucleophiles are negatively charged or neutral electron-rich entities with lone pairs of electrons that are attracted to an electron-deficient center. The nucleophiles participate in the chemical reactions by donating an electron pair to form a bond with the electrophilic reactant. Now since the nucleophiles donate electrons, they are seen as Lewis bases.

4. Leaving Group: The part of the substrate molecule that gets detached is typically called as the leaving group. The leaving groups with electron-pair and without electron pairs as labeled as nucleofuges and electrofuges, respectively.

5. Reaction intermediates: The chemical species that are formed somewhere during the course of a chemical reaction are called as reaction intermediates. These are actual molecules that are short-lived and unstable. Sometimes they are called temporary reactants or products because they are neither present in actual reactants nor the actual products.

6. Transition states: The transition state of an organic reaction is a specific configuration along the reaction coordinate and corresponds to the highest potential energy along with this reaction coordinate. Unlike "reaction intermediate", the transition state is not an actual molecule that can be isolated, and therefore it is often marked with the double dagger $\ddagger$ symbol to differentiate.

Now depending upon the bonds-breaking, the reaction mechanism of organic compounds can be divided basically into three basic types as given below.

> ### 1. Polar Mechanism (Bond Heterolysis)

Polar or ionic or bond-heterolysis mechanisms have the characteristic feature of the electron-pair migrating from a well-defined source (such as lone pair or a nucleophilic bond) to a well-defined sink (some electrophilic center having low-lying antibonding molecular orbitals). This mechanism results from the heterolytic or unsymmetrical breaking of the covalent bond in which the bonding pair goes with one of the fragments.

The participating atoms in this type of mechanism undergo a change in their charges (formally as well as in reality). Most of the organic reactions fall into this category. For instance, the hydrolytic cleavage of tert-butyl bromide generates a carbocation and bromide ion.

Similarly, the acetone's reaction with halogen in the presence of base occurs through the formation of carbonian.

In an ionic mechanism, it's the electronegativity of the binding atom that primarily determines the fragment with bonding pair of electrons. The bond pair goes with the more electronegative atom and generates the anion. Now owing to the very small value of carbon's electronegativity, the heterolytic cleavage gives rise to carbocation most of the time; and therefore, carbocations are much more common than the carbanions.

➢ *2. Radical Mechanism (Bond Homolysis)*

Radical or bond-homolysis mechanisms have the characteristic feature of generating species with unpaired electrons (radicals) and the migration of single electrons. Furthermore, the radical reactions can also be divided into chain and nonchain processes. This mechanism results from the homolytic or symmetrical breaking of the covalent bond in which the bonding pair is distributed equally to the fragments.

$$: \overset{..}{A} \cdot \quad | \quad \cdot \overset{..}{B} : \quad \longrightarrow \quad : \overset{..}{A} \cdot \; + \; \cdot \overset{..}{B} :$$

For instance, consider the photochlorination of methane in which the first step is the homolytic bond breaking of Cl_2 i.e.

$$Cl_2 \quad \xrightarrow{\;h\nu\;} \quad : \overset{..}{Cl} \cdot \; + \; \cdot \overset{..}{Cl} :$$

$$CH_4 \; + \; Cl_2 \quad \xrightarrow{\;h\nu\;} \quad CH_3Cl \; + \; HCl$$

Both Cl atoms generated carry one unpaired electron each and are neutral. The chlorine atoms produced this way are called free radicals and are extremely reactive due to the availability of unpaired electrons.

➢ *3. Pericyclic Mechanism*

Pericyclic reactions involve a concerted cyclic shift of electrons to yield new chemical bonds via a ring transition state. Though the electron pairs are involved formally, no true source or sink can be assigned due to the cyclic movement. These reactions need a continuous overlap of participating orbitals and are dictated by orbital symmetry considerations; so, it is impossible to label them as homolytic or heterolytic cleavage.

Heterolysis

Homolysis

In these reactions, bond breaking and bond making takes place simultaneously and these are said to occur via a concerted route.

❖ Types of Reactions

Although the number of possible organic reactions and corresponding mechanisms is very large, certain patterns can still be observed which are used to describe numerous useful organic reactions. Each type of reaction has a stepwise route which explains how it occurs, though the actual picture of these steps cannot always be visualized only by looking at the reactants' list. Furthermore, despite their basic classification, many organic reactions may fall into more than one category. For instance, some of the substitution reactions follow an addition-elimination route. Hence, this classification doesn't mean to include all the organic reactions but most of them for general studies.

➢ *Substitution Reactions*

A substitution or single displacement reaction may simply be defined as a chemical change where one functional group in an organic compound is displaced by another functional group.

Unimolecular Aliphatic Nucleophilic Substitution Reactions

These types of organic reactions can be classified primarily into three categories; electrophilic substitution, nucleophilic substitution, and radical substitution depending upon the type and nature of the attacking reagent involved. Further classification is also possible by considering whether the reactive intermediate is a carbanion, a carbocation, or a free radical; or if the substrate is aliphatic or aromatic.

➢ *Addition Reaction*

The addition reaction in organic chemistry may simply be defined as a chemical change where two or more molecular entities combine to give rise to a bigger molecule (i.e., the adduct). Also, since the incoming group needs to bind to substrate, addition reactions are pretty much limited to organic compounds with multiple bonds, like molecules with carbon-carbon double or triple bonds (alkenes), and compounds that possess rings in them (i.e. also a kind of unsaturation). Furthermore, besides alkenes, alkynes, or ring structures, the organic compound can also have carbon-hetero multiple double bonds like imine (C=N) groups or carbonyl (C=O) groups; and therefore, are also capable of undergoing addition reactions.

The addition reaction can also be treated as the opposite of an elimination reaction. For example, the alkene's hydration to alcohol is the opposite of the dehydration reaction. The addition reactions can primarily be classified into two types: polar addition and non-polar addition. Polar additions are further divided into electrophilic addition and nucleophilic additions; whereas the non-polar addition reactions can be subdivided into free-radical addition and cycloaddition types.

Nucleophilic Addition Reaction

Finally, it is also worthy to note that addition reactions are also found in polymerization processes and are typically labeled as addition polymerization.

> ### *Elimination Reactions*

An elimination reaction in organic chemistry may simply be defined as a chemical change where two substituents are detached from a molecule in either a one- or two-step pathway. The one-step pathway is abbreviated as the E_2 mechanism, whereas the two-step pathway is abbreviated as the E_1 mechanism. Hence, the subscript in E_1 or E_2 reactions does not represent the number of steps involved, but the kinetics followed; E_1 is unimolecular (first-order) while E_2 is bimolecular (second-order).

E1 Elimination Reaction

Furthermore, if the molecule can stabilize an anion but does have a poor leaving group, the third kind of mechanism, called E_1CB, also exists. Lastly, a fourth kind, called the internal elimination (E_i), also exists which is generally followed by pyrolysis of xanthate and acetate esters.

> ### *Rearrangement Reactions*

A rearrangement reaction in organic chemistry may simply be defined as a chemical change where the carbon skeleton of an organic compound rearranges itself to give rise to a structural isomer. Generally, a group moves from one atom to another atom within the same molecule.

Now although the domain of rearrangement reactions is extremely wide, these changes can still primarily be classified into four categories; 1, 2 rearrangements, metathesis reactions, sigmatropic rearrangements, and electrocyclic reactions. A fifth kind called group transfer reactions also exist but are far less important than what we have mentioned. One of the most common examples of rearrangement reactions is the 'Cope rearrangement' which is a 1, 3-sigmatropic rearrangement involving the movement R group from 1st carbon to 3rd carbon in the same molecule.

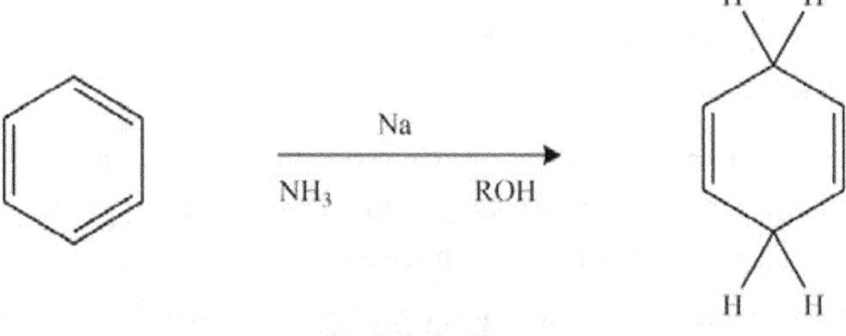

Cope Rearrangement

Besides intramolecular rearrangements, intermolecular rearrangements are also possible in many organic compounds; the group transfer reaction we mentioned is an example. It is also worthy to note that a rearrangement cannot be represented by the simple and discrete electron in a very good manner. For instance, in Wagner-Meerwein rearrangement, the actual mechanism of alkyl group migration involves the transfer of the alkyl group fluidly along with a bond, and not the typical bond breaking-making. Similarly, the pericyclic rearrangements are explained in terms of orbital interactions rather than discrete electron transfers. Nevertheless, it is quite possible to draw the curved arrows mechanism for rearrangement reaction for simple and fast understanding, even if these are not realistic necessarily, excepting in allylic rearrangement.

> ### Redox Reactions

Redox reactions in organic chemistry may simply be defined as the chemical changes where the reduction or oxidation of organic compounds occurs to give rise to new products. It is also worthy to note that the meaning of oxidations and reductions in organic chemistry is different from simple redox reactions because numerous reactions bear the label but do not include the actual electron transfer in the electrochemical context. In its place, organic oxidation is the gain of oxygen or loss of hydrogen; whereas organic reduction means the gain of hydrogen or the loss of oxygen. Nevertheless, simple functional groups can still be organized in the ascending order of oxidation states for approximation.

Birch Reduction

Finally, we need to remember that the reactant can undergo both oxidation and reduction in the same chemical reaction to give rise to two separate compounds (disproportionation reactions).

❖ Thermodynamic and Kinetic Requirements

In this section, we will discuss a brief idea of the thermodynamic and kinetic requirements of an organic reaction.

➢ *Thermodynamic Requirements for a Reaction*

A reaction takes place spontaneously if the magnitude of ΔG is negative i.e. the free energy of the reactants must be higher than the free energy of the products and vice-versa. From thermodynamics, we know that

$$\Delta G = \Delta H - T\Delta S \qquad (1)$$

$$\Delta G = -RT \ln K \qquad (2)$$

Where $\Delta H = \Sigma H_{products} - \Sigma H_{reactants}$, $\Delta S = \Sigma S_{products} - \Sigma S_{reactants}$ and T is the temperature. The symbols R and K represent the gas constant and equilibrium constant, respectively. Now, as the entropy changes for most of the organic reactions is quite small, the free energy change is primarily governed by the value of ΔH. Furthermore, it follows from equation (1) that ΔG will become more negative with a more negative value of ΔH, which in turn would increase the value of equilibrium constant K. Hence, more and more product will be formed as the enthalpy of the system decreases.

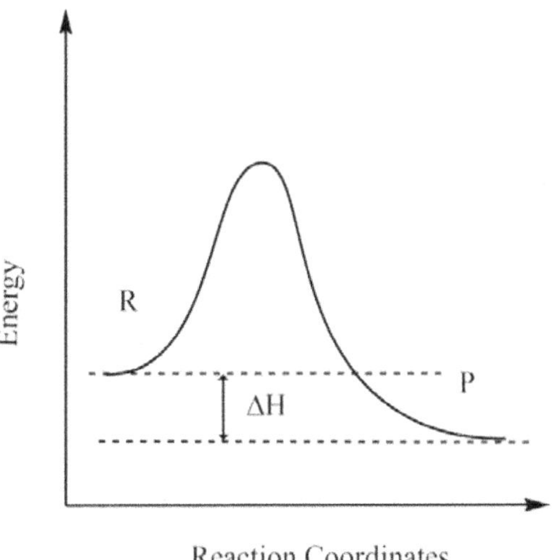

Figure 1. Energy profile diagram for thermodynamics of a typical organic reaction.

Furthermore, it is also very obvious from equation (1) that entropy becomes more important as the temperature of the reaction vessel is raised.

> ### *Kinetic Requirements for a Reaction*

The kinetic requirements of an organic reaction depend upon the activation energy of the same. If the activation energy barrier is low, the reaction will take place at a higher speed. If the activation energy barrier is high, the substance will react slowly and will take a very long time to complete.

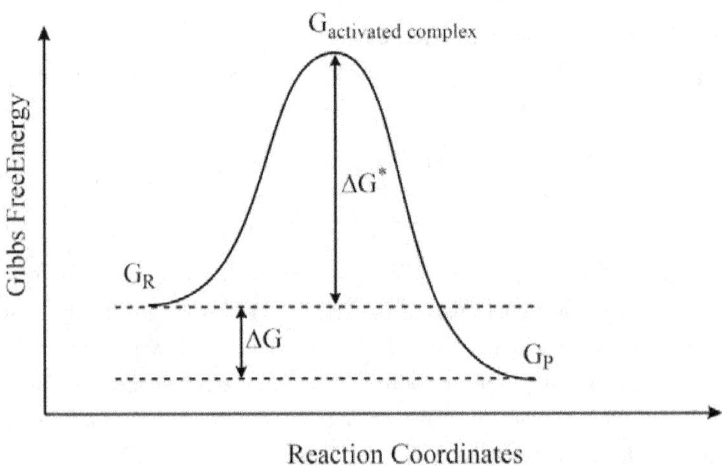

Figure 2. Energy profile diagram for kinetics of a typical organic reaction.

Since the kinetics of organic reactions is generally studied in the framework of "Activated complex Theory", which states that the rate constant for a typical reaction is

$$k = \frac{RT}{Nh}e^{-\frac{\Delta G^*}{RT}} = \frac{RT}{Nh} \times e^{\frac{\Delta S^*}{R}} \times e^{-\frac{\Delta H^*}{RT}} \tag{3}$$

Where ΔG^*, ΔH^* and ΔS^* are free energy change, enthalpy change, and the entropy change of the activation step, respectively. As far as equation (3) is concerned, it can easily be seen that as the free energy change of the activation step increases, the rate constant would decrease.

However, if we look at the simplified form i.e. equation (3), we find three factors; one is RT/Nh which is constant if the temperature is kept constant. The second factor involves ΔS^*, and therefore, we can conclude that the reaction rate would show an exponential increase if the entropy of activation increases. The third factor includes ΔH^*, and therefore, we can conclude that the reaction rate would show an exponential decrease if the enthalpy of activation increases. It is also worthy to note that the first two terms collectively make the frequency factor.

❖ Kinetic and Thermodynamic Control

The thermodynamic or the kinetic control in a chemical reaction can dictate the composition in the final product mixture when two or more competing pathways giving rise to different products and the selectivity or stereoselectivity is influenced by the reaction conditions. Consider a typical reaction in which reactant A can transform into product B and C i.e.

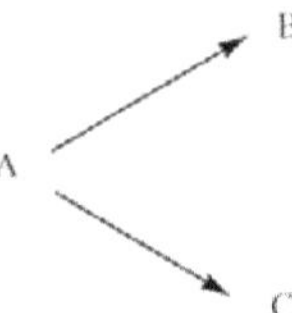

Now, will discuss the two cases of thermodynamic and kinetic control of the overall reaction depending upon the nature of the product.

➤ *Different Products from Thermodynamic and Kinetic Control*

The difference will be obvious when product C forms at a faster rate than product B because the activation energy for product C is lower than that for product B (even if B is more stable than A). In such a scenario, C is favored kinetically and called as the kinetic product; whereas B is favored thermodynamically controlled and is called as the thermodynamic product. The exact nature of the final product (thermodynamic or kinetic) will be dictated by many reaction conditions like temperature, pressure, solvent, or any other factor which might affect the reaction pathway. Also, it must be kept in mind that the above-mentioned statement is true only if the activation energy of the two routes are different; with dominating yield for the pathway having lower activation energy (E_a) than the other.

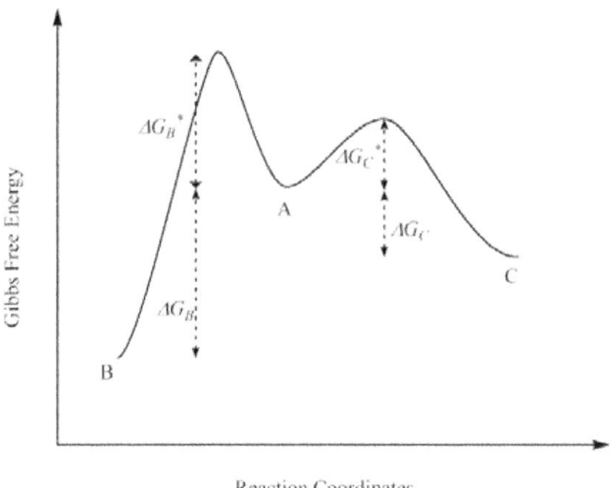

Figure 3. Energy profile diagram for kinetic versus thermodynamic product reaction.

> ### *Same Products from Thermodynamic and Kinetic Control*

In this case, product B forms at a faster rate than product C because the activation energy for product B is lower than that for product C. Also, since $\Delta G_B > \Delta G_C$, B is more stable thermodynamically than C. Therefore, the thermodynamic as well kinetic control of the reaction will favor the formation of product B. The activation of thermodynamic or kinetic control dictates the composition of the end product when these competing pathways give rise to different products influencing the selectivity of the reaction.

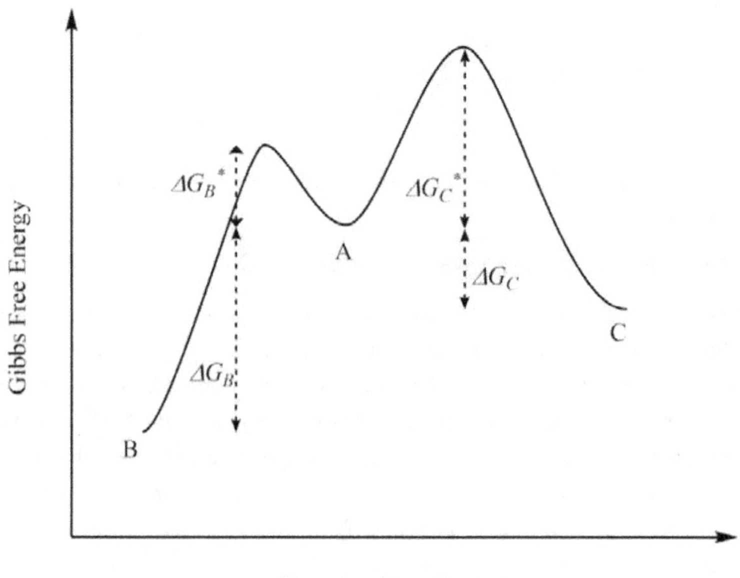

Figure 4. Energy profile diagram for kinetic versus thermodynamic product reaction.

The selectivity arising from kinetic vs thermodynamic control is extremely important in the case of asymmetric synthesis. This can be attributed to the fact that the pairs of enantiomers have the same Gibbs free energy; and therefore, the thermodynamic control will yield a racemic mixture by obligation. So, any catalytic reaction that gives the product with enantiomeric excess different than zero must be under some kind of kinetic control. Several stoichiometric asymmetric syntheses occur giving enantiomeric products due to a chiral substrate i.e., the reaction is actually a diastereoselective type; and therefore, even if such transformations are kinetically controlled, thermodynamic control is at least theoretically possible.

❖ Hammond's Postulate

Hammond's postulate states when a transition state gives rise to an unstable reaction intermediate (or product) it will have comparable energy to that intermediate, and both will be interconvertible by a very minute reorganization of molecular structure.

This hypothesis was first proposed by George Hammond in 1955 to correlate the transition states with intermediates and products. Nevertheless, besides Hammond's postulate, another chemist J. E. Leffler also gave an assumption that the transition state will have a greater resemblance to the less stable species (i.e., reactant, or intermediate, or product). Since many teachers or textbooks state 'Leffler assumption' but refer to the statement as 'Hammond's postulate' due to their similar arguments, it is better to use 'Hammond- Leffler postulate' to give the credit where it is due.

Therefore, the geometric structure of a state can be predicted by comparing its energy to the species neighboring it along with the reaction coordinate. For instance, the geometry of the transition state resembles reactants in an exothermic reaction; whereas the endothermic reactions will have transition states closer to products. This kind of assessment is very useful especially when most transition states cannot be analyzed experimentally.

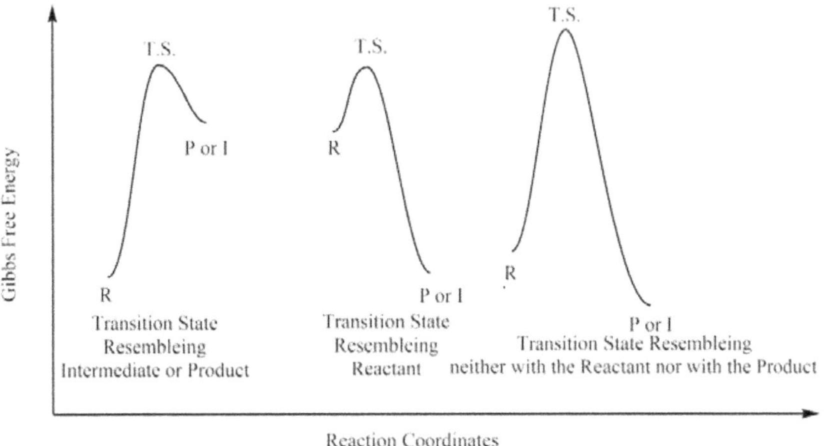

Figure 5. Three cases of according to Hammond's postulate.

Hammond's postulate can also be used to rationalize the Bell-Evans-Polanyi principle describes the experimental observation that the rate of a reaction (and so the activation energy) is influenced by the enthalpy of that reaction. Hammond's postulate describes this observation by explaining how does enthalpy variation results in the structure change of the transition state; which in turn, would change the energy of the transition state, and so the reaction rate as well (via changing activation energy).

❖ Curtin-Hammett Principle

The Curtin–Hammett principle in chemical kinetics was proposed by David Yarrow Curtin and Louis Plack Hammett to predict the relative ratio of the products obtained from a conformational equilibrium.

It states that the relative ratio of the products obtained from a conformational equilibrium is independent of the relative concentration of participating conformers and is a function of the free-energy-gap of the corresponding transition states provided that the rate of conformational-interconversion is much larger than the rate of formation of products.

To understand the Curtin-Hammett principle in perspective, consider a typical reaction in which products C and D are formed from two conformers A and B as

$$C \xleftarrow{k_C} A \underset{k_A}{\overset{k_B}{\rightleftharpoons}} B \xrightarrow{k_D} D \tag{4}$$

Since the enthalpy of activation for most of the organic reactions is greater than the equilibrium between participating conformers ($\Delta G_A^* > \Delta G^o$ and $\Delta G_B^* > \Delta G^o$), the reactants A and B are will remain abundant.

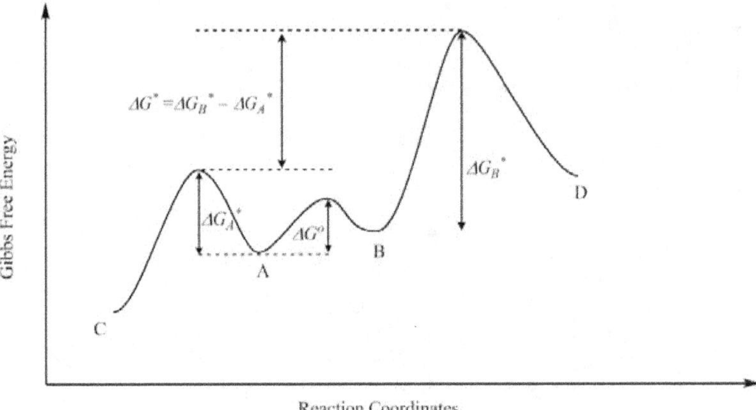

Figure 6. Potential energy diagram for Curtin-Hammett principle.

For the conformational equilibrium, the equilibrium constant (K) will be

$$K = \frac{B}{A} \tag{5}$$

The formation rate for C can be given as

$$\frac{dC}{dt} = k_C[A] \tag{6}$$

$$\frac{dD}{dt} = k_D[B] \tag{7}$$

Now considering the reaction is first-order or pseudo-first-order in nature, the ratio of the products can simply be obtained by diving equation (7) by equation (6) i.e.

$$\frac{dD}{dt} \times \frac{dt}{dC} = \frac{k_D[B]}{k_C[A]} \tag{8}$$

$$\frac{dD}{dC} = \frac{k_D}{k_C}\frac{[B]}{[A]} \tag{9}$$

Using the value of $[B]/[A]$ from equation (5), we get

$$\frac{dD}{dC} = \frac{k_D}{k_C}K \tag{10}$$

But we know from chemical kinetics that

$$K = e^{-\Delta G^o/RT} \tag{11}$$

$$k_C = \frac{RT}{Nh}e^{-\Delta G_A^*/RT} \tag{12}$$

$$k_D = \frac{RT}{Nh}e^{-\Delta G_B^*/RT} \tag{13}$$

Using equations (11–13) in equation (10), we get

$$\frac{dD}{dC} = \frac{\frac{RT}{Nh}e^{-\Delta G_B^*/RT}}{\frac{RT}{Nh}e^{-\Delta G_A^*/RT}}e^{-\Delta G^o/RT} \tag{14}$$

$$\frac{dD}{dC} = Products\ Ratio = e^{-\Delta G_B^*/RT}e^{+\Delta G_A^*/RT}e^{-\Delta G^o/RT} \tag{15}$$

$$Products\ Ratio = e^{-\frac{\Delta G_B^* + \Delta G_A^* - \Delta G^o}{RT}} \tag{16}$$

However, it is obvious from Figure 5 that $-\Delta G_B^* + \Delta G_A^* - \Delta G^o = -\Delta G^*$; therefore, we have

$$Products\ Ratio = e^{-\frac{\Delta G^*}{RT}} \tag{17}$$

Hence, the product ratio is governed by the energy gap between the transition states rather than ΔG^o, which is the famous Curtin-Hammett principle.

❖ Potential Energy Diagrams: Transition States and Intermediates

The potential energy diagrams or the reaction progress curves are nothing but a visual representation of the energy changes that occur during a chemical reaction. The energy of various species participating in the reaction is plotted on the y-axis (ordinate) whereas the progress of the reaction on the x-axis (abscissa). Now since a reaction can have transition states and intermediates (in addition to the reactants and final product), we must discuss the corresponding potential energy diagrams.

➤ *Potential energy Diagram of Reactions with Reactant, Product and Transition State*

The transition state of an organic reaction is a specific configuration along the reaction coordinate and corresponds to the highest potential energy along with this reaction coordinate. Furthermore, the transition state is not an actual molecule that can be isolated, and therefore it is often marked with the double dagger or star symbol to differentiate. The typical potential energy diagrams of endothermic and exothermic reactions with reactants, products, and transition states are shown below.

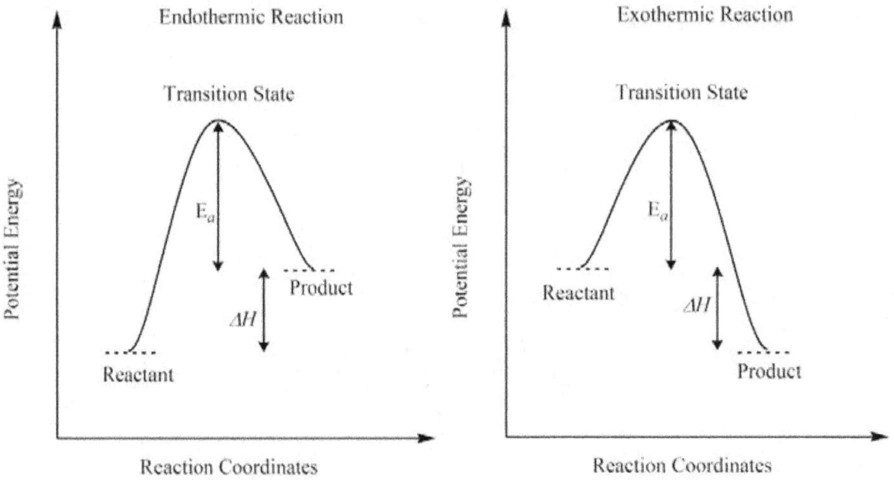

Figure 7. Typical potential energy curves for exothermic and endothermic reactions.

It is obvious from the potential energy diagrams that the enthalpy change is positive for an endothermic reaction ($\Delta H > 0$) and negative for an exothermic reaction ($\Delta H > 0$). For exothermic reactions, the potential energy of the reacting species first increases, attains a maximum at the transition state, and then decreases even more than the reactants. For endothermic reactions, the potential energy of the reacting species first increases, attains a maximum at the transition state, and then decreases less than the reactants.

Furthermore, it should also be noted that the energy gap between the transition state and reactant is the activation energy (E_a) which is responsible for the speed of the reaction. The higher activation energy leads to slow transformation and vice-versa.

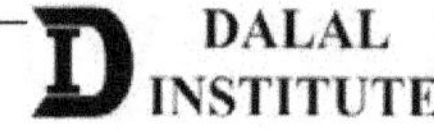

> ➤ *Potential energy Diagram of Reactions with Reactant, Product, Transition State, and Intermediates*

The chemical species that are formed somewhere during the course of a chemical reaction are called as reaction intermediates. Unlike transition states, these are actual molecules that are short-lived and unstable. Sometimes they are called temporary reactants or products because they are neither present in actual reactants nor the actual products. The typical potential energy diagram of a typical exothermic reaction with a reactant, product, intermediate, and transition state is shown below.

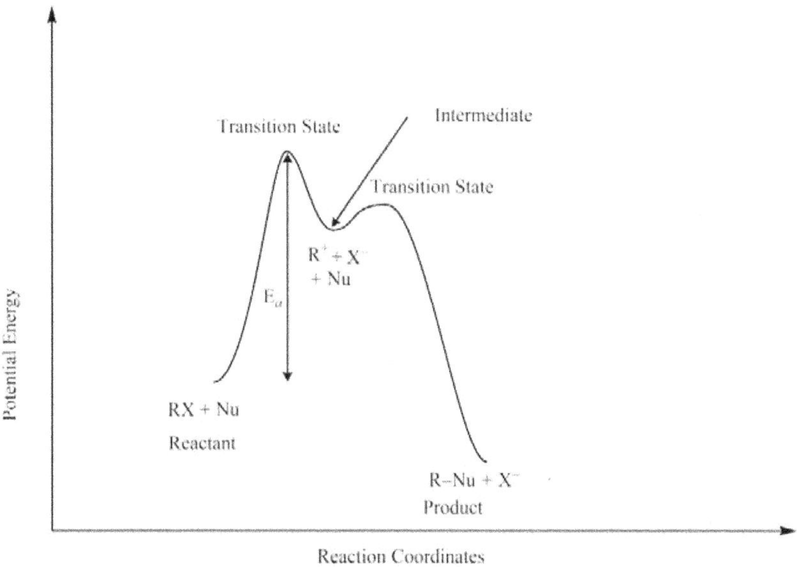

Figure 8. Potential energy diagram for the SN_1 reaction.

It is obvious from the potential energy diagrams that the intermediates are situated at the minima whereas the transition states are shown at the maxima. Also, just like the previous case, the energy gap between the transition state and reactant is the activation energy (E_a) which is responsible for the speed of the reaction i.e. the higher activation energy leads to slow transformation and vice-versa.

Reaction coordinate diagrams also give information about the equilibrium between a reactant or a product and an intermediate. If the barrier energy for going from intermediate to product is much higher than the one for the reactant to intermediate transition, it can be safely concluded that a complete equilibrium is established between the reactant and intermediate. Nevertheless, if the reactant-to-intermediate barrier is almost equal in energy to intermediate-to-product change, then no full equilibrium is set and steady-state approximation becomes activated to dictate the kinetic rate expressions.

❖ Methods of Determining Mechanisms

The reaction mechanism of organic compounds can be determined via several individual routes; however, it is quite common to use multiple methods for confirmatory results. Some of the major methods to determine the organic reaction mechanism are given below.

> *1. Product Identification*

The proposed mechanism must account for all the products in the experimental reaction including by-products. Therefore, the theoretically proposed mechanisms are shortlisted depending upon the nature and number of the products. This can be illustrated by the hydrolysis of isomeric allyl chlorides. Since the hydrolysis of 3-chloro-3-methylbut-1-en or 4-chloro-2-methylbut-2-en give rise to 15% of primary alcohol and 85% tertiary alcohol in each case, a common intermediate must be proposed to rationalize the results.

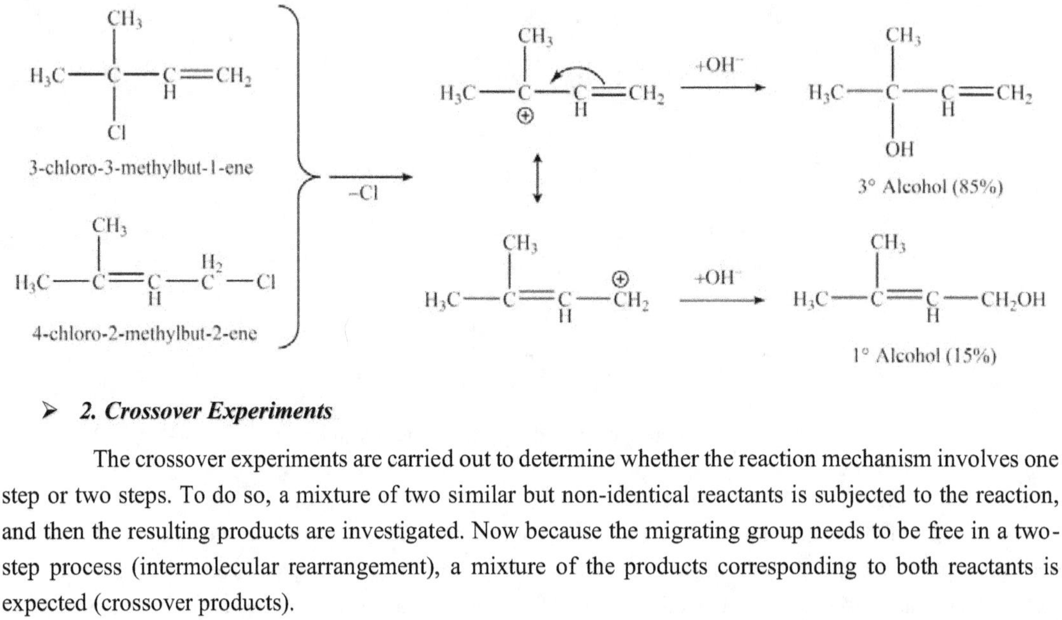

3-chloro-3-methylbut-1-ene

4-chloro-2-methylbut-2-ene

3° Alcohol (85%)

1° Alcohol (15%)

> *2. Crossover Experiments*

The crossover experiments are carried out to determine whether the reaction mechanism involves one step or two steps. To do so, a mixture of two similar but non-identical reactants is subjected to the reaction, and then the resulting products are investigated. Now because the migrating group needs to be free in a two-step process (intermolecular rearrangement), a mixture of the products corresponding to both reactants is expected (crossover products).

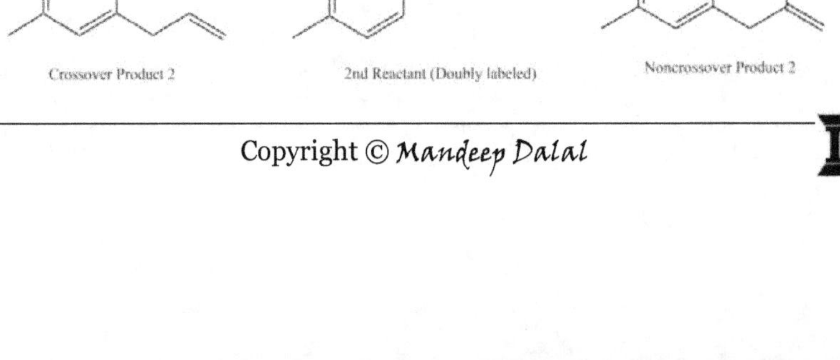

Crossover Product 1

1st Reactant (Not labeled)

Noncrossover Product 1

+

+

Crossover Product 2

2nd Reactant (Doubly labeled)

Noncrossover Product 2

On the other hand, if the process is single-step (intramolecular rearrangement) we will not get any crossover products. This can be illustrated by the Claisen-rearrangement in which allyl aryl ethers are transformed into allyl phenols.

The concept underlying the crossover experiment is a basic one: provided that the labeling method chosen does not affect the way a reaction proceeds, a shift in the labeling as observed in the products can be attributed to the reaction mechanism. The most important limitation in crossover experiment design is therefore that the labeling does not affect the reaction mechanism itself.

> ### ➢ *3. Isolation, Detection, and Trapping of Intermediates*

Several organic reactions proceed via the generation of intermediates. If these intermediates are identified experimentally, the writing of a reasonable mechanism becomes very easy. These intermediates are usually detected by spectrophotometric methods directly; however, some unstable intermediates are first isolated by trapping them with reactive substrates.

i) Isolation: In some cases, the isolation of an intermediate from the reaction mixture is possible either under mild conditions or by ending the reaction after some time. For instance, three intermediates can be isolated in the transformation of primary amides to primary amines (Hofmann rearrangement).

$$R-\underset{\underset{1^\circ\ \text{amide}}{}}{\overset{\overset{O}{\|}}{C}}-NH_2 \xrightarrow[-KBr,\ -H_2O]{Br_2/KOH} R-\overset{\overset{O}{\|}}{C}-\overset{H}{\underset{\cdot\cdot}{N}}-Br \xrightarrow[-H_2O]{KOH} K^+\left[R-\overset{\overset{O}{\|}}{C}-\overset{\ominus}{\underset{\cdot\cdot}{N}}-Br\right]$$

$$\downarrow -KBr$$

$$K_2CO_3 \ +\ R-NH_2 \xleftarrow{\quad 2KOH \quad} R-N{=}C{=}O$$

$$\underset{1^\circ\ \text{amine}}{} \qquad\qquad\qquad\qquad \underset{\text{Isocynate}}{}$$

ii) Detection: If the intermediate formed in the course of the chemical reaction is very unstable, it is better to use spectrophotometric methods to detect them. For instance, the carbocation formed in the E_1 and SN_1 pathways can be detected via NMR spectroscopy confirming the route.

iii) Trapping: In some reactions, the intermediate formed is so unstable and short-lived that the conventional techniques cannot be used to identify them. In such cases, it is better to mix the reactant with a special reactive agent which can react with the intermediate to form detectable species. For instance, the formation of carbene during the decomposition of diazomethane can be detected by trapping them via cyclohexene.

bicyclo[4.1.0]heptane

> ### 4. Isotopic Labelling

Isotopic labeling is a technique that is used to track the passage of reactants through a chemical reaction. The reactant is 'labeled' by displacing specific atoms by the corresponding isotope. The reactants are then permitted to undergo the chemical change, and the sites of the isotopes in the products are analyzed to find the sequence of the isotopes followed in the reaction mechanism. For instance, the mechanism of ester hydrolysis can be understood by replacing the normal water with H_2O^{18}.

It is obvious from the above reaction that it is the acyl-oxygen bond that gets broken rather than the alkyl-oxygen bond.

> ### 5. Kinetic Studies

The kinetic of an organic reaction is very useful in identifying the correct reaction mechanism. The rate-determining step (slowest step) infers about the reaction route from the overall rate law. One of the most popular examples where the reaction kinetics has been found to be exceptionally useful to determine reaction mechanism aliphatic nucleophilic substitution.

For instance, the hydrolysis of methyl bromide follows second order kinetics i.e.

$$-\frac{d[CH_3Br]}{dt} = k[CH_3Br][OH^-] \tag{18}$$

This is possible only if both reactants undergo a transition state instated of an intermediate. On the other hand, the hydrolysis of tert-butyl bromide follows first-order kinetics i.e.

$$-\frac{d[CH_3Br]}{dt} = k[(CH_3)_3CBr] \tag{19}$$

This is possible only if the substrate forms an intermediate instated transition state.

> ### 6. Stereochemical Analysis

In some cases, the stereochemistry of the reactants and products can also be used to identify the pathway of an organic reaction. For instance, the optical purity of the resulting product in aliphatic nucleophilic substitution in chiral compounds can infer whether it occurs via SN_1 or SN_2.

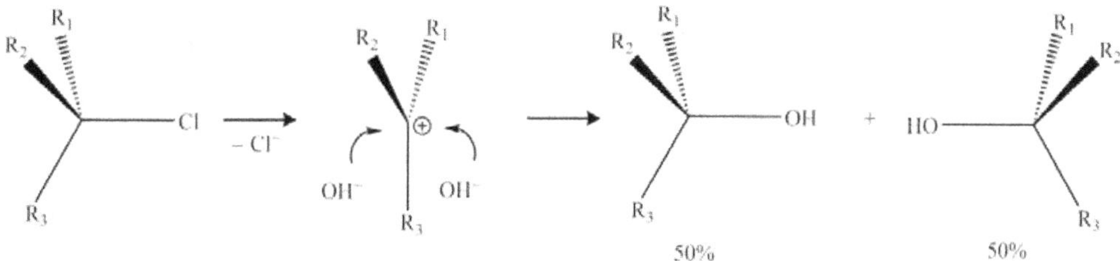

The racemization of the product showed that the reaction has proceeded via a dissociative pathway where the attack on the two faces of the carbocation has happened. On the other hand, we get a mostly optically pure compound, which suggests that the reaction has proceeded via a dissociative pathway where the inversion has happened through a transition state.

❖ Isotope Effects

In the previous section of this chapter, we studied isotopic labeling as a tool to detect the reaction mechanism. There we assumed that the isotopic substitution has no effect on the chemical profile of the reaction whatsoever which is far from the truth because the isotopic substitution can change the rate of chemical reactions in a significant manner.

The changes in the reaction rate when a particular atom in the reactant molecule is replaced by one of its isotopes are termed as isotopic effects.

Formally, it is called as "kinetic isotope effect", and it is calculated rate constants for the reactions with the lighter isotope (k_L) by the rate constant for reaction with the heavier isotope (k_H) i.e.

$$KIE = \frac{k_L}{k_H} \qquad\qquad (20)$$

This variation in the rate of reaction because the heavier isotopologues have lower vibrational frequencies relative to their lighter isotopes, and can easily be rationalized quantum mechanically. In general, this means that greater energy is needed for heavier isotopologues to cross the transition state (or the dissociation limit in rare cases), and therefore, a lower reaction rate is observed. The kinetic isotopic effect can primarily be classified into two categories; primary and secondary isotope effects.

➢ *Primary Isotope Effects*

The primary kinetic isotope effect is observed if a bond to the isotopically-labeled atom is being broken or formed. Now depending on the nature of the route followed to study kinetic isotope effect (parallel measurement of rates vs intramolecular competition vs intermolecular competition), we can find if the bond-formation or bond-breaking has happened at the rate-determining step, or in the subsequent product-determining step. Furthermore, it is also worthy to note that many textbooks are propagating the misconception that a primary kinetic isotope effect must echo bond breaking or formation to the isotope at the rate-determining step always.

For nucleophilic substitutions mentioned earlier, the primary kinetic isotope effects have been studied for the attacking nucleophiles, leaving groups, and the α-carbon at which the displacement occurs. The interpretation of the kinetic isotope effects as a function of the leaving group had been very complicated at first because of the significant contributions from temperature-independent factors. Though the primary kinetic isotope effect is less sensitive than the ideal, and the contribution from non-vibrational factors; the kinetic isotope effects at the α-carbon are employed to advance the understanding of the transition state's symmetry of the SN_2 pathway.

➢ *Secondary Isotope Effects*

We will get a secondary kinetic isotope effect if no bond formation or breaking happens to the isotopically-labeled atom in the reactant. Also, the secondary kinetic isotope effects are much smaller in magnitude than primary kinetic isotope effects; though the secondary isotope effects of deuterium can reach

up to 1.4 per atom. Furthermore, many experimental techniques have been invented and refined to measure heavy-element isotope effects with extremely good precision; and therefore, these kinds of effects are still very valuable for mechanisms-elucidating of many organic reactions.

In the case of many nucleophilic substitution reactions, secondary isotope effects of H at α-carbon offer a direct route to differentiate between SN_2 and SN_1 type reactions. It is observed that SN_1 reactions generally give rise to great secondary isotope effects (up to a theoretical maximum of 1.22); whereas SN_2 reactions usually have primary isotope effects which are close to or less than one in most cases. Kinetic isotope effects having a value greater than unity are typically labeled as 'normal kinetic isotope effects', whereas isotope effects having a value less than unity are called as 'inverse kinetic isotope effects'. On the whole, transition states with smaller force constants are expected to have a normal kinetic isotope effect; and transition states with larger force constants are expected to have an inverse kinetic isotope effect when stretching vibrational contributions control the isotope effect.

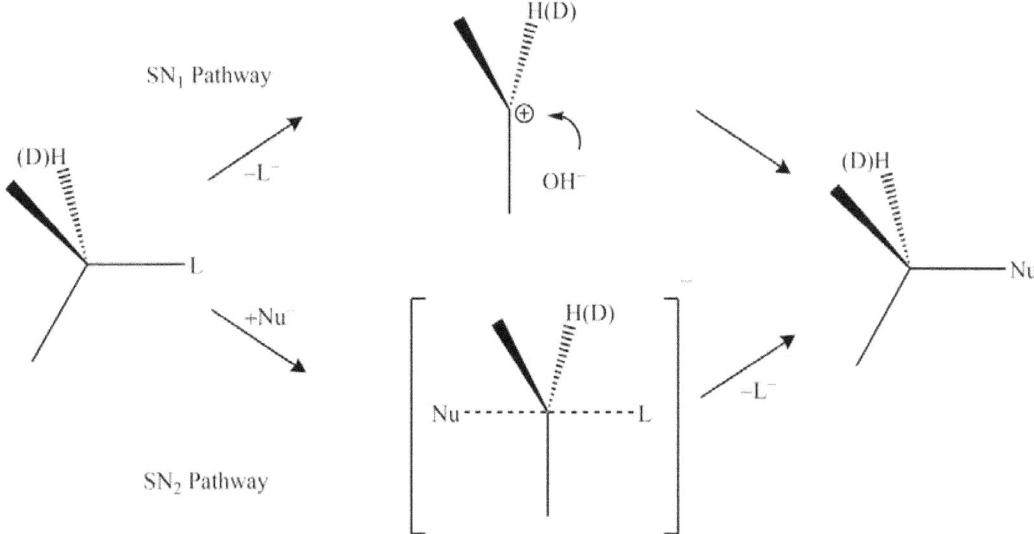

The Cα-H(D) vibration at the α-carbon dictates the magnitudes of these secondary isotope effects. Now because the carbon is converted into an *sp²* hybridized carbocation during an SN_1 reaction, the transition state for the rate-limiting step will have an increase in Cα-H(D) bond order; and therefore, an inverse kinetic isotope effect would be expected if only the stretching vibrations were the chief factors. On the other hand, large normal kinetic isotope effects are observed which are induced by significant out-of-plane bending vibrational contributions if we go from the reactants to the transition state of carbocation formation. In the case of SN_2 reactions, bending vibrations will still play a vital role in the kinetic isotope effect, but stretching vibrational contributions are also of significant importance; and therefore, the resulting isotope effect may be inverse or normal depending upon the dominating factor.

❖ Hard and Soft Acids and Bases

Most of the Lewis acids can be distributed into two categories; class a and class be acids. Generally speaking, class a acids prefer to bind with Lewis bases with donor atom from second row non-metal (i.e. F, O or N); whereas class b acids prefer to bind with Lewis bases with donor atom coming from third or below row (i.e. Br, I, S or P). An American chemist Ralph Pearson then developed a model where Lewis acids and bases can be characterized by strength and softness. To understand the concept, consider a typic acid-base reaction i.e.

$$A + \ddot{B} \rightleftharpoons A{:}B \tag{21}$$

Where A is the acid, $\ddot{B}$ is the base and $A{:}B$ represents the complex. The equilibrium constant for the reaction (21) can be written as

$$\log K = S_A S_B + \sigma_A \sigma_B \tag{22}$$

Where S_A and S_B represent the strength of acid A and base B, respectively. The symbol σ_A and σ_B represent the softness of acid A and base B, respectively. The experimental data showed that class a metals form stable complexes with ligands having N, O, or F as the donor site; whereas class b metals form stable complexes with ligands having P, S, or Cl like donor atoms. Therefore, Ralph Pearson articulated his observation as given below.

This trend is explainable by the hard-soft acid-base principle which states that hard acid prefers a hard base while soft acid prefers a soft base for binding to yield stable systems.

The typical classification of different species based on hardness or softness is quite useful in the understanding of the chemical reactivity of organic reactions.

➢ *Hard and Soft Acids*

Metals such as Li^{1+}, Ba^{2+}, $Mg^{2+,}$ and Al^{3+}, which have large negative reduction potential have a lesser tendency to attract electrons, and hence, form stable complexes with highly electronegative groups like N, F, or O so that they become unable to draw the unwanted electron density due to polarization. However, Metals like Pd^{2+} or Pt^{2+} which have large positive reduction potential have a greater tendency to accept electrons and hence form stable complexes with less electronegative groups like P so that they can easily grab the electron density by polarizing the surrounding.

A typical hard acid has a small size, high polarizing power, less distortable outer cloud, and high positive oxidation sometimes. On the other hand, a typical soft acid has a large size, low polarizing power, a highly distortable outer cloud, and zero or low positive oxidation sometimes.

Borderline cases are also identified: borderline acids are trimethylborane, sulfur dioxide and ferrous Fe^{2+}, cobalt Co^{2+} cesium Cs^+ and lead Pb^{2+} cations. It is also very important to mention that the borderline list may vary from book to book it is also medium-dependent.

Table 1. The class *a*, class *b*, and borderline metals.

Hard acid	Soft acid	Borderline acids
H^+, Li^+, Na^+, K^+	Cu^+, Ag^+, Au^+, Tl^+	
Be^{2+}, Mg^{2+}, Ca^{2+}, Sr^{2+}	Hg^{2+}, Pd^{2+}, Pt^{2+}	Mn^{2+}, Fe^{2+}, Co^{2+}, Sr^{2+} Ni^{2+}, Cu^{2+}, Zn^{2+}
Al^{3+}, Ga^{3+}, In^{3+} Cr^{3+}, Mn^{3+}, Fe^{3+}, Co^{3+}, La^{3+}, Ce^{3+}, Gd^{3+}	Tl^{3+}	
In^{4+}, Zr^{4+}, Hf^{4+}, Th^{4+}, U^{4+}, Pu^{4+}	Pt^{4+}	

➤ *Hard and Soft Bases*

The base becomes more and more soft as the donor atom is selected from the lower part of a particular group in the periodic table. For instance, F^- is the hardest, and I^- is the softest base in its group.

A typical soft base has the donor atom of low polarizability, high electronegativity, high negative charge density, and tightly held electron cloud. On the other hand, a typical soft base the donor atom of high polarizability, low electronegativity, less negative charge density, and loosely held electron cloud.

Borderline cases are also identified: borderline bases are aniline, pyridine, nitrogen N_2, and the azide, chloride, bromide, nitrate, and sulfate anions. It is also very important to mention that the borderline list may vary from book to book it is also medium-dependent.

Table 2. The hard, soft, and borderline bases.

Hard bases	Soft bases	Borderline bases
F^-, OH^-, F^-, CH_3COO^-, H_2O, SO_4^{2-}	RSH, R_2S, R_3P	$C_6H_5NH_2$
NO_3^-, Cl^-, ROH, R_2O, RO^-	RS^-, I^-, SCN^-, H^-, $(RO)_3P$, CN^-, Alkenes, R^-	N_3^-, Br^-, NO_2^-, N_2
N_2H_4, NH_3, RLi, $RMgX$		SO_2^{2-}
$RCOOR$, RNH_2		

It is also worthy to mention that hard nucleophiles or hard bases have the highest occupied molecular orbitals (HOMO) of higher energy whereas soft nucleophiles or soft bases have HOMOs of lower energy.

❖ Generation, Structure, Stability and Reactivity of Carbocations, Carbanions, Free Radicals, Carbenes and Nitrenes

A number of chemical reactions proceed via the formation of certain chemical species which are formed somewhere during the overall pathway. These species are called as reaction intermediates and are actual molecules that are short-lived and unstable. Sometimes they are called temporary reactants or products because they are neither present in actual reactants nor the actual products. Here, we will study the generation, structure, stability, reactivity of carbocations, carbanions, free radicals, carbenes, and nitrenes.

➢ *Carbocations*

The term carbocations in organic chemistry may simply be defined as the chemical species that carries a positive charge on the carbon with only six valence electrons.

Since the carbon in carbocations has only six electrons, it is electron deficient; and therefore, acts as an electrophile in chemical reactions.

1. Generation of carbocations: The heterolytic cleavage of the covalent bond is responsible for the generation of most of the carbocation species. Some reactions involving the production of carbocations are given.

i) Ionization of alkyl halides in polar solvents:

$$H_3C-\underset{\underset{CH_3}{|}}{\overset{\overset{CH_3}{|}}{C}}-Cl \longrightarrow H_3C-\underset{\underset{CH_3}{|}}{\overset{\overset{CH_3}{|}}{C}}^{\oplus} + Cl^-$$

ii) Protonation of alcohols followed by dehydration:

$$R-\ddot{O}H + H^+ \longrightarrow R-\overset{\oplus}{\underset{\cdot\cdot}{O}}H_2 \longrightarrow R^+ + H_2O$$

iii) Protonation of unsaturated systems:

$$R-\underset{H}{C}=CH_2 + H^+ \longrightarrow R-\overset{H}{\underset{\oplus}{C}}-CH_3$$

iv) Action of super acids on alkyl fluorides:

$$R-F + SbF_5 \xrightarrow{FSO_3H} R^{\oplus} + SbF_6^{\ominus}$$

v) Deamination of primary aliphatic amines by nitrous acid:

$$R-NH_2 + HNO_2 \xrightarrow[-2H_2O]{} R-\overset{\oplus}{N}\equiv N \longrightarrow R^{\oplus} + N_2$$

2. Structure of carbocations: It has been experimentally found that the carbocations are trigonal planar around the carbon bearing positive charge. Now valence bond theory, as well as molecular orbital theory, easily accounted for such structure, it is more comfortable to discuss the valence bond approach. The carbon with a positive charge is in sp^2 hybridization with three hybrid orbitals oriented at $120°$ in a plane with an empty p_z orbital at the perpendicular.

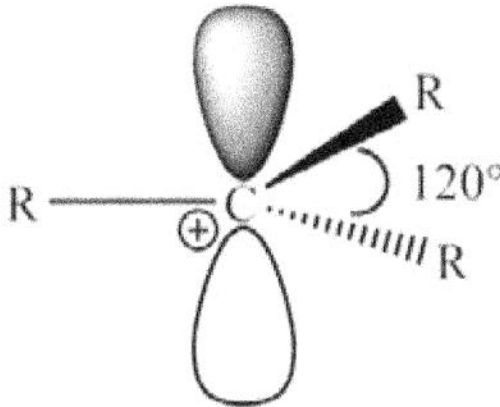

Figure 9. Orbital structure of carbocation.

3. Stability of carbocations: Before we discuss the stability of carbocations, we need to classify them on the basis of saturation. The first case is alkyl carbocations which are given below.

Methyl carbocation	Ethyl carbocation (1°)	Isopropyl carbocation (2°)	tert-butyl carbocation (3°)

The second case is of unsaturated carbocations where the carbon bearing positive charge is directly connected to a carbon participating in multiple bond i.e.

Allyl carbocation	Benzyl carbocation	Diphenylmethyl carbocation	Triphenylmethyl carbocation

Since the carbon in carbocations has only six electrons, it is electron deficient; and therefore, any effect that can compensate for the deficiency will stabilize the carbocation.

i) Stability of alky carbocations on the basis of inductive effect:

Since the alkyl group has an electron donating effect (+I), the stability of the carbocation will increase as the number of donating ability of the attached group increases. The stability order of alky carbocations on the basis of inductive effect is given below.

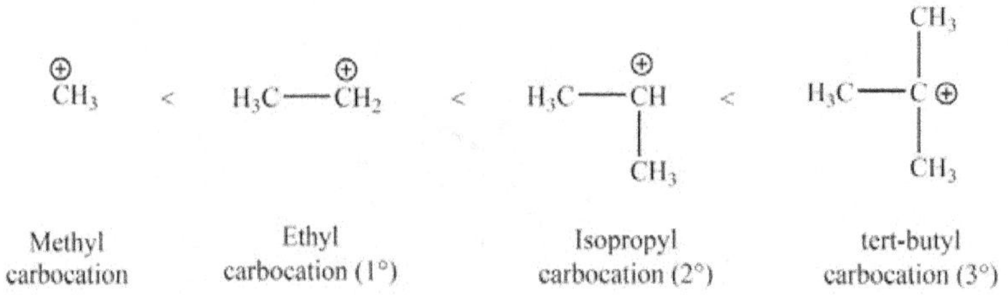

$$\overset{\oplus}{CH_3} \quad < \quad H_3C-\overset{\oplus}{CH_2} \quad < \quad H_3C-\overset{\oplus}{\underset{CH_3}{CH}} \quad < \quad H_3C-\overset{CH_3}{\underset{CH_3}{\overset{|}{\underset{|}{C}}}}\oplus$$

| Methyl | Ethyl | Isopropyl | tert-butyl |
| carbocation | carbocation (1°) | carbocation (2°) | carbocation (3°) |

ii) Stability of alky carbocations on the basis of hyperconjugation:

The existence of the hyperconjugation effect can be used to rationalize the relative stability of different carbocations as shown below.

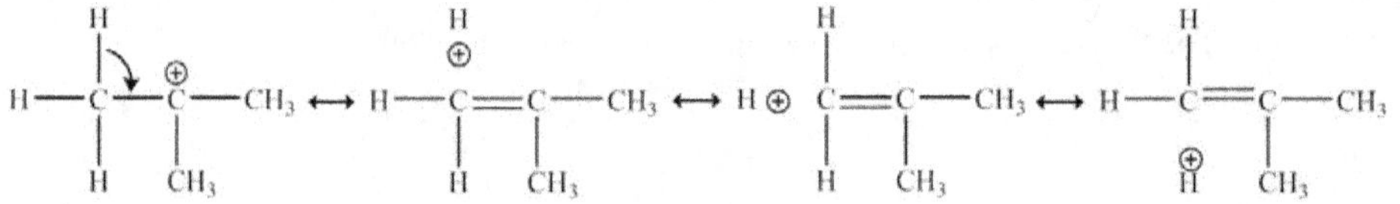

tert-butyl carbocation

$$\left(\begin{array}{l}\text{3 hyperconjugative structures for one methyl group}\\ \text{Total hyperconjugative structures = 9}\end{array}\right)$$

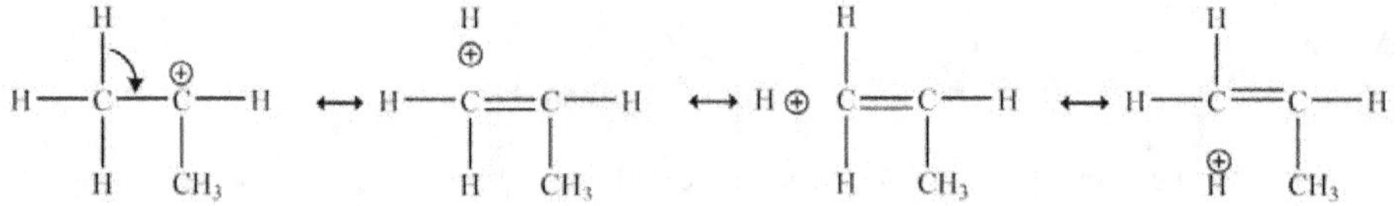

iso-propyl carbocation

$$\left(\begin{array}{l}\text{3 hyperconjugative structures for one methyl group}\\ \text{Total hyperconjugative structures = 6}\end{array}\right)$$

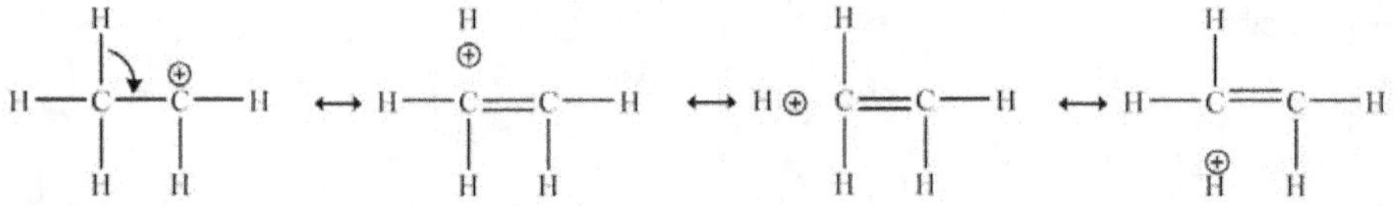

ethyl carbocation

$$\left(\begin{array}{l}\text{3 hyperconjugative structures for one methyl group}\\ \text{Total hyperconjugative structures = 3}\end{array}\right)$$

Hence, as far as the number of possible hyper-conjugative structures possible is concerned, tertiary carbocation should be more stable than secondary, which in turn should be more stable than primary.

iii) Stability of alky carbocations on the basis of steric effect:

Since the alkyl carbocations are primarily obtained from alkyl halides with tetrahedral geometry, a link between the steric relief and carbocation formed can be established. During the formation of carbocations in such cases, the carbon-carbon bond angles change from 109°28' to 120°. Therefore, the carbon with bulky groups around is expected to get more relief from this carbocationic conversion. The stability order of alky carbocations on the basis of steric effect is given below.

$$CH_3X \qquad (CH_3)CH_2X \qquad (CH_3)_2CHX \qquad (CH_3)_3CX$$

$$\overset{\oplus}{C}H_3 \quad < \quad H_3C\!-\!\overset{\oplus}{C}H_2 \quad < \quad H_3C\!-\!\underset{CH_3}{\overset{\oplus}{C}H} \quad < \quad H_3C\!-\!\underset{CH_3}{\overset{CH_3}{\overset{|}{C}}}\!\oplus$$

iv) Stability of ally and benzyl carbocations:

The stability of the carbocations in which the carbon bearing positive charge is adjacent to the double or triple bond can be rationalized in terms of resonance effect. First of all, let us draw the resonance structures allyl and benzyl carbocations.

$$H_2C\!=\!\underset{H}{C}\!-\!\overset{\oplus}{C}H_2 \quad \longleftrightarrow \quad \overset{\oplus}{H_2C}\!-\!\underset{H}{C}\!=\!CH_2$$

Resonance stablized allyl cation

Resonance stablized benzyl cation

Now, as the number of phenyl groups attached to carbon bearing positive charge increases, the number of resonating structures will also increase, and hence the stability.

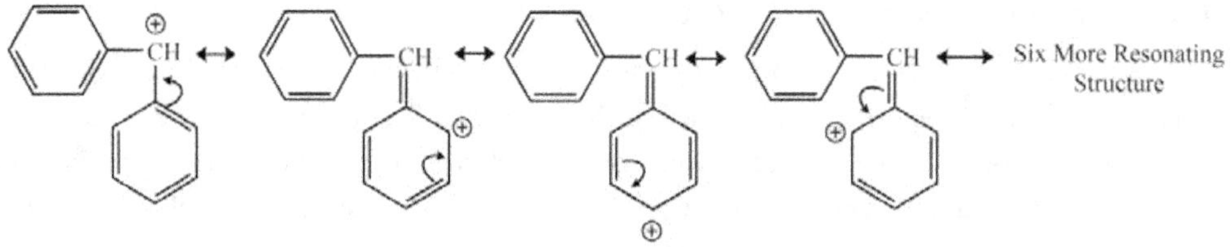

Resonance stablized diphenylmethyl carbocation

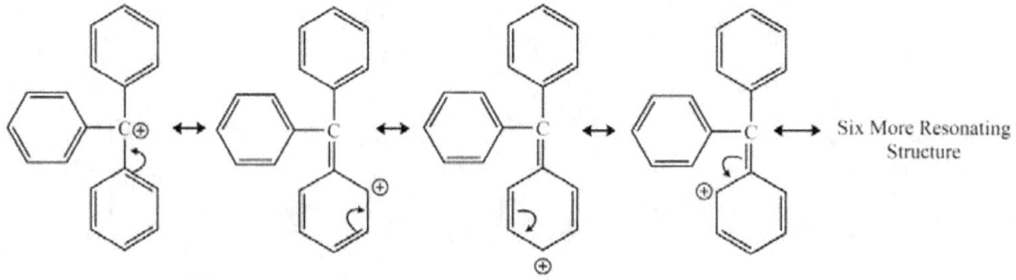

Resonance stablized triphenylmethyl carbocation

Therefore, the expected order of the stability of unsaturated systems with carbon bearing positive charge should be as given below.

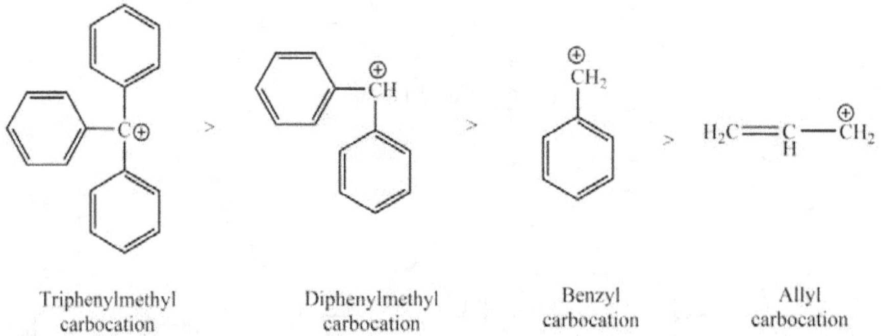

| Triphenylmethyl carbocation | Diphenylmethyl carbocation | Benzyl carbocation | Allyl carbocation |

Similarly, the order of stability in phenylcyclopropenyl, diphenylcyclopropenyl and triphenylcyclopropenyl should follow the following order.

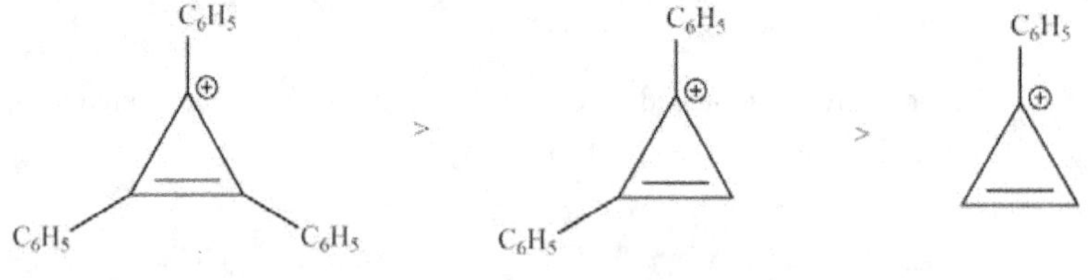

v) Stability of substituted benzyl carbocations: Since the carbon is electron bearing positive charge is electron deficient in nature, any group with +R effect will stabilize the system and vice-versa. The order of stability of some typically substituted carbocations is given below.

(4-methoxyphenyl)methylium   phenylmethylium   (4-nitrophenyl)methylium

vi) Stability of tropylium ion: The cycloheptatrienyl cation or tropylium ion is exceptionally stable due to its aromatic character (planar and $4n+2$ π electrons). According to molecular orbital theory, its delocalization energy is significantly greater than the delocalization energy of its acyclic counterpart. Similarly, the valence bond theory can also explain its exceptional stability of the basis of resonance as given below.

vii) Instability of cyclopentadienyl cation: The cyclopentadienyl cation is very unstable due to its antiaromatic character (planar and $4n$ π electrons). According to molecular orbital theory, its delocalization energy is significantly less than the delocalization energy of its acyclic counterpart.

$4n+2$ π electrons
(antiaromatic)

viii) Stability of alkoxyalkyl cation: if the positive charge bearing carbon in the carbocationic species is connected to a hetero atom with lone pair of electrons, the resonance will get it stabilized.

ix) Stability of acyl cation: Just like alkoxyalkyl cation, the resonance will also stabilize the acyl cation as shown below.

$$R—C\equiv\overset{\oplus}{O} \longleftrightarrow R—\overset{\oplus}{C}=\overset{..}{\underset{..}{O}}$$

x) Instability of phenyl and vinyl cation: If the positive charge is on the double-bonded carbon atom, the system cannot be stabilized because the *sp²* orbital carrying positive charge will be perpendicular to the orbital of the double bond.

$$H_2C=\overset{\oplus}{CH}$$

4. Reactivity of carbocations: The principal routes by which the carbocations can react to give rise to stable products are given below.

i) Nucleophilic attack: In these types of reactions, a carbocation may combine with a species by accepting an electron pair. Furthermore, it should also be noted that if all the three groups on the carbocation are different, a racemic mixture will be obtained.

ii) Proton removal: In these types of reactions, a carbocation may result in the removal of a proton from the adjacent atom forming a double bond.

iii) Rearrangement reaction: 1-2 methyl shit or 1-2 hydride shifts are very common in carbocation chemistry to attain a more stable counterpart. For instance, a primary carbocation will prefer to rearrange itself into a more stable tertiary carbocation.

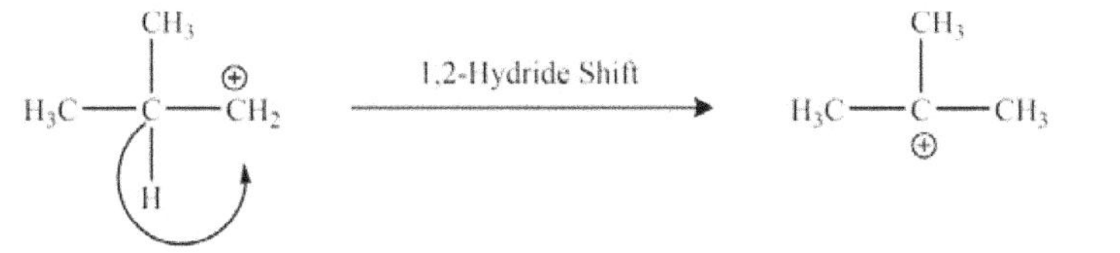

$1°$ carbocation (less stable) $2°$ carbocation (less stable)

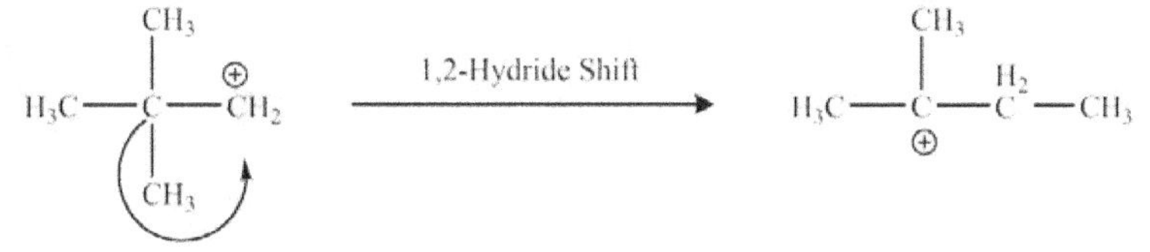

$1°$ carbocation (less stable) $3°$ carbocation (less stable)

iv) Addition reactions: A carbocation may attack at the triangular face of a double bond to create a new positively charged center as shown below.

> ➢ **Carbanions**

The term carbanions in organic chemistry may simply be defined as the chemical species that carries a negative charge on the carbon with only eight valence electrons.

Since the carbon in carbanions has its octet complete, it is electron-rich; and therefore, acts as a nucleophile in chemical reactions.

1. Generation of carbanions: The heterolytic cleavage of the covalent bond is responsible for the generation of most of the carbanions species. Some reactions involving the production of carbanions are given.

i) Hydrogen abstraction by a strong base from the carbon alpha to cyano, nitro, or carbonyl groups:

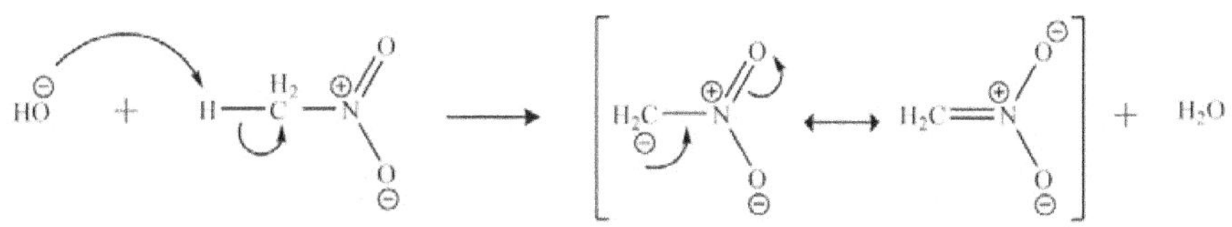

Resonance stablized ion

ii) Nucleophilic addition to α, β-unsaturated species:

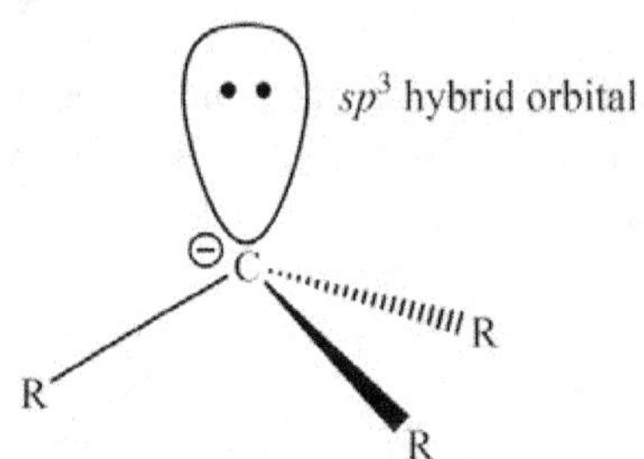

Resonance stablized ion

iii) Abstraction of terminal hydrogen from acetylene:

2. Structure of carbanions: It has been experimentally found that the carbanions are trigonal pyramidal around the carbon bearing negative charge. Now valence bond theory, as well as molecular orbital theory, easily accounted for such structure, it is more comfortable to discuss the valence bond approach. The carbon with a negative charge is in sp^3 hybridization with three hybrid orbitals forming bonds and the fourth hybrid orbital containing lone pair of electrons.

Figure 10. Orbital structure of carbonation.

Furthermore, it should also be kept in mind that if the carbon bearing negative charge is adjacent to multiple bonds, the carbanions will adopt a planar structure to get stable by dispersing the negative charge.

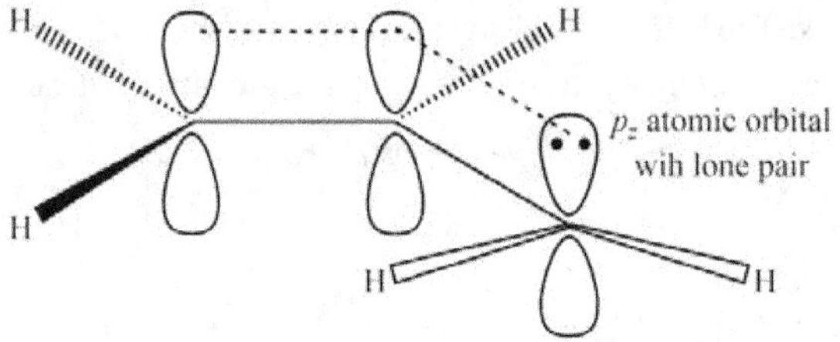

Figure 11. The planar structure of allyl carbanion.

3. Stability of carbanions: Before we discuss the stability of carbanions, we need to classify them on the basis of saturation. The first case is alkyl carbanions which are given below.

Methyl carbation	Ethyl carbation (1°)	Isopropyl carbation (2°)	tert-butyl carbation (3°)

The second case is of unsaturated carbanions where the carbon bearing negative charge is directly connected to a carbon participating in multiple bonds i.e.

Allyl carbation	Benzyl carbation	Diphenylmethyl carbation	Triphenylmethyl carbation

Since the carbon in carbanions has eight valence electrons, it is electron-rich; and therefore, any effect that can compensate the electron density accumulation will stabilize the carbanions.

i) Stability of alky carbanions on the basis of inductive effect:

Since the alkyl group has an electron donating effect (+I), the stability of the carbocation will increase as the number of donating ability of the attached group increases. The stability order of alky carbocations on the basis of inductive effect is given below.

Methyl carbation	Ethyl carbation (1°)	Isopropyl carbation (2°)	tert-butyl carbation (3°)

ii) Stability of carbanions on the basis of hybridization:

The nature and type of hybridization effect can be used to rationalize the relative stability of different carbanions as shown below.

$$HC \equiv C:^{\ominus} \quad > \quad H_2C = \overset{\ominus}{C}H \quad > \quad H_3C - \overset{\ominus}{C}H_2$$

50% *s*-character   33% *s*-character   25% *s*-character
(*sp*-hybridization)   (*sp²*-hybridization)   (*sp³*-hybridization)

Hence, we can say that as the s-character of carbon bearing negative charge increases, the lone pair gets better stabilization; and therefore, overall carbanionic stability also increases.

iii) Stability of ally and benzyl carbanions:

The stability of the carbanions in which the carbon bearing negative charge is adjacent to the double or triple bond can be rationalized in terms of resonance effect. First of all, let us draw the resonance structures of allyl and benzyl carbanions.

$$H_2C - \underset{H}{C} = CH_2 \quad \longleftrightarrow \quad H_2C = \underset{H}{C} - CH_2$$

Resonance stablized allyl anion

Resonance stablized benzyl anion

Now, as the number of phenyl groups attached to carbon bearing negative charge increases, the number of resonating structures will also increase, and hence the stability.

Three more resonating structure

Resonance stablized diphenylmethyl anion

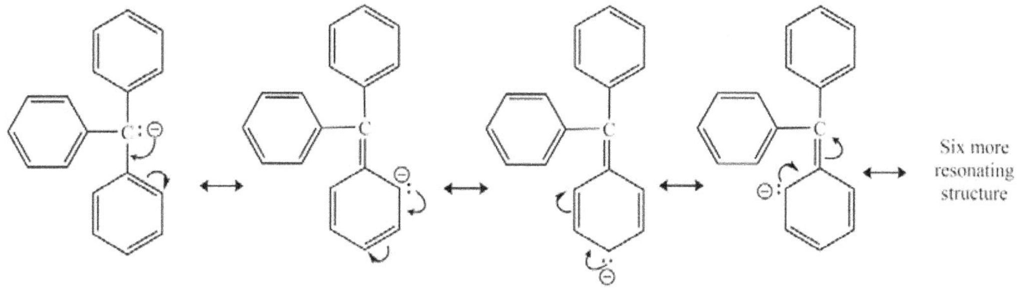

Resonance stablized triphenylmethyl anion

Therefore, the expected order of the stability of unsaturated systems with carbon bearing negative charge should be as given below.

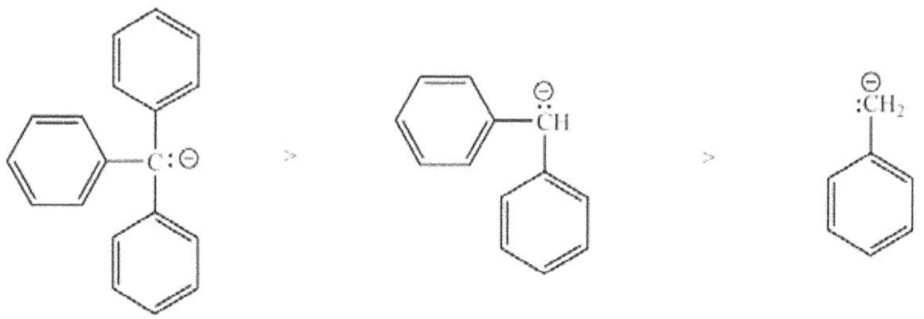

iv) Stability of substituted benzyl carbanions: Since the carbon is electron bearing negative charge is electron-rich in nature, any group with −R effect will stabilize the system and vice-versa. The order of stability of some typically substituted carbanions is given below.

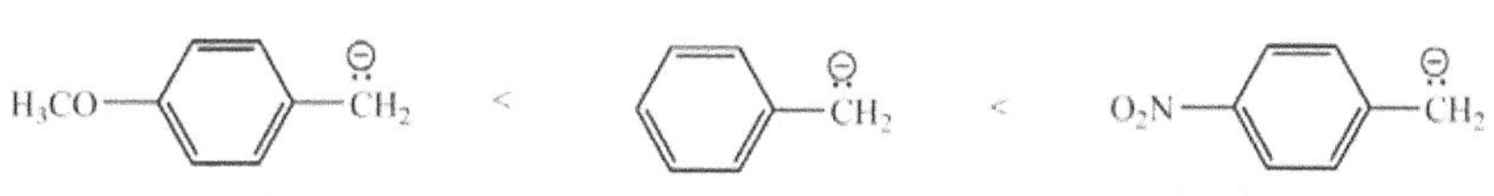

(4-methoxyphenyl)methanide   <   phenylmethanide   <   (4-nitrophenyl)methanide

v) Instability of cyclopentadienyl anion: The cyclopentadienyl anion is very unstable due to its aromatic character (planar and $4n+2$ π electrons). According to molecular orbital theory, its delocalization energy is significantly greater than the delocalization energy of its acyclic counterpart.

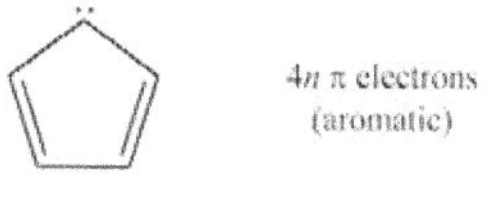

$4n$ π electrons
(aromatic)

vi) Stability of carbanions *with electron-withdrawing groups:* The presence of electron-withdrawing group will distribute the charge over a wider range; and therefore, will result in a greater stabilization.

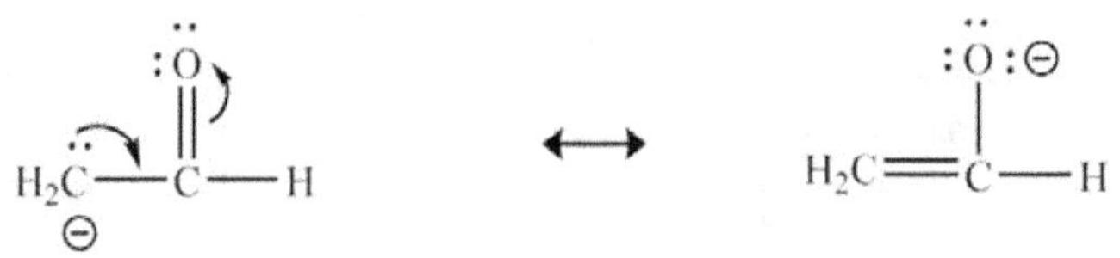

Resonance stablization in acetaldehyde anion

4. Reactivity of carbanions: The principal routes by which the carbanions can react to give rise to stable products are given below.

i) Lone pair donation: One of the most common pathways of carbanions reaction is the donation of electrons to a positive species like proton, or some species with an empty orbital.

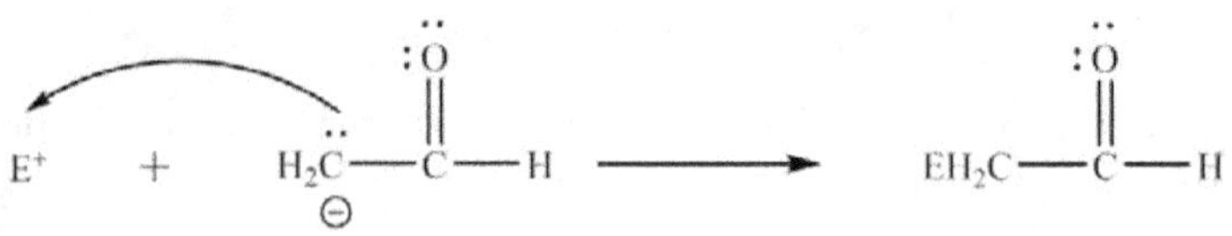

ii) Associative reaction: The carbanions can bond with a tetra-coordinated carbon followed by displacement of one of the previously attached groups.

iii) Rearrangement reaction: Like carbocations, the carbanions can also undergo rearrangement reaction although it is not very common.

iv) Addition reactions: A carbonation may attack at the triangular face of a double bond to create a new negatively charged center as shown below.

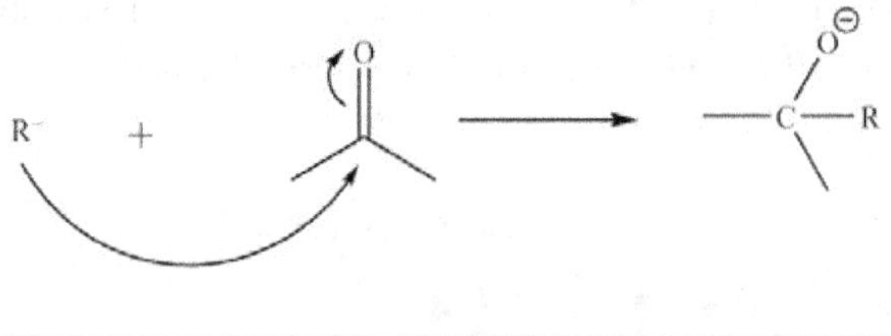

> *Free Radicals*

The term free radicals in organic chemistry may simply be defined as the chemical species that carries odd or unpaired electrons on the carbon with only seven valence electrons.

Since the carbon in free radicals has only seven electrons, it is electron deficient; and therefore, acts as an electrophile in chemical reactions.

1. Generation of free radicals: The homolytic cleavage of the covalent bond is responsible for the generation of most of the free radicals species. Some reactions involving the production of free radicals are given below.

i) Thermal cleavage:

$$R-N{=}N-R \quad \xrightarrow{\ \Delta\ } \quad 2\ \dot{R} \quad + \quad N_2$$

$$Me_4Pb \quad \xrightarrow{\ \Delta\ } \quad 4\ \dot{C}H_3 \quad + \quad PbH2$$

ii) Photochemical Cleavage:

$$Cl_2 \quad \xrightarrow{\ h\nu\ } \quad Cl\cdot \quad + \quad Cl\cdot$$

2. Structure of free radicals: It has been experimentally found that the free radicals are trigonal planar around the carbon bearing odd electron. Now valence bond theory, as well as molecular orbital theory, easily accounted for such structure, it is more comfortable to discuss the valence bond approach. The carbon with the odd electron is in sp^2 hybridization with three hybrid orbitals oriented at $120°$ in a plane perpendicular to p_z orbital occupied by the odd electron.

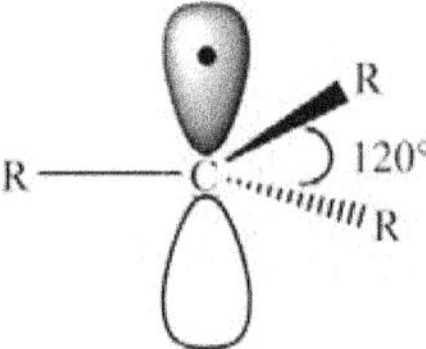

Figure 12. Orbital structure of free radical.

3. Stability of free radicals: Before we discuss the stability of free radicals, we need to classify them on the basis of saturation. The first case is alkyl free radicals which are given below.

$\dot{C}H_3$	$H_3C \!-\! \dot{C}H_2$	$H_3C \!-\! \dot{C}H \,|\, CH_3$	$H_3C \!-\! \dot{C} \cdot$ with CH_3
Methyl free radical	Ethyl free radical (1°)	Isopropyl free radical (2°)	tert-butyl free radical (3°)

The second case is of unsaturated free radicals where the carbon bearing odd electron is directly connected to a carbon participating in multiple bonds i.e.

$H_2C \!=\! \underset{H}{C} \!-\! \dot{C}H_2$	$\dot{C}H_2$ (phenyl)	(phenyl)$\!-\!\dot{C}H$ (phenyl)	(phenyl)$\!-\!\dot{C}\!-\!$(phenyl) with (phenyl)
Allyl free radical	Benzyl free radical	Diphenylmethyl free radical	Triphenylmethyl free radical

Since the carbon in free radicals has only seven electrons, it is electron deficient; and therefore, any effect that can compensate for the deficiency will stabilize the carbocation.

i) Stability of alky free radicals on the basis of inductive effect:

Since the alkyl group has an electron-donating effect (+I), the stability of the free radicals will increase as the number of donating groups attached increases. The stability order of alky free radicals on the basis of inductive effect is given below.

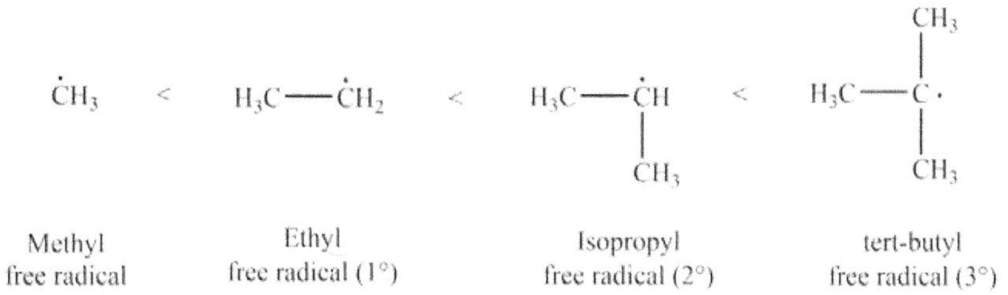

$$\dot{C}H_3 \quad < \quad H_3C \!-\! \dot{C}H_2 \quad < \quad H_3C \!-\! \dot{C}H \!-\! CH_3 \quad < \quad H_3C \!-\! \overset{CH_3}{\underset{CH_3}{\overset{|}{\underset{|}{C}}}}\! \cdot$$

Methyl free radical   Ethyl free radical (1°)   Isopropyl free radical (2°)   tert-butyl free radical (3°)

ii) Stability of alky free radicals on the basis of hyperconjugation:

The existence of the hyperconjugation effect can be used to rationalize the relative stability of different free radicals as shown below.

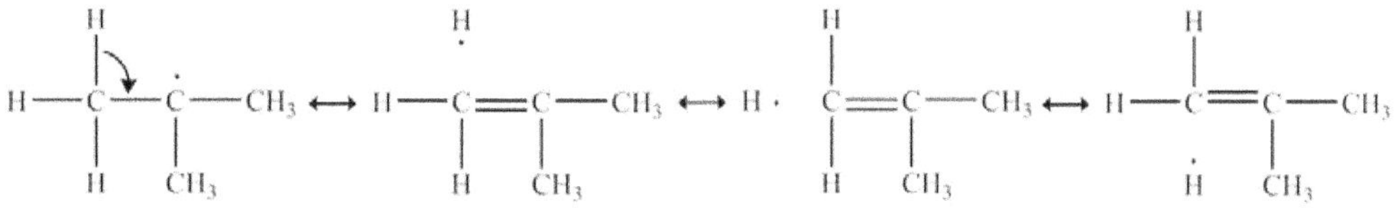

tert-butyl carbocation

$$\left(\begin{array}{c}\text{3 hyperconjugative structures for one methyl group}\\ \text{Total hyperconjugative structures} = 9\end{array}\right)$$

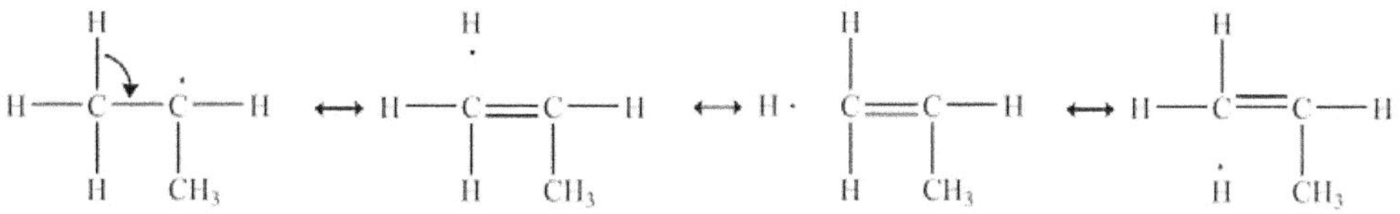

iso-propyl carbocation

$$\left(\begin{array}{c}\text{3 hyperconjugative structures for one methyl group}\\ \text{Total hyperconjugative structures} = 6\end{array}\right)$$

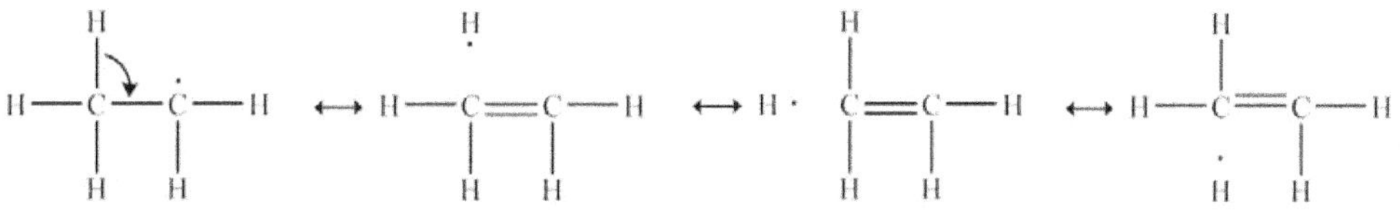

ethyl carbocation

$$\left(\begin{array}{c}\text{3 hyperconjugative structures for one methyl group}\\ \text{Total hyperconjugative structures} = 3\end{array}\right)$$

Hence, as far as the number of possible hyper-conjugative structures possible is concerned, tertiary free radicals should be more stable than secondary, which in turn should be more stable than primary.

iii) Stability of ally and benzyl free radicals:

      The stability of the free radicals in which the carbon bearing odd electron is adjacent to the double or triple bond can be rationalized in terms of resonance effect. First of all, let us draw the resonance structures of allyl and benzyl free radicals.

Resonance stablized allyl radical

Resonance stablized benzyl radical

Now, as the number of phenyl groups attached to carbon bearing electron increases, the number of resonating structures will also increase, and hence the stability.

Resonance stablized diphenylmethyl free radical

Resonance stablized triphenylmethyl free radical

Therefore, the expected order of the stability of unsaturated systems with carbon bearing odd electron should be as given below.

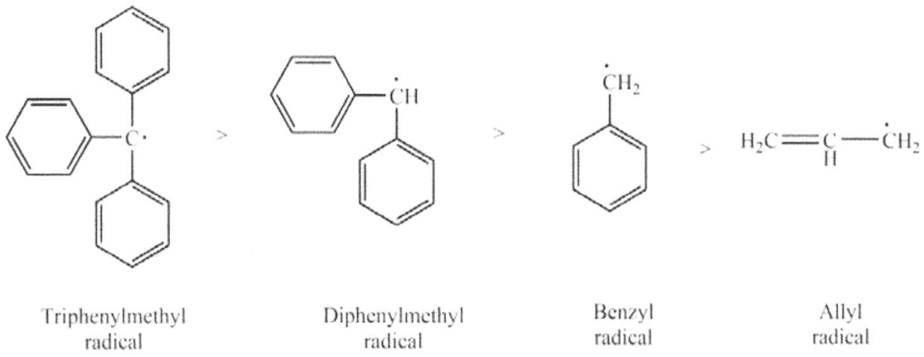

Triphenylmethyl radical > Diphenylmethyl radical > Benzyl radical > Allyl radical

4. Reactivity of free radicals: The principal routes by which the free radicals can react to give rise to stable products are given below.

i) Decomposition reactions: One of the most common examples of this type of process is the decomposition of the benzoxy radical.

$$ C_6H_5\text{-}C(=O)\text{-}O\cdot \longrightarrow \overset{\cdot}{C_6H_5} + CO_2 $$

ii) Hydrogen abstraction: In these types of reactions, a free radical may result in the removal of a proton from the same (intermolecular process) or another molecule (intramolecular process).

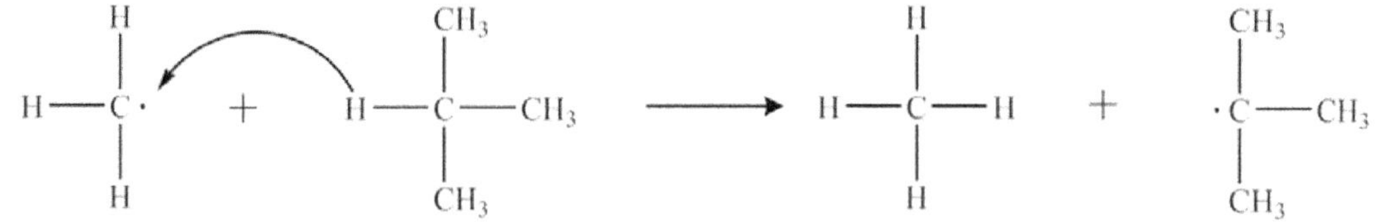

$$ CH_3\cdot + (CH_3)_3CH \longrightarrow CH_4 + \cdot C(CH_3)_3 $$

iii) Rearrangement reaction: Free radicals may undergo rearrangement reactions to yield different but stable free radical counterparts.

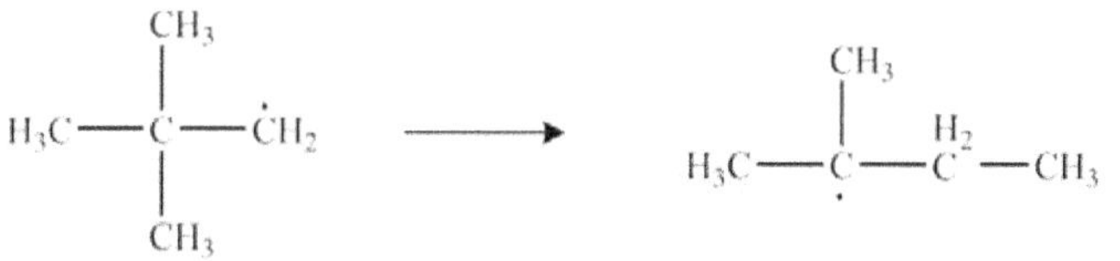

$$ (CH_3)_3C\text{-}\overset{\cdot}{C}H_2 \longrightarrow (CH_3)_2\overset{\cdot}{C}\text{-}CH_2\text{-}CH_3 $$

Copyright © *Mandeep Dalal*

iv) Addition reactions: The free radical obtained from an alkene may attack the triangular face of the double bond of another alkene molecule, and the process continues to yield polymers.

➢ **Carbenes**

The term carbenes in organic chemistry may simply be defined as the chemical species that carries two non-bonding electrons (paired or unpaired) on the carbon with a total of six valence electrons.

Since the carbon in carbenes has only six electrons, it is electron deficient; and therefore, acts as an electrophile in chemical reactions.

1. Generation of carbenes: Some of the most common pathways involving the production of carbenes are given below.

i) Decomposition of ketones or diazoalkanes:

ii) Thermal or photolytic cleavage of cyclopropanes and oxiranes:

iii) Attack of strong base on chloroform:

2. Structure of carbenes: Before we discuss the structure of carbenes, it is better to understand how carbenes can be classified on the basis of the distribution of non-bonding electrons. There are primarily two types carbenes; singlet carbenes and triplet carbenes.

It has been experimentally found that the singlet carbenes are V-shaped and are derivatives of trigonal planar geometry. Now valence bond theory, as well as molecular orbital theory, easily accounted for such structure, it is more comfortable to discuss the valence bond approach. The central carbon is in sp^2 hybridization with three hybrid orbitals oriented at 120° in a plane; two half-filled orbitals participating in bonding whilst the third hybrid orbital contains the lone pair. The p_z orbital remains empty and is perpendicular to the molecular plane.

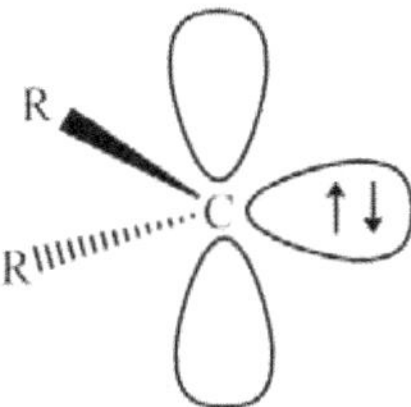

Figure 13. Orbital structure of singlet carbenes.

The triplet carbenes are either linear or V-shaped (depending upon the reaction requirement). Now valence bond theory, as well as molecular orbital theory, easily accounted for such structure, it is more comfortable to discuss the valence bond approach. The central carbon in linear triplet carbene is in sp hybridization with two hybrid orbitals oriented at 180° along the z-axis participating in bonding; whilst the atomic p_x and p_y atomic orbital containing the unpaired electrons.

Figure 14. Orbital structure of triplet carbenes.

The central carbon in bent triplet carbene is in sp^2 hybridization with three hybrid orbitals oriented at 120° in a plane; two half-filled orbitals participating in bonding whilst the third hybrid orbital contains an unpaired electron. The other unpaired electron is in the atomic p_z orbital perpendicular molecular plane.

3. Stability of carbenes:

In the case of simple hydrocarbons, triplet carbenes typically have energies 8 kcal/mol less than singlet carbenes due to Hund's rule of maximum multiplicity; and therefore, as a whole, we can conclude that the triplet is the ground state and singlet one is the excited state entities. Also, groups that can donate electron pairs can stabilize the singlet carbene by delocalizing the pair into an empty p_z orbital. Furthermore, the singlet state can become the ground state if its energy is significantly reduced. However, triplet carbenes cannot be stabilized by this strategy. A carbene 9-fluorenylidene is found to exist in a rapid equilibrating mixture of triplet and singlet states with an energy difference of roughly 1.1 kcal/mol. Nevertheless, it is disputed if diaryl carbenes like fluorene carbene are true carbenes since the electrons can be delocalized to such a level that they become biradicals in nature. The experimental studies have suggested that triplet carbenes can be stabilized thermodynamically with heteroatoms of electropositive nature (like in silyl and silyloxy carbenes).

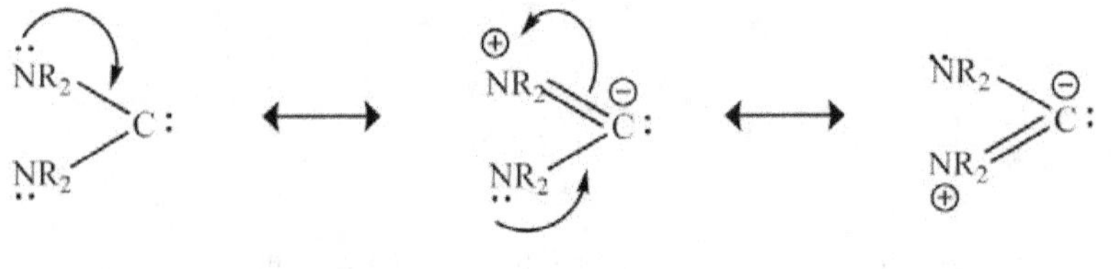

4. Reactivity of carbenes: The principal routes by which the carbenes can react to give rise to stable products are given below.

i) Addition reactions: besides the carbon-carbon double bonds (including aromatic systems), carbenes may also add to carbon-heteroatom multiple bonds.

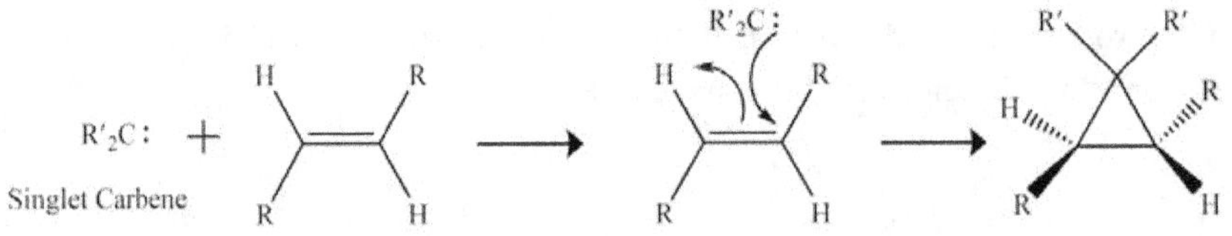

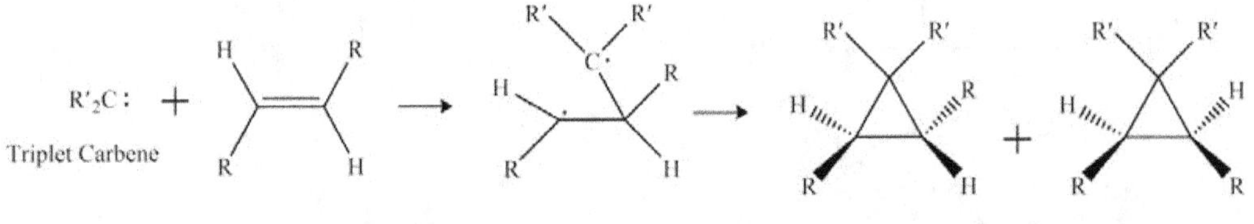

ii) Insertion reaction: In these reactions, carbenes get inserted into CH bonds to give stable products.

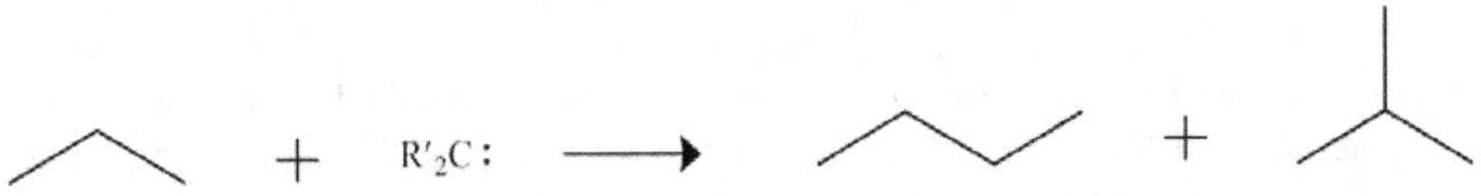

iii) Dimerization reaction: Carbenes may undergo dimerization to form an alkene; however, it is more likely to arise from the attack by a carbene on a molecule of a carbene precursor.

$$R_2C: \quad + \quad R_2CN_2 \quad \longrightarrow \quad R_2C\!=\!CR_2 \quad + \quad N_2$$

iv) Rearrangement reactions: The carbenes can also undergo rearrangement reactions to yield very stable products as given below.

v) Fragmentation reactions: many substitutions and elimination products are obtained from the fragmentation reactions of alicyclic oxychlorocarbenes as given below.

vi) Rearrangement reactions: Triplet carbenes are also able to abstract hydrogen or any other groups or atoms to yield free radicals' products as given below.

$$R_2C: \ + \ H_3C\!-\!CH_3 \quad \longrightarrow \quad R_2HC^{\bullet} \ + \ H_2C^{\bullet}\!-\!CH_3$$

> ➤ **Nitrenes**

    The term nitrenes in organic chemistry may simply be defined as the chemical species that carries four non-bonding electrons (paired or unpaired) on the nitrogen with a total of six valence electrons.

    Since the nitrogen in carbenes has only six electrons, it is electron deficient; and therefore, acts as an electrophile in chemical reactions.

1. Generation of nitrenes: Some of the most common pathways involving the production of nitrenes are given below.

i) Photochemical or thermal decomposition of isocyanates or azides:

$$R\!-\!N\!=\!C\!=\!O: \quad \longrightarrow \quad R\!-\!N: \quad + \quad :CO$$

Alkyl isocyanate   Alkyl nitrene

$$R\!-\!N\!=\!\overset{\oplus}{N}\!=\!\overset{\ominus}{N}: \ \longleftrightarrow \ R\!-\!\overset{\ominus}{N}\!-\!\overset{\oplus}{N}\!\equiv\!N: \ \xrightarrow{h\nu \text{ or } \Delta} \ R\!-\!N: \ + \ N\!\equiv\!N$$

Alkyl azide   Alkyl nitrene

ii) Elimination of sulphonate ion from certain compounds:

2. Structure of nitrenes: Before we discuss the structure of nitrenes, it is better to understand how nitrenes can be classified on the basis of the distribution of non-bonding electrons. There are primarily two types of nitrenes; singlet nitrenes and triplet nitrenes.

It has been experimentally found that the singlet nitrenes are linear and are derivatives of trigonal planar geometry. Now valence bond theory, as well as molecular orbital theory, easily accounted for such structure, it is more comfortable to discuss the valence bond approach. The nitrogen is in sp^2 hybridization with three hybrid orbitals oriented triangularly in a plane; one half-filled hybrid orbital participating in bonding with carbon whilst the second and third hybrid orbitals contain the lone pairs. The p_z orbital remains empty and is perpendicular to the above-mentioned triangular plane.

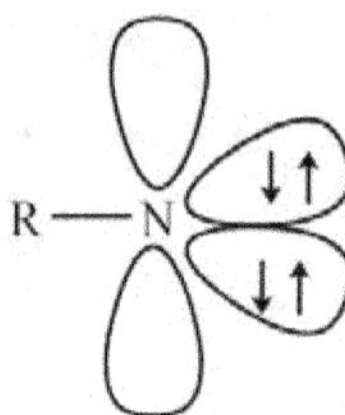

Figure 15. Orbital structure of singlet nitrenes.

The triplet nitrenes are also linear. The nitrogen in triplet nitrenes is in sp hybridization with two hybrid orbitals oriented at 180° along the z-axis; one hybrid orbital (half-filled) participating in bonding whilst the other hybrid orbital contains a lone pair. The atomic p_x and p_y atomic orbital containing the unpaired electrons.

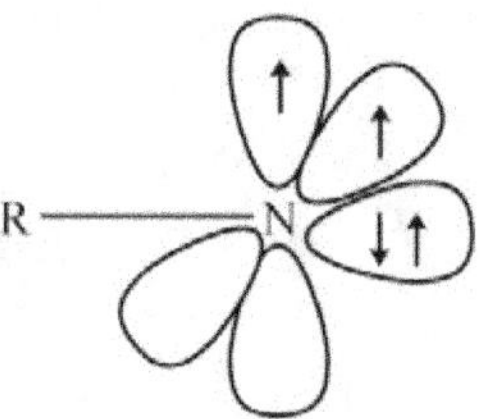

Figure 16. Orbital structure of triplet nitrenes.

3. Stability of nitrenes:

Although nitrenes are too reactive to isolate under normal conditions, in 2019, an authentic triplet nitrene was isolated by Betley and Lancaster, stabilized by coordination to a copper center in a bulky ligand. Furthermore, triplet nitrenes are thermodynamically more stable but react stepwise allowing free rotation and thus producing a mixture of stereochemistry. They are usually detected by adding carbon monoxide as it can form isocyanates with nitrenes which can be isolated easily.

$$C_6H_5 \!-\! \overset{\ominus}{\underset{..}{\overset{..}{N}}} \!-\! N\!\equiv\!N: \quad \xrightarrow{\;\Lambda\;} \quad C_6H_5 \!-\! \underset{..}{\overset{..}{N}}: \quad \xrightarrow{\;CO\;} \quad C_6H_5 \!-\! \overset{..}{N}\!=\!C\!=\!\underset{..}{\overset{..}{O}}:$$

Phenylazide   Phenylnitrene   Phenylisocyanate

4. Reactivity of nitrenes:

The principal routes by which the nitrenes can react to give rise to stable products are given below.

i) Addition reactions: Nitrenes may add to carbon-carbon multiple bonds to give rise to some stable product.

$$R\!-\!\underset{..}{N}: \;\; + \;\; H_2C\!=\!CH_2 \;\longrightarrow\; \triangleright\!N\!-\!R$$

ii) Insertion reaction: In these types of reactions, nitrenes get inserted into CH bonds to give stable products.

$$\underset{R'}{\overset{O}{\underset{\diagdown}{\overset{\|}{C}}}}\!\!\underset{N:}{} \;\; + \;\; R_3C\!-\!H \;\longrightarrow\; \underset{R'}{\overset{O}{\overset{\|}{C}}}\!\!\underset{\underset{H}{N}}{}\!\!-\!CR_3$$

iii) Dimerization reaction: Nitrenes may undergo dimerization to form diamide.

$$2\,Ar\!-\!\underset{..}{N}: \;\longrightarrow\; Ar\!-\!N\!=\!N\!-\!Ar$$

iv) Rearrangement reactions: The nitrenes can also undergo rearrangement reactions to yield very stable products as given below.

$$\underset{H}{\overset{R}{\diagdown}}\!CHN \;\longrightarrow\; RHC\!=\!NH$$

v) Hydrogen abstraction: Nitrenes are also able to abstract hydrogen or any other groups or atoms to yield free radicals' as given below.

$$R\!-\!\underset{..}{N}: \;\; + \;\; R\!-\!H \;\longrightarrow\; R\!-\!\overset{.}{N}\!-\!H \;\; + \;\; R^{\cdot}$$

❖ Effect of Structure on Reactivity

When chemical equations are used to represent chemical reactions, functional groups are usually represented as a condensed formula because all the compounds with a particular functional group give more or less the same products. It enables us to categorize a large number of reactions in the same class for a better understanding and memorization. However, it must be kept in mind that two compounds with the same functional group may react in an entirely different manner because the structure of the rest of the molecule affects the overall reactivity at the corresponding functional group. Moreover, even if the two compounds with the same functional group undergo the same reaction, their rate rates may be slightly or largely different. The effects of molecular structure on the overall reactivity can be fragmented into three main classes; resonance or mesomeric effect, inductive effect, and steric effect. Now although we see a combined result of two or all three phenomena in most of the cases; it is still possible to extract information about their individual effects. A brief idea of all the three effects of molecular structure on reactivity is discussed below.

➤ *Impact of Resonance Effect on Reactivity*

The resonance or mesomeric effect in organic compounds may affect the reactivity up to a great as it can produce polarity by creating centers of high and low electron density. For instance, +R groups increase the electron density at o- and p- positions making them more susceptible to attacking electrophile.

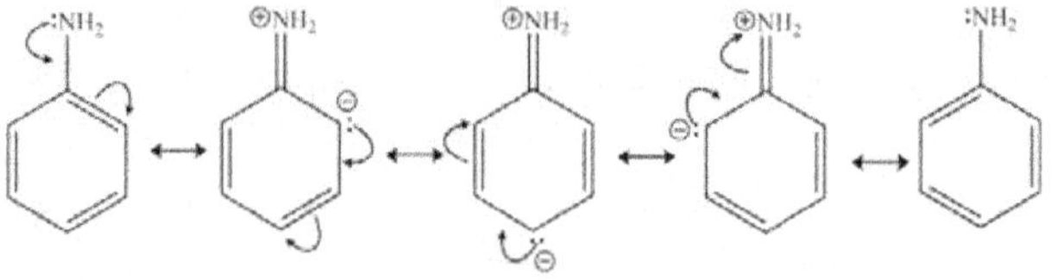

It is also worthy to mention that groups showing the +R effect are having lone pair of electrons that can be put into conjugation with the double of the chain or ring to which it gets attached with. Some of the typical groups showing +R effect are $-O^- > -NH_2 > -NHR > -OR > -NHCOR > -OCOR > -Ph > -F > -Cl > -Br > -I$

Similarly, groups with the −R effect tends to decrease the electron density at o- and p- positions in the benzene ring making the m-position more susceptible towards attacking electrophile.

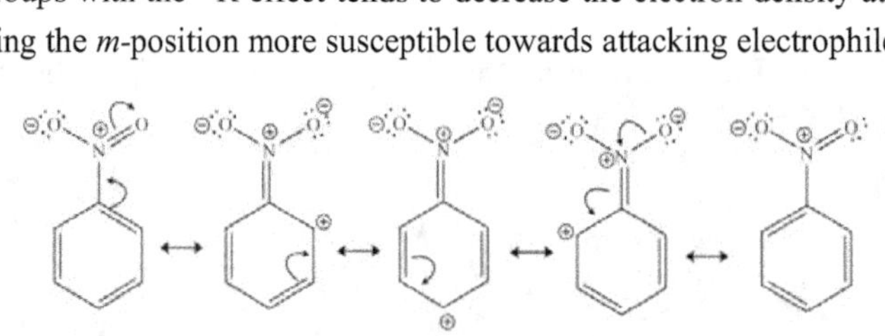

It is also worthy to mention that groups showing the −R effect are having a double bond that can be put into conjugation with the double of the chain or ring to which it gets attached with. Some of the typical groups showing −R effect is $-NO_2 > -CN > -S(=O)_2-OH > -CHO > -C=O > -COOCOR > -COOR > -COOH > -CONH_2 > -COO^-$. It is also worthy to recall that the alkyl groups, which do not have multiple bonds or unshared pairs, can also show the +M effect due to hyperconjugation.

> ### *Impact of Inductive Effect on Reactivity*

The inductive effect in organic compounds may affect the reactivity up to a greater as it can produce polarity by its electron-donating or electron-withdrawing ability. For instance, groups with the +I effect tend to increase the electron density and making them less or more stable or susceptible to attacking electrophile. Relative inductive effects have been experimentally measured with reference to hydrogen, in increasing order of +I effect or decreasing order of -I effect, as follows:

$-NH_3^+ > -NO_2 > -SO_2R > -CN > -SO_3H > -CHO > -CO > -COOH > -COCl > -CONH_2 > -F > -Cl > -Br > -I > -OR > -OH > -NR_2 > -NH_2 > -C_6H_5 > -CH=CH_2 > -H$ and $C-H < C-D < C-T$ in increasing order of +I effect, where H is Hydrogen and D or T are hydrogen's isotopes.

17.6   1   0.28

Relative rates of reaction of RBr with ethanol

The inductive effect is extremely useful to describe the molecular stability depending on the sign and magnitude of the charge present on the atom and the substituent that is attached to this atom. For instance, if an atom carries a positive charge and binds to a group with −I effect, its charge becomes 'boosted' and the molecule tends to be less stable. Likewise, if an atom carries a negative charge and binds to a group with +I effect, its charge will also get 'amplified' making it less stable. Conversely, if an atom has a positive charge and binds to a +I effect, its charge will get 'de-boosted' and therefore, the molecule will become more stable. Similarly, if an atom has a negative charge and is attached to a −I group its charge becomes 'de-amplified' and the molecule will get more stable. The attribution can be found for such behavior in the fact more charge on an atom cuts stability and less charge on an atom raises the stability.

The basicity and acidity of a molecular species are also affected by the inductive effect. If substituents with +I inductive effect are attached to a molecule, the electron density at the donor site increases, which in turn, makes it more basic. Likewise, if substituents with −I inductive effect are attached to a molecule, the electron density at the acceptor site decreases, which in turn, makes it more acidic. Furthermore, the acidity also increases as the number of −I groups attached to a molecule increases; and the same is true for bases since the increased number of +I groups on a molecule increases its basicity.

It is also worthy to recall that some groups can also affect the reactivity in the same way as an effect but through space i.e. field effect.

> ➤ *Impact of Steric Effect on Reactivity*

The steric effect in organic compounds may affect the reactivity up to a greater extent by affecting the stability of reactants, intermediates, or transition states. These effects are primarily of nonbonding interactions that affect the reactivity and conformation of various ions and molecules. Also, these effects also couple with electronic effects which usually govern reactivity and shape. Steric effects primarily arise from repulsive interactions between overlapping electronic clouds. These effects are largely employed in academic and applied chemistry for many purposes. For instance, the reaction-rate increases with the increase in bulky groups in the SN_1 pathway.

Steric effects also give rise to steric hindrance which slows down the chemical reactions due to bulk presence. Nevertheless, it should be kept in mind that steric hindrance is primarily an intermolecular phenomenon, whilst the dialogue of steric effects usually emphasizes intramolecular interactions. The steric hindrance is mainly employed to dictate the reaction selectivity like slowing down unwanted concurrent reactions.

Also, the steric hindrance between cis substituents can disturb the torsional bond angles; and is responsible for the observed profile of rotaxanes and the small rates of racemization of compounds like 2,2'-disubstituted biphenyls and their derivatives. The rate of reaction also increases with the decrease in the number size of bulky groups in the SN_2 pathway.

Comparative reaction rates give a very useful vision into the effects of the steric bulk of different groups. When standard experimental conditions were used, the solvolysis of methyl bromide is 10^7 times faster than what in neopentyl bromide. The variance shows the oppose of attack on the substrate with the sterically bulky $(CH_3)_3C$ substituent. These values (resulting from equilibrium measurements of monosubstituted cyclohexanes) provide another measure of the bulk of various substituents. Also, the bulk of a substituent can be measured by the extent that it favors the equatorial position.

❖ The Hammett Equation and Linear Free Energy Relationship

In this section, we will discuss the quantitative treatments of the effect of structure on reactivity i.e. how the resonance effect, field-effect, and steric effect impact the reaction rate in measurable numbers.

➤ *The Hammett Equation*

Consider an organic reaction is carried out on a substrate which can be denoted as XRY, X a variable substituent and Y is the reaction spot, and R represents the basic substrate structure. In this type of case, replacing X = H with X = CH_3 results in an increment in the rate of reaction up ten times. However, it is still a mystery what part of the rate enhancement comes from resonance effect, field-effect, or steric effect. To do so, it is reasonable to use compounds where one or two effects are so small that they simply can be neglected. Although it is the oversimplification of the problem, quantitative results can still be obtained. The Hammett equation is the first attempt to give numerical values for the quantitative treatment of structure on reactivity. Hammett proposed the equation for the cases of *m*- and *p*-XC_6H_4Y as given below.

$$\log \frac{k}{k_0} = \sigma\rho \tag{23}$$

where k and k_0 are the constant for the group X ≠ H and X = H; ρ and σ are the constants for reaction conditions and substituent X, respectively.

➤ *Derivation of Hammett Equation*

To derive the Hammett equation, we need to recall the quantitative relationship between the structure and reactivity first. To do so, we need to find some mathematical parameter that can be used to represent the combined magnitude of inductive and resonance effects of different substituents. This can be achieved by considering the hydrolysis of a series of different benzoic acids as given below.

$$XC_6H_4COOH + H_2O \;\overset{K_a}{\rightleftharpoons}\; XC_6H_4COO^- + H_3O^+ \tag{24}$$

Where X is a substituent at the *m*- or *p*-position and K_a is the dissociation constant. As expected, the dissociation constant was found to be different for differently substituted substrates.

Since an electron-withdrawing group will better stabilize the conjugate base (i.e., $XC_6H_4COO^-$), resulting in a larger magnitude of K_a (lower pK_a). On the other hand, an electron-donating group will destabilize the conjugate base (i.e., $XC_6H_4COO^-$), resulting in a smaller magnitude of K_a (higher pK_a). Therefore, we can say that the electronic effect (inductive plus mesomeric effect) of a substituent can be represented as the difference between the pK_a value of its benzoic acid derivative and the pK_a value of benzoic acid itself; mathematically, we can say

$$\sigma_X = \log(K_a) - \log(K_a)_0 = -p(K_a) + p(K_a)_0 \tag{25}$$

Where the parameter σ_X (or simply σ) is called as substituent constant; and was found for several different groups just subtracting its benzoic acid derivative's pK_a value from pK_a value of benzoic acid.

Table 1. pK_a values and substituent constants for XC_6H_6COOH using benzoic acids $p(K_a)_0 = 4.21$.

Substituent	$p_m(K_a)$	$p_p(K_a)$	$\sigma_m = p(K_a)_0 - p_m(K_a)$	$\sigma_p = p(K_a)_0 - p_p(K_a)$
NO_2	3.50	3.43	0.71	0.78
CH_3	4.28	4.38	−0.07	−0.17
OCH_3	4.09	4.48	0.12	−0.27
$CH(CH_3)_2$	4.28	4.36	−0.07	−0.15
F	3.87	4.15	0.34	0.06
Br	3.82	3.98	0.39	0.23
Cl	3.84	3.98	0.37	0.23
I	3.86	3.93	0.35	0.28
$COCH_3$	3.83	3.71	0.38	0.50

Using $\log m - \log n = \log m/n$, equation (36) can also be written as

$$\log \frac{(K_a)}{(K_a)_0} = \sigma \tag{26}$$

Now if we plot a curve between $\log(K_a)/(K_a)_0$ vs σ, we will definitely get a straight line with a slope = 1.

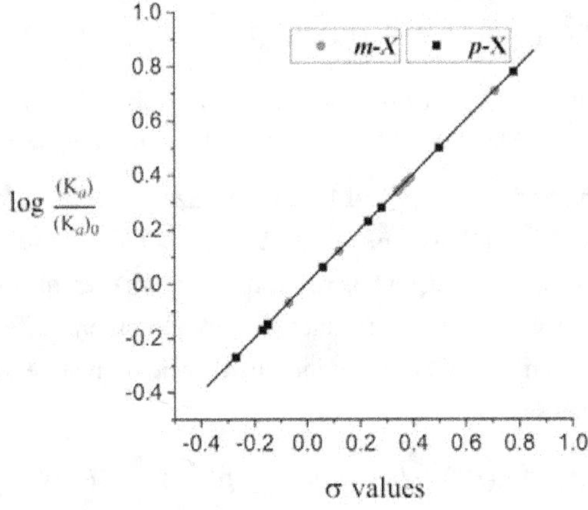

Figure 17. Variation of $\log(K_a)/(K_a)_0$ vs σ for substituted benzoic acids.

Now we need to check if these σ values (i.e., of substituted benzoic acids) can also be used for other meta- or para-substituted benzene derivatives. To do so, consider two series of reactions; the first one is the acid dissociation of phenyl phosphonic acid, and the second one is the base hydrolysis of substituted ethyl benzoate. Here we will find if different substituents affect their dissociation constants or rates in the same manner as affected in the case of substituted benzoic acid. Also, we did not use ortho-substituents or substituents in the aliphatic system because they also contain steric factors and don't not linear variation.

The experimental $\log(K_a)/(K_a)_0$ for the reaction-I and experimental $\log k/k_0$ for reaction-II are given below.

Table 2. Experimental values of $\log(K_a)/(K_a)_0$ and $\log k/k_0$ for the acid dissociation of phenyl phosphonic acid and base hydrolysis of substituted ethyl benzoates, respectively.

Substituent	meta-$\log(K_a)/(K_a)_0$	para-$\log(K_a)/(K_a)_0$	meta-$\log k/k_0$	para-$\log k/k_0$
NO_2	0.53	0.59	1.83935	2.06423
Br	0.29	0.23	–	–
Cl	0.28	0.17	0.88536	0.63347
CH_3	–	−0.15	−0.16115	−0.34679
OCH_3	–	–	–	−0.67923

When plotted the experimental $\log(K_a)/(K_a)_0$ for the reaction-I and experimental $\log k/k_0$ for reaction-II vs the substituent constants obtained for the substituted benzoic acids, we get the following curves.

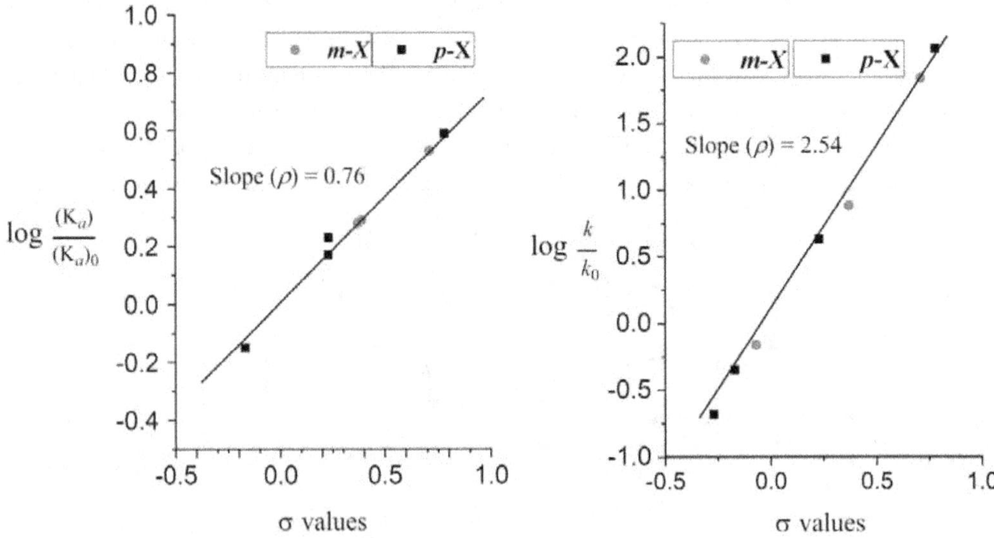

Figure 18. $\log(K_a)/(K_a)_0$ and $\log k/k_0$ vs σ for reaction-I and reaction-II.

It is obvious that the plots are still linear like in Figure 3 but the slope has changed. This implies that the order and relative effects for different substituents on both reactions remain the same though the magnitude has been changed which can be attributed to the different nature reaction considers from 'base reaction'.

Therefore, our aim, in this case, should be the determination of the slope (let us say ρ). Since on the vertical side we have '$[\log(K_a)/(K_a)_0]_{sppa}$' for reaction-I (acid dissociation of phenyl phosphonic acid) and on the horizontal side we have 'σ or $[\log(K_a)/(K_a)_0]_{sba}$' for base reaction (hydrolysis of substituted benzoic acid), the slope should be

$$\rho = \frac{[\log(K_a)/(K_a)_0]_{seb}}{[\log(K_a)/(K_a)_0]_{sba}} \tag{27}$$

or

$$\rho[\log(K_a)/(K_a)_0]_{sba} = [\log(K_a)/(K_a)_0]_{seb} \tag{28}$$

But from equation (26), we know that $[\log(K_a)/\log(K_a)_0]_{sba} = \sigma$; and therefore, equation (40) takes the form

$$\left[\log\frac{(K_a)}{(K_a)_0}\right]_{seb} = \rho\sigma \tag{29}$$

For any reactions,

$$\log K_a = \rho\sigma - \log (K_a)_0 \tag{30}$$

Similarly, on the vertical side we have '$\log k/k_0$' for reaction-II (base hydrolysis of substituted ethyl benzoate) and on the horizontal side we have 'σ or $[\log(K_a)/(K_a)_0]_{sba}$' for base reaction (hydrolysis of substituted benzoic acid), the slope should be

$$\rho = \frac{[\log k/k_0]_{seb}}{[\log(K_a)/(K_a)_0]_{sba}} \tag{31}$$

or

$$\rho[\log(K_a)/(K_a)_0]_{sba} = [\log k/k_0]_{seb} \tag{32}$$

But from equation (26), we know that $[\log(K_a)/\log(K_a)_0]_{sba} = \sigma$; and therefore, equation (32) takes the form

$$\left[\log\frac{k}{k_0}\right]_{sppa} = \rho\sigma \tag{33}$$

For any reactions,

$$\log k = \rho\sigma - \log k_0 \tag{34}$$

The results given by equation (29, 30, 33, 34) are called as Hammett's equations; which shows that the rates of ortho and para-substituted benzene derivatives can be obtained if the substituent contents for substituted benzoic acid are known. Now we will discuss the substituent and reaction constants in more detail

> ### ➤ *Linear Free Energy Relationship (LFER)*

The Hammett equation is a linear free energy relationship that can be proved for any group X by recalling the kinetics of organic reactions is in the framework of "Activated complex Theory", which states that the rate constant (k) for a typical reaction is

$$k = \frac{RT}{Nh}e^{-\frac{\Delta G^*}{RT}} \tag{35}$$

Where ΔG^* is the free energy change of the activation step at temperature T. The symbols R, N, and h are the gas constant, Avogadro number, and Planck's constant, respectively. Similarly, for k_0 we have

$$k_0 = \frac{RT}{Nh}e^{-\frac{\Delta G_0^*}{RT}} \tag{36}$$

After putting the value of equation (35) and equation (36) in Hammett equation (23), we get

$$\log\frac{\dfrac{RT}{Nh}e^{-\frac{\Delta G^*}{RT}}}{\dfrac{RT}{Nh}e^{-\frac{\Delta G_0^*}{RT}}} = \sigma\rho \tag{37}$$

$$\log \frac{e^{-\frac{\Delta G^*}{RT}}}{e^{-\frac{\Delta G_0^*}{RT}}} = \sigma\rho \tag{38}$$

Multiplying both sides by 2.303, we have

$$2.303 \, \log \frac{e^{-\frac{\Delta G^*}{RT}}}{e^{-\frac{\Delta G_0^*}{RT}}} = 2.303 \, \sigma\rho \tag{39}$$

$$\ln \frac{e^{-\frac{\Delta G^*}{RT}}}{e^{-\frac{\Delta G_0^*}{RT}}} = 2.303 \, \sigma\rho \tag{40}$$

$$\ln e^{-\frac{\Delta G^*}{RT}} - \ln e^{-\frac{\Delta G_0^*}{RT}} = 2.303 \, \sigma\rho \tag{41}$$

or

$$\left(-\frac{\Delta G^*}{RT} \ln e\right) - \left(-\frac{\Delta G_0^*}{RT} \ln e\right) = 2.303 \, \sigma\rho \tag{42}$$

or

$$-\frac{\Delta G^*}{RT} + \frac{\Delta G_0^*}{RT} = 2.303 \, \sigma\rho \tag{43}$$

Which implies

$$\frac{\Delta G_0^*}{RT} - \frac{\Delta G^*}{RT} = 2.303 \, \sigma\rho \tag{44}$$

or

$$\frac{\Delta G_0^* - \Delta G^*}{RT} = 2.303 \, \sigma\rho \tag{45}$$

or

$$\Delta G_0^* - \Delta G^* = 2.303 RT \, \sigma\rho \tag{46}$$

or

$$-\Delta G^* = 2.303 RT \, \rho\sigma - \Delta G_0^* \tag{47}$$

Hence, the variation of negative of the free energy of activation varies linearly with slope $2.303 RT\rho$ and $-\Delta G_0^*$ as intercept.

❖ Substituent and Reaction Constants

To discuss the substituent and reaction constants, we need to recall the quantitative relationship between the structure and reactivity first. To do so, we need to find some mathematical parameter that can be used to represent the combined magnitude of inductive and resonance effects of different substituents. This can be achieved by considering the hydrolysis of a series of different benzoic acids as given below.

$$XC_6H_4COOH + H_2O \overset{K_a}{\rightleftharpoons} XC_6H_4COO^- + H_3O^+ \tag{48}$$

Where X is a substituent at the *m*- or *p*-position and K_a is the dissociation constant. As expected, the dissociation constant was found to be different for differently substituted substrates.

Since an electron-withdrawing group will better stabilize the conjugate base (i.e., $XC_6H_4COO^-$), resulting in a larger magnitude of K_a (lower pK_a). On the other hand, an electron-donating group will destabilize the conjugate base (i.e., $XC_6H_4COO^-$), resulting in a smaller magnitude of K_a (higher pK_a). Therefore, we can say that the electronic effect (inductive plus mesomeric effect) of a substituent can be represented as the difference between the pK_a value of its benzoic acid derivative and the pK_a value of benzoic acid itself; mathematically, we can say

$$\sigma_X = \log(K_a) - \log(K_a)_0 = -p(K_a) + p(K_a)_0 \tag{49}$$

Where the parameter σ_X (or simply σ) is called as substituent constant; and was found for several different groups just subtracting its benzoic acid derivative's pK_a value from pK_a value of benzoic acid. Using $\log m - \log n = \log m/n$, equation (49) can also be written as

$$\log\frac{(K_a)}{(K_a)_0} = \sigma \tag{50}$$

Now if we plot a curve between $\log(K_a)/(K_a)_0$ vs σ, we will definitely get a straight line with a slope which is equal to unity.

Now we need to check if these σ values (i.e., of substituted benzoic acids) can also be used for other meta- or para-substituted benzene derivatives. To do so, consider two series of reactions; the first one is the acid dissociation of phenyl phosphonic acid, and the second one is the base hydrolysis of substituted ethyl benzoate.

Here we will find if different substituents affect their dissociation constants or rates in the same manner as affected in the case of substituted benzoic acid. Also, we did not use ortho-substituents or substituents in the aliphatic system because they also contain steric factors and don't not linear variation.

Reaction-I

Reaction-II

When plotted the experimental $\log(K_a)/(K_a)_0$ for the reaction-I and experimental $\log k/k_0$ for reaction-II vs the substituent constants obtained for the substituted benzoic acids, we get the following curves. It is obvious that the plots will still be linear like but the slope will be changed. This implies that the order and relative effects for different substituents on both reactions remain the same though the magnitude has been changed which can be attributed to the different nature reaction considers from 'base reaction'.

Therefore, our aim, in this case, should be the determination of the slope (let us say ρ). Since on the vertical side we have '$[\log(K_a)/(K_a)_0]_{sppa}$' for reaction-I (acid dissociation of phenyl phosphonic acid) and on the horizontal side we have 'σ or $[\log(K_a)/(K_a)_0]_{sba}$' for base reaction (hydrolysis of substituted benzoic acid), the slope should be

But from equation (50), we know that $[\log(K_a)/\log(K_a)_0]_{sba} = \sigma$; and therefore, we get

$$\left[\log\frac{(K_a)}{(K_a)_0}\right]_{seb} = \rho\sigma \tag{51}$$

For any reactions,

$$\log K_a = \rho\sigma - \log (K_a)_0 \tag{52}$$

Similarly, on the vertical side we have '$\log k/k_0$' for reaction-II (base hydrolysis of substituted ethyl benzoate) and on the horizontal side we have 'σ or $[\log(K_a)/(K_a)_0]_{sba}$' for base reaction (hydrolysis of substituted benzoic acid), the slope should be

But from equation (50), we know that $[\log(K_a)/\log(K_a)_0]_{sba} = \sigma$; and therefore, we get

$$\left[\log\frac{k}{k_0}\right]_{sppa} = \rho\sigma \tag{53}$$

Equation (51, 53) are Hammett equations where ρ and σ are the substituent and reaction constants.

> ➤ *Substituent Constants (σ)*

From the derivation of Hammet's equation, we know that the substituent constants can be collected by finding the change in the pKa value of substituted benzoic acid in water at 25 °C. The reaction constant for this 'base reaction" will simply be equal unity.

Since an electron-withdrawing group will better stabilize the conjugate base (i.e., $XC_6H_4COO^-$), resulting in a larger magnitude of K_a (lower pK_a). On the other hand, an electron-donating group will destabilize the conjugate base (i.e., $XC_6H_4COO^-$), resulting in a smaller magnitude of K_a (higher pK_a). Therefore, we can say that the electronic effect (inductive plus mesomeric effect) of a substituent can be represented as the difference between the pK_a value of its benzoic acid derivative and the pK_a value of benzoic acid itself.

Various substituent effects can be concluded by looking at σ values displayed in Table 1. With $\rho = 1$, the group of substituents with increasing positive values (such as nitro) makes the equilibrium constant increase relative to the hydrogen as the substituent, which in turn, means that the acidity of the benzoic acid has been increased. This is because the substituents like NO_2 stabilize the negative charge on the carboxylate ion by an inductive effect (−I) and also by a negative resonance effect (−R).

The second kind of substituents is the halo- groups, for which the substituent effect is modestly positive. This can be attributed to the fact that even though the inductive effect is still negative, the resonance effect is positive (+R), canceling the former partially. Experimental data also demonstrated that for these substituents, the *m*-effect is much bigger than the *p*-effect, because the resonance effect is largely reduced in the *m*-substituent. In the case of *m*-substituted substrates, a C atom with the negative charge is further away from the COOH group.

The behavior of resonance effect in the perspective of substituent constants can be understood by the given below, where, in a *p*-substituted arene (*1a*), one resonance structure (*1b*) is a quinoid with the positive charge on the substituent X, freeing electrons and so destabilizing the Y group. This kind of destabilizing outcome is not likely to happen when X is at the *m*-site.

Furthermore, groups such as ethoxy and methoxy can even show opposite signs for the substituent constant due to the opposing nature of the inductive and resonance effects. It's just the aryl and alkyl substituents like methyl are electron-donating in w.r.t. inductive as well as resonance effects. Finally, If the reaction constant's sign is negative, substituents with a negative substituent constant will raise K_a values.

Now although the substituent constants derived from substituted benzoic acid were quite accurate in predicting $\log(K_a)/(K_a)_0$ or $\log k/k_0$ for several reactions, cases where the rates or dissociations constants predicted were not in line if the substituent is either strongly electron-withdrawing or strongly electron-donating in nature. For instance, $\log(K_a)/(K_a)_0$ of substituted phenols for p-CN and p-NO$_2$ are above the line indicating that systems with these substituents act as stronger acids than expectations. This is because if electron-withdrawal arising from mesomeric effects is extended to the reaction site via 'through conjugation', the conjugated acid will exceptionally stable. Since the substituent is developing a negative charge during this process, the modified substituent constant will be labeled as σ_p^-. Similarly, if the electron-donating arising from mesomeric effects is extended to the reaction site via 'through conjugation, the conjugated acid will less stable. Since the substituent is developing a positive charge during this process, the modified substituent constant will be labeled as σ_p^+.

SN$_1$ hydrolysis of substituted phenyldimethyl carbinyl chloride (σ^+)

Dissociation of substituted para-nitro phenol (σ^-)

The magnitude by which $\log(K_a)/(K_a)_0$ or $\log k/k_0$ deviate from σ value is added to produce a new scale of substituent constants. The same is true for the m-site excepting the fact that values of σ_m^+ will be the same as the σ_m values.

Table 3. Substituent constants: para and meta substituted benzene rings.

Group	σ_p	σ_m	σ_p^+	σ_m^+	σ_p^-
COOH	0.44	0.35	0.42	0.32	0.73
COOR	0.44	0.35	0.48	0.37	0.48
CN	0.70	0.62	0.66	0.56	1.00
NO$_2$	0.81	0.71	0.79	0.73	1.27
Cl	0.24	0.37	0.11	0.40	–

> ➤ *Reaction Constants (ρ)*

After knowing the values of substituent constants, the reaction constant (ρ) can be obtained for an extensive range of reactions. The 'prototype or base' reaction is the alkaline hydrolysis of ethyl benzoate in a water mixture at 25 °C. For instance, plotting experimental values of $\log(K_a)/(K_a)_0$ and $\log k/k_0$ for the acid dissociation of phenyl phosphonic acid and base hydrolysis of substituted ethyl benzoates vs substituent constants yielded the ρ values equal to 0.76 and 2.4, respectively.

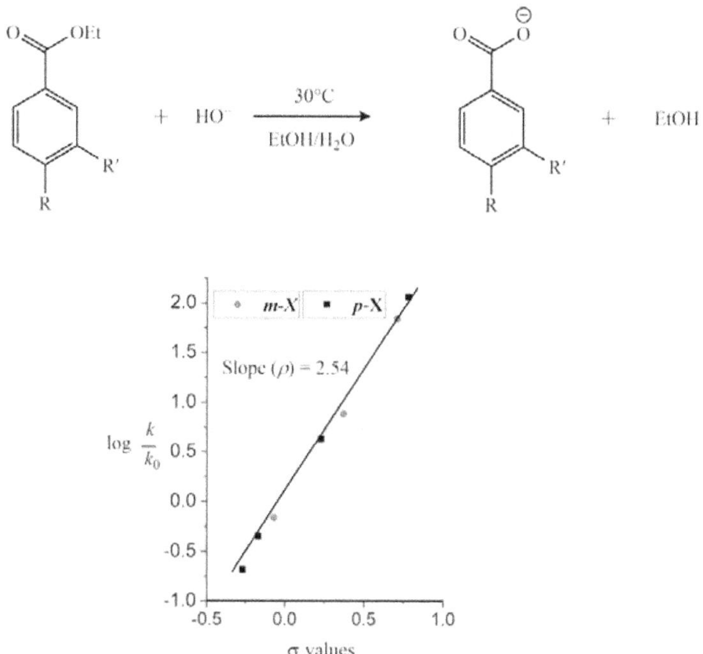

Figure 19. $\log k/k_0$ vs σ for alkaline hydrolysis of substituted ethyl benzoate.

Reaction constants or ρ values for many other reactions and equilibria have been obtained over years. P-values for some of the important reactions provided by Hammett himself are given below.

i) Hydrolysis of substituted cinnamic acid ester in water /ethanol (+1.267).

ii) Acid-catalyzed esterification of substituted benzoic esters in C_2H_5OH (−0.085).

iii) Ionization of substituted phenols in H_2O (+2.008).

iv) Substituted benzyl chlorides' hydrolysis in H_2O-acetone at 69.8 °C (−1.875).

v) The acid-catalyzed bromination of substituted acetophenones in CH_3COOH (acetic acid) or water or hydrochloric acid (+0.417).

Now, as far as the significance is concerned, the sensitivity constant (i.e. reaction constant ρ), defines the reaction's susceptibility to different substituents, relative to the ionization of benzoic acid; and is equal to the slope of the Hammett's equation or plot. The reaction's information and the mechanism involved can be found using the value of ρ as given below.

Case 1: if $\rho > 1$, the reaction has a greater sensitivity to substituents than the benzoic acid and a negative charge will accumulate (or a loss of positive charge) in the course of the reaction.

Case 2: if $0 < \rho < 1$, the reaction will be less sensitive to substituents than the benzoic acid and a negative will accumulate (or a loss of positive charge).

Case 3: if $\rho = 0$, the reaction will show no sensitivity to substituents, and no charge will be lost or built.

Case 4: if $\rho < 0$, a positive will accumulate (or a loss of negative charge) during the reaction.

The correlations given above can be used to explain the mechanism of an organic reaction. Since the ρ-value is connected to the charge in the course of the rate-limiting step, the mechanism involved can be developed using the data obtained. For instance, if an aromatic compound's reaction is believed to happen via one of two routes, the organic compound can simply be modified with substituents with dissimilar σ values and then the shortlisting can is done by taking kinetic measurements. After the measurements we mentioned, the Hammett plot can be raised to find the ρ value. Now, if the mechanisms we believe to be true encompass the charge formation, the ρ value will easily confirm our predictions. On the other hand, if the Hammett plot demonstrates that no charge is created during the reaction (i.e., slope or $\rho = 0$), the mechanism with the charge development can simply be neglected.

It is also worthy to note that the Hammett plots may not always be flawlessly linear. For example, a plot may have an unexpected or rapid change in the ρ value or slope. A case like this means that the mechanism responsible for the reaction has simply been changed due to the addition of different substituents. Some other kinds of deviations from linear variation may be attributed to a change in the site of the transition state. A situation like this means that certain substituents may cause the transition state to form later (or earlier) during the mechanism involved.

Table 4. The reaction constants (ρ-values) for a relative analysis for some typical organic chemical reaction types.

Reaction type	ρ-value
Ionization of acids	1.464
Alkaline hydrolysis of ethyl esters	2.494
Acids with diphenyldiazomethane	0.937
Acid dissociation of phenyl phosphonic acid	0.76

❖ Taft Equation

The Taft equation, just like the Hammett equation, is also a linear free energy relationship (LFER) that is employed in physical organic chemistry in the analysis of reaction mechanisms and to develop the quantitative relationships between structure and activity for organic species. Robert W. Taft developed this equation in 1952 as an amendment to the Hammett equation. Nevertheless, unlike the Hammett equation (accounts only for inductive, field, and resonance effects on rate of reaction), the Taft equation also explains the steric effects of a substituent. The mathematical form of the Taft equation is given below.

$$\log \frac{k}{k_0} = \sigma^* \rho^* + \delta E_s \tag{56}$$

where k and k_0 are the constant for the group $X \neq H$ and $X = H$, respectively; ρ^* and σ^* are the modified reaction constants (sensitivity factor for the reaction to polar effects) and substituent X (describes the field and inductive effects), respectively. The symbol δ is the sensitivity factor for the reaction to steric effects, and E_s is the steric substituent constant.

➤ *Polar Substituent Constants (σ^*)*

The polar substituent constants explain the way a substituent affects a reaction pathway via polar (field, inductive, and mesomeric effect) influences. Taft examined the hydrolysis of methyl esters to get σ* values. The idea of using rates of ester hydrolysis to study polar effects was initially proposed by Ingold in early 1930. Esters hydrolysis can proceed via either acid- or base-catalyzed pathway, and both routes involve a tetrahedral intermediate species. During the base-catalyzed pathway, the reactant transforms from a neutral entity to a negatively charged intermediate in the rate-limiting step; whereas in the acid-catalyzed pathway, a positively charged reactant transforms to an intermediate species with a positive charge.

Base Catalyzed Ester Hydrolysis

Acid Catalyzed Ester Hydrolysis

Owing to the same nature of intermediates (tetrahedral), Taft suggested that any steric factors under identical conditions should be approximately the same for the two pathways; and so would not affect the rates' ratio. Nevertheless, since a charge difference is built up in the rate-limiting steps it was suggested that polar effects would only affect the rate of reaction for base-catalyzed transformation because a new charge was created. The mathematical formulation of polar substituent constant (σ*) can be written as given below.:

$$\sigma^* = \frac{1}{2.48\rho^*}\left[\left(\log\frac{k}{k_0}\right)_B - \left(\log\frac{k}{k_0}\right)_A\right] + \delta E_s \tag{57}$$

Where $(\log k/k_0)_B$ represents the ratio of the base-catalyzed reaction rate compared to the reference transformation; whilst $(\log k/k_0)_A$ represents the ratio of the acid-catalyzed reaction rate relative to the reference transformation. The symbol ρ^* shows the reaction constant which explains the sensitivity of the series of reactions. For the base reaction series, we use $\rho^* = 1$ and R = CH_3 is set as the reference transformation with $\sigma^* = 0$. The incorporation of 1/2.48 is to make the magnitude σ^* equal σ values given by Hammett.

> ➤ *Steric Substituent Constants (E_s)*

Though the base- and acid-catalyzed esters' hydrolysis yield transition states for the rate-limiting steps which have different densities of charge, their molecular structures differ by 2 H atoms only. Therefore, Taft thought that the steric effects should affect both pathways by equal extent. Owing to this fact, the magnitude of E_s (steric substituent constant) can be obtained from purely the acid-catalyzed pathway since polar effects would be excluded this way. E_s was defined as:

$$E_s = \frac{1}{\delta}\log\frac{k}{k_0} \tag{58}$$

Where symbol δ represents the reaction constant which explains the reaction's susceptibility to steric factor. The values $\delta = 1$ and $E_s = 0$ were used for the definition reaction series. The equation (49) can be combined with equation (48) to write the complete form of the Taft equation. Comparing E_s values for CH_3, C_2H_5, isopropyl, and tert-butyl; it is obvious that it increases with growing steric bulk. Nevertheless, E_s values can deviate from expectation due to steric interactions. For instance, the phenyl's E_s is larger than tert-butyl; however, if we compare these groups using another measure, the tert-butyl will come out to be dominant.

Table 5. Constants used in the Taft equation.

Group	E_s	σ^*
–H	1.24	0.49
–CH_3	0	0
–CH_2CH_3	–0.07	–0.1
–$CH(CH_3)_2$	–0.47	–0.19
–$C(CH_3)_3$	–1.54	–0.3
–CH_2Ph	–0.38	0.22
–Ph	–2.55	0.6

> *Sensitivity Factors*

A brief discussion on the nature and significance of all the sensitivity factors used in the Taft equation is given below.

1. Polar sensitivity factor (ρ^*): Just like Hammett's ρ-values, Taft's ρ^*-values describe the reaction's susceptibility to polar effects. If the steric effects of substituents do not influence the rate of reaction significantly, the Taft equation will reduce to Hammett equation as given below.

$$\log\frac{k}{k_0} = \sigma^*\rho^* \tag{59}$$

The ρ^* value can be found by plotting the log of the ratio of the experimental rates (k) to the reference reaction (k_0) vs the σ^* values of different substituents. The slope of such plot will be equal to ρ^*. Just like Hammett's ρ-value:

i) If $\rho^* > 1$, a negative charge will accumulate in the transition state during the reaction, and the reaction will be accelerated by electron-withdrawing substituents.

ii) If $1 > \rho^* > 0$, a negative charge will accumulate in the transition state during the reaction, and the reaction will show mild sensitivity to polar effects.

iii) If $\rho^* = 0$, the reaction will simply not get influenced by polar effects.

iv) If $0 > \rho^* > -1$, a positive charge will accumulate in the transition state during the reaction, and the reaction will show mild sensitivity to polar effects.

v) If $-1 > \rho^*$, a positive charge will accumulate in the transition state during the reaction, and the reaction will be accelerated by electron-donating substituents.

2. Steric sensitivity factor (δ): Just like the polar sensitivity factor, δ or the steric sensitivity factor of reaction explains to what extent the rate of reaction is affected by steric effects. If the polar effects of substituents do not influence the rate of reaction significantly, the Taft equation will reduce to the equation as given below.

$$\log\frac{k}{k_0} = \delta E_s \tag{60}$$

By plotting the log of the ratio rates vs the E_s value of different substituents, we get a straight line with a slope equal to δ. Similarly to Hammett's ρ-value, the magnitude of δ gives the extent of steric effects:

i) If the δ-value is very high, the reaction will be extremely sensitive to steric effects, whereas a smaller δ-value implies that the reaction has little to no sensitivity to steric influence.

Also, owing to larger and negative E_s-values for bulkier substituents, we may conclude that the flowing point about the reaction profile.

i) If $\delta > 1$, the raise in steric bulk cuts the rate of reaction and steric effects will be higher in the transition state.

ii) If $\delta < 1$, the raise in steric bulk will raise the rate and steric effects will be reduced in the transition state.

3. Reactions influenced by polar and steric effects: If the steric, as well as polar effects, influence the rate of chemical reaction, the Taft equation can be employed to evaluate ρ^*- and δ-values via the use of standard least-squares fitting for getting a bivariant regression plane. The application of this technique was outlined by Taft in 1957 as a demonstration of accuracy.

❖ Problems

Q 1. What is the reaction mechanism? How it can be used as a basis to classify different types of organic reactions?

Q 2. What are the thermodynamic and kinetic requirements of an organic reaction to occur? Explain with a suitable example.

Q 3. What do you mean by kinetically controlled product? How is it different from a thermodynamically controlled reaction?

Q 4. State and explain Hammond's postulate.

Q 5. Derive the mathematical formulation of the Curtin-Hammett principle.

Q 6. Draw and explain the potential energy diagrams of different kinds of organic reactions.

Q 7. Discuss the methods of determining the reaction mechanism with special reference to the detection of reaction intermediates.

Q 8. Define isotope effect.

Q 9. State and explain Pearson's hard-soft-acid-base principle.

Q 10. What are the differences and similarities in carbocations and carbanions?

Q 11. How does the structure of the substrate affect the reaction rate?

Q 12. Give Hammett equation. How it can be used to prove the linear free energy relationship?

Q 13. Write down the Taft equation. Discuss its scope and physical significance.

❖ Bibliography

1. C. A. Coulson, B. O'Leary, R. B. Mallion, *Hückel Theory for Organic Chemists*, Academic Press, Massachusetts, USA, 1978.

2. C. D. Johnson, *The Hammett Equation*, Cambridge University Press, London, UK, 1973.

2. M. B. Smith, *March's Advanced Organic Chemistry: Reactions, Mechanisms, and Structure*, John Wiley & Sons, Inc., New Jersey, USA, 2013.

3. D. Klein, *Organic Chemistry*, John Wiley & Sons, Inc., New Jersey, USA, 2015.

4. H. Zimmerman, *Quantum Mechanics for Organic Chemists*, Academic Press, New York, USA, 1975.

5. J. Clayden, N. Greeves, S. Warren, *Organic Chemistry*, Oxford University Press, Oxford, UK, 2012.

6. R. L. Madan, O*rganic Chemistry*, Tata McGraw Hill, New Delhi India, 2013.

7. R. W. Taft, *Linear Free Energy Relationships from Rates of Esterification and Hydrolysis of Aliphatic and Ortho-substituted Benzoate Esters*, Am. Chem. Soc. 74 (1952) 2729–2732.

CHAPTER 4

Carbohydrates

❖ Types of Naturally Occurring Sugars

It is quite a well-known fact that carbohydrates can primarily be classified into three categories; monosaccharides, oligosaccharides, and polysaccharides. The monosaccharides are the simplest carbohydrates that cannot be further hydrolyzed to simpler molecules. The general formula of monosaccharides is $(CH_2O)_n$ where n = 3–8. The oligosaccharides are the carbohydrate molecules that can produce 2–10 molecules of monosaccharides. Polysaccharides are carbohydrate molecules that can produce a very large number of monosaccharides' molecules upon hydrolysis.

Furthermore, in addition to the number of hydrolysis produce, the carbohydrates can also be classified on the basis of their taste. It has been found that all the monosaccharides and oligosaccharides (di-, tri-, tetra-saccharides, etc.) are crystalline compounds, soluble in water and sweet in taste; and typically labeled as sugars. On the other hand, the polysaccharides are amorphous compounds, insoluble in water, and don't have any taste; and therefore, these carbohydrates are typically called as non-sugars. In this section, we will discuss the different types of naturally occurring sugars.

➤ *D-(+)-Glucose or Dextrose or Grape Sugar ($C_6H_{12}O_6$)*

The D-(+)-glucose or dextrose is the most abundant monosaccharide in nature; and is also found in the combined state in many disaccharides, polysaccharides, and glycosides. The name grape sugar comes from the fact that D-(+)-Glucose is found in very large amounts in ripe grapes.

Properties: *i)* It is a white solid with crystalline nature that melts at 419K. It is not soluble in ether but may dissolve to some extent in alcohol. However, the sweet solid is highly dissolvable in water. Furthermore, as the name suggests D-(+)-glucose is optically active and is dextrorotatory in nature. The dextrose possesses 75% sweetness to that of table sugar.

ii) Glucose shows most of the aldehydic reactions but does not respond to Schiff's reagent test and is unable to yield addition compounds with sodium bisulfite.

iii) Glucose also reacts with hydroxyl groups. For instance, it reacts with acetic anhydride and methanol to yield glucose penta-acetate and α- or β-methylglucosides, respectively.

iv) Dextrose doesn't react with dilute acids but can give 5-hydroxymethylfurfural upon heating with concentrated HCl solution.

v) Upon treatment with concentrated alkali solution, glucose first turns yellow and then brown resinous mass. On the other hand, reaction with dilute alkali solution gives rise to an equilibrium mixture of glucose, fructose, and mannose.

vi) Glucose gets fermented to ethanol when mixed with yeast due to enzyme zymase.

Structure: The terminal aldehydic carbon in open-chain glucose molecule may participate in hemiacetal formation by using the hydroxyl group of 4th and 5th carbon in open-chain glucose molecule, giving rise to a five-membered furan-like and six-membered pyran-like ring structure, respectively. In the solution phase, the open-chain type of glucose (either "L-" or "D-") happens to be in equilibrium with numerous cyclic isomers, where each contains a cycle of carbons closed by one O atom. Nevertheless, in an aqueous phase, greater than 99% of glucose amount, at any given time, exists as the pyranose form; on the other hand, furanose form exists in negligible concentration with the open-chain type is restricted to 0.25% only.

i) Pyranose form: The terminal aldehydic carbon participate in hemiacetal formation by using the hydroxyl group of 5th carbon in open-chain glucose molecule to give pyranose form.

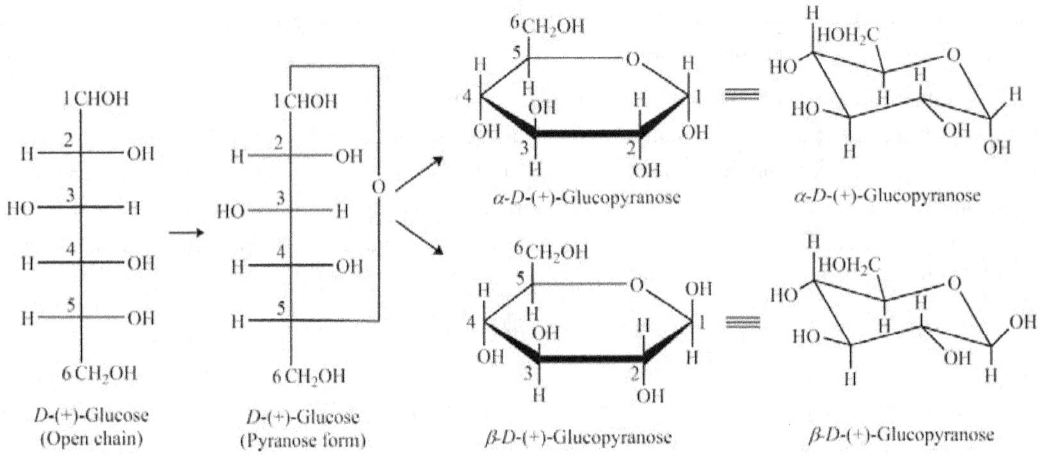

ii) Furanose form: The terminal aldehydic carbon participate in hemiacetal formation by using the hydroxyl group of 4th carbon in open-chain glucose molecule to give furanose form.

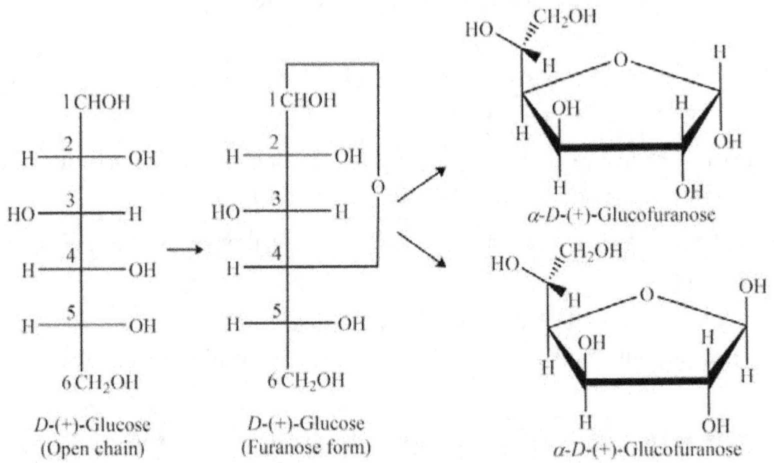

> ### D-(−)-Fructose or Laevulose or Fruit Sugar ($C_6H_{12}O_6$)

The D-(−)-fructose or laevulose is the most important ketoses monosaccharide in nature. It exists freely in honey and is also found in the combined state in many disaccharides, polysaccharides, and glycosides. The name fruit sugar comes from the fact that D-(−)-fructose is found in very large amounts in sweet fruits.

Properties: *i)* It is a white solid with crystalline nature that melts at 375K. It has a higher solubility in water and alcohol than glucose. Furthermore, as the name suggests D-(−)-fructose is optically active and is laevorotatory in nature.

ii) Fructose gives rise to most of the typical ketonic chemical reactions including oxidation and reduction types as well.

iii) Fructose also gives many typical reactions of hydroxyl groups like acetylation or the formation of fructosates etc.

iv) Fructose doesn't react with dilute acids but can give laevulinic acid upon heating with concentrated HCl acid solution.

v) Upon treatment of fructose with dilute alkali solution, we get an equilibrium mixture of glucose, fructose, and mannose.

vi) Like glucose, the fructose also gets fermented to ethyl alcohol when mixed with yeast due to enzyme zymase.

Structure: The ketonic carbon in open-chain fructose molecule may participate in hemiketal formation by using the hydroxyl group of 5th and 6th carbon in open-chain glucose molecule, giving rise to a five-membered furan-like and six-membered pyran-like ring structure, respectively.

i) Pyranose form: The ketonic carbon participate in hemiketal formation by using the hydroxyl group of 6th carbon in open-chain glucose molecule to give pyranose form.

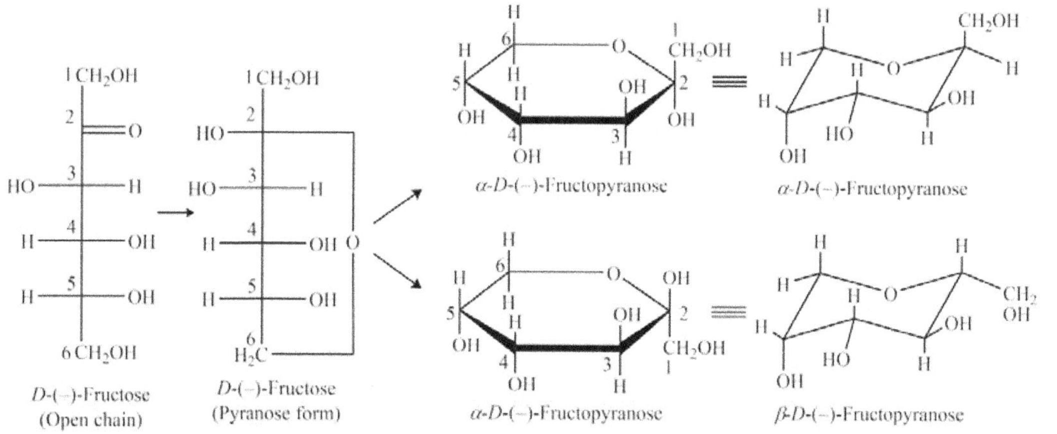

i) Furanose form: The ketonic carbon participate in hemiketal formation by using the hydroxyl group of 6th carbon in open-chain glucose molecule to give furanose form.

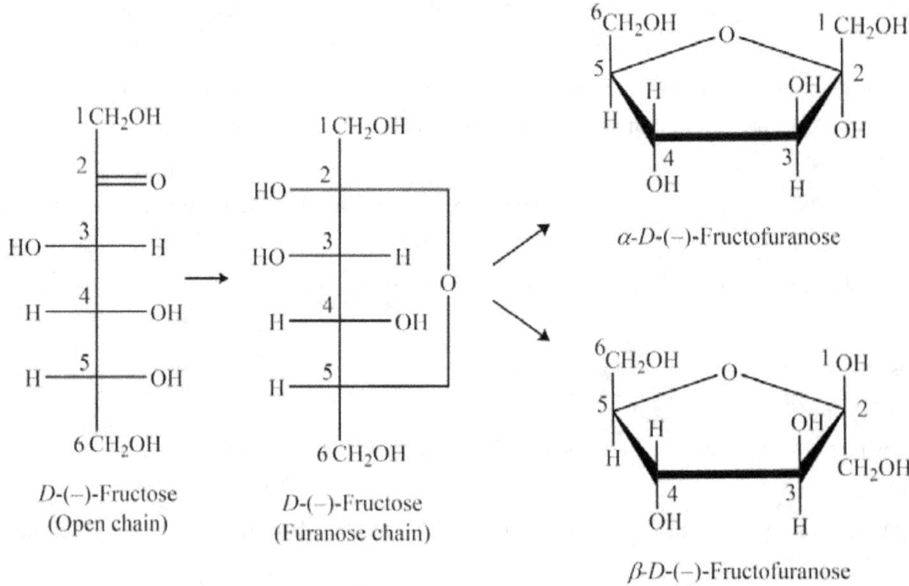

> **(+)-Sucrose or Cane-Sugar or Table Sugar ($C_{12}H_{22}O_{11}$)**

Sucrose is the most important disaccharide in nature and is the most widely produced pure chemical. The name cane sugar comes from the fact that (+)-sucrose is found in very large amounts in sugar cane and sugar beets.

Properties: *i)* It is a white solid with crystalline nature that melts at 453K. It is not soluble in ether and alcohol. However, the sweet solid is highly dissolvable in water. Furthermore, as the name suggests (+)-sucrose is optically active and is dextrorotatory in nature.

ii) Upon heating above its melting point, it gets converted to caramel, a brown amorphous solid which is beverage coloring and confectionery. The further heating of the same produces charring with burnt sugar's smell.

iii) Upon treating with yeast, sucrose yields an equimolar mixture of D-(+)-glucose and D-(–)-fructose which is due to the enzyme invertase.

iv) Sucrose reacts with acetic anhydride to give sucrose octaacetate.

v) Sucrose yields oxalic acid when treated with concentrated HCl.

vi) It gives sugar charcoal when treated with concentrated sulphuric acid with a large amount of SO_2 and CO_2 release.

Structure: The sucrose is made up of one glucose and one fructose unit which are joined together by the glycosidic linkage.

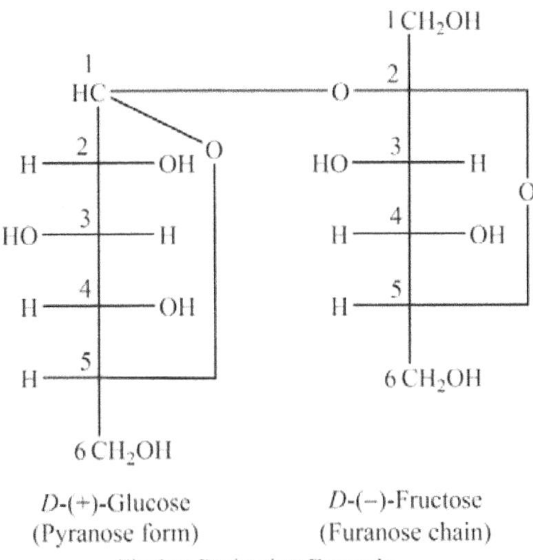

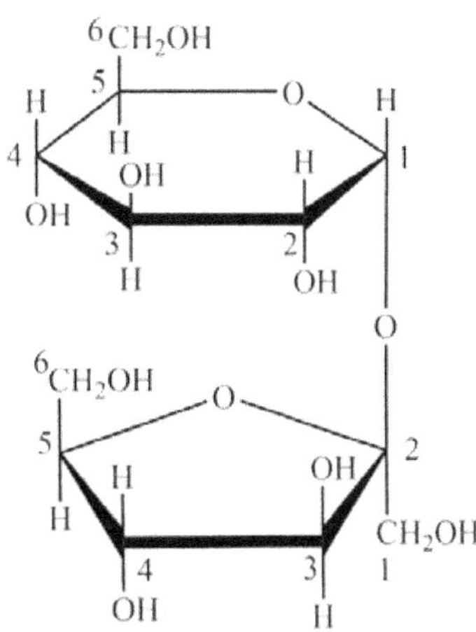

> ➤ *(+)-Maltose or Malt Sugar ($C_{12}H_{22}O_{11}$)*

Maltose in nature is primarily present in germinating seeds especially cereals. Originally, Augustin-Pierre Dubrunfaut discovered Maltose; nevertheless, his finding was well accepted in 1872 after the confirmation by Irish brewer and chemist Cornelius O'Sullivan. The name maltose comes from malt, combined with the suffix '-ose' which is used in sugars' nomenclature.

Properties: *i)* It is a white solid with crystalline nature that melts at 438K. It is not soluble in ether and alcohol. However, the sweet solid is well dissolvable in water. Furthermore, as the name suggests (+)-maltose is optically active and is dextrorotatory in nature.

ii) Upon treating with Br_2/H_2O, maltose yields maltobionic acid, an organic compound with the same number of carbon atoms as maltose.

iii) Upon treating with dilute acids or yeast, maltose yields to moles of D-(+)-glucose which is due to the enzyme maltase.

iv) Just like the case of (+)-sucrose (cane sugar) Maltose reacts with acetic anhydride to give maltose octaacetate.

v) Tollens' reagent and Fehling's solution are well reduced by maltose.

vi) The maltose molecules react with hydroxylamine to yield phenylhydrazine or oxime to form phenylhydrazone.

Structure: The maltose is made up of two glucose units, one reducing and one none reducing, with are joined together by the glycosidic linkage.

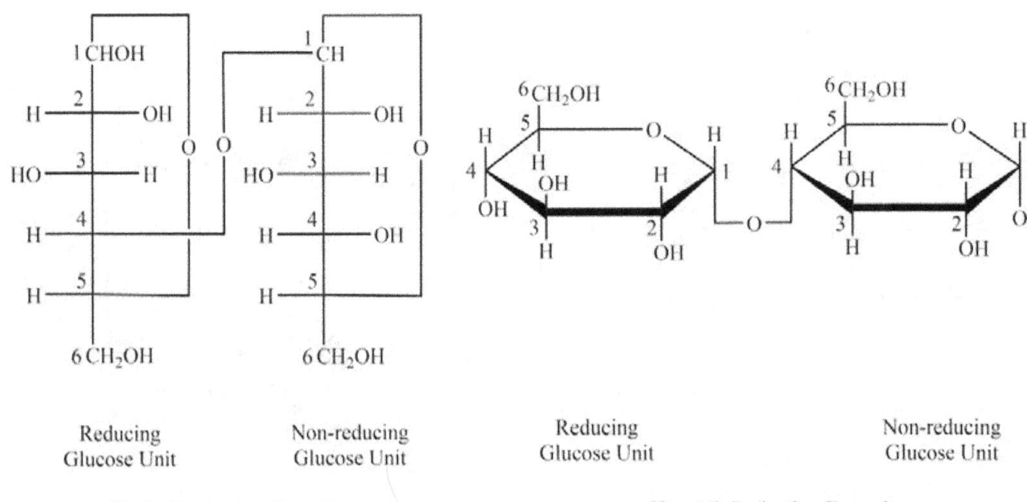

> ### *(+)-Lactose or Milk Sugar ($C_{12}H_{22}O_{11}$)*

Lactose is a disaccharide which is a sugar composed of galactose and glucose subunits and has the molecular formula $C_{12}H_{22}O_{11}$. Lactose is mainly found in mammals' milk; and therefore, it is also called as milk sugar. The milk gets sour if the bacterial action turns (+)-lactose into lactic acid. Lactose makes up around 2–8% of milk (by weight).

Properties: *i*) Lactose is a mildly sweet, non-hygroscopic, water-soluble, white solid with α- and β- forms which melt at 496K and 525K, respectively. Furthermore, as the name suggests (+)-lactose is optically active and is dextrorotatory in nature.

ii) Upon treating with dilute acid or yeast, lactose yields an equimolar mixture of D-(+)-glucose and D-(+)-galactose which is due to the enzyme lactase.

iii) Tollens' reagent and Fehling's solution are well reduced by lactose. This confirms the reducing nature of lactose like maltose.

iv) The lactose molecules or milk sugar react with hydroxylamine to yield oxime, and with phenylhydrazine gives osazone.

vi) Upon treating with Br_2/H_2O, lactose yields lactobionic acid, an organic compound with the same number of carbon atoms as lactose.

vii) Lactose has relatively low cariogenicity among sugars.

viii) Undigested lactose acts as dietary fiber.

Structure: The sucrose is made up of one glucose and one galactose unit with are joined together by the glycosidic linkage.

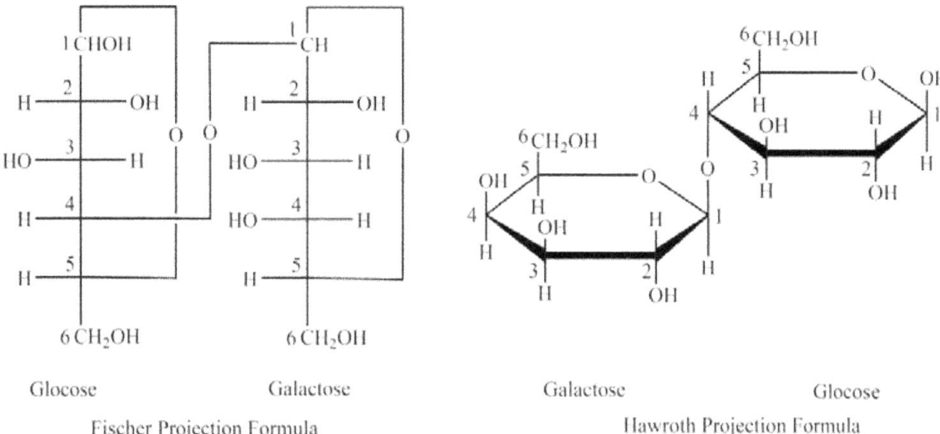

❖ Deoxy Sugars

Deoxy sugars may simply be defined as the sugars in which at least one hydroxyl group is replaced by a hydrogen atom.

The common nomenclature of deoxy sugars is carried out by adding the prefix "deoxy" along with the location of carbon at which the displacement has taken place, followed by the common name of parent aldoses or ketoses. On the other hand, the IUPAC nomenclature is done by adding two prefixes before the IUPAC name of parent aldoses or ketoses; the first prefix "deoxy" (along with the location of substituted carbon) and the at which the displacement has taken place followed by the configurational prefix of chiral carbon atoms.

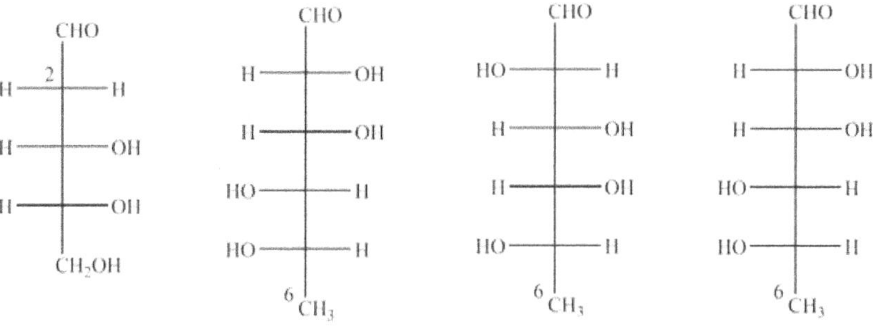

Common Name:	2-Deoxy-D-(–)-Ribose	6-Deoxy-L-(–)-Mannose	6-Deoxy-L-(–)-Galactose	2, 6-Dideoxy-D-(+)-Allose
IUPAC Name:	2-Deoxy-D-erythropentose	6-Deoxy-L-Mannohexose	6-Deoxy-L-Galactohexose	2, 6-Dideoxy-L-Ribohexose

The most abundant deoxy sugar is deoxyribose which is a major constituent of DNA. Furthermore, it is also worthy to note that although the replacement of the hydroxyl group (by the hydrogen atom) is feasible at any carbon, most of the deoxy sugars are 6-deoxy type.

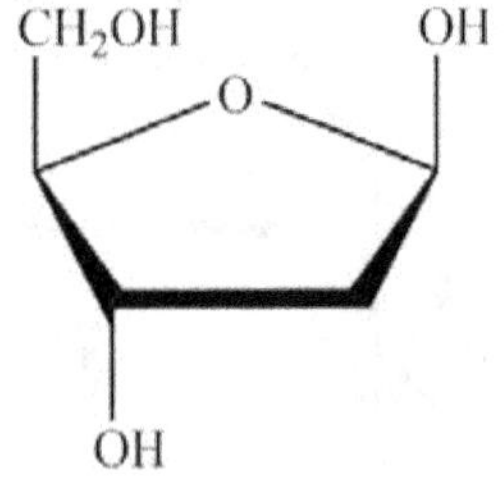

(I)

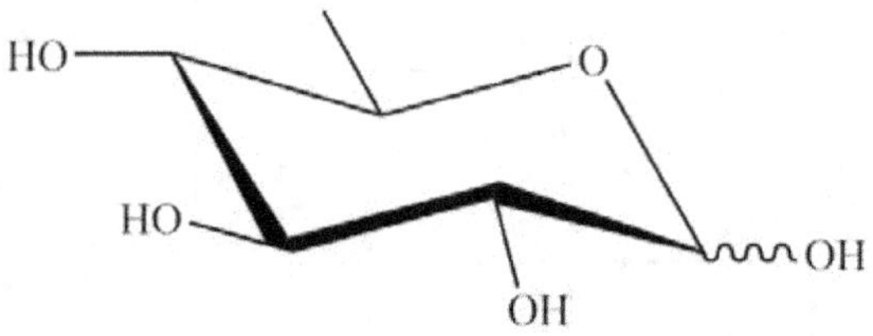

(II)

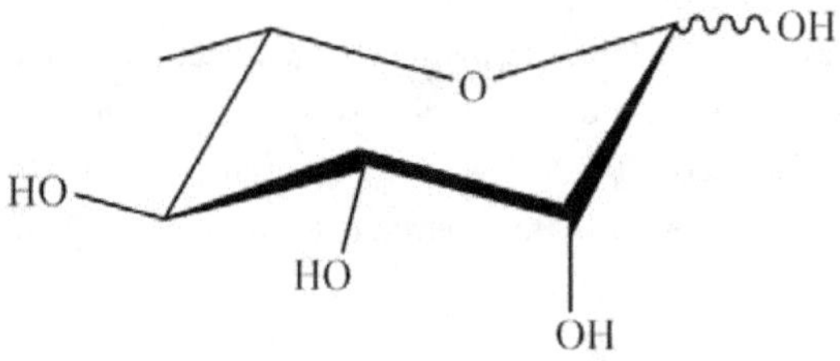

(III)

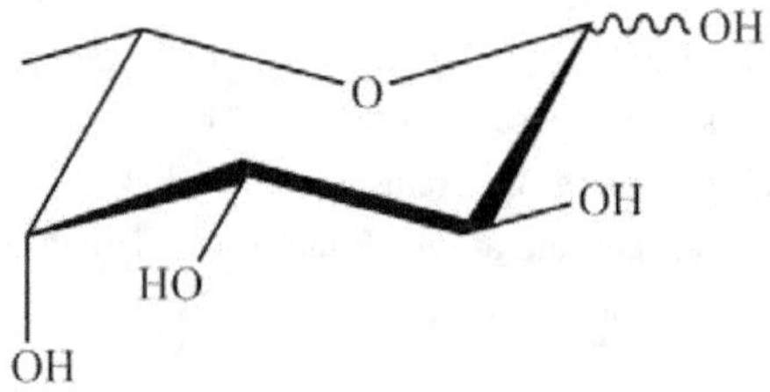

(IV)

It seems that the deoxygenation at C2 in β-D-ribofuranose enhances the overall stability of the consequential nucleic acid; however, the better hydrophobic interaction at C6 (methyl group) with appropriate receptor sites appears to be the dominant factor in the other three molecules.

❖ Amino Sugars

Amino sugars may simply be defined as the sugars in which at least one hydroxyl group is replaced by an amine group.

Since amino sugars play key roles in the structure and proper functioning of glycoproteins many biologically significant polysaccharides, these types of sugars are absolutely vital components for the well-functioning of humans and other organisms. The primary cause of such behavior lies in the good polarity to the molecule due to the positively charged ammonium ion or the acetamido group which can be created from the protonation of a free amine in a monosaccharide residue. Two of the most popular members of amino sugars are given below.

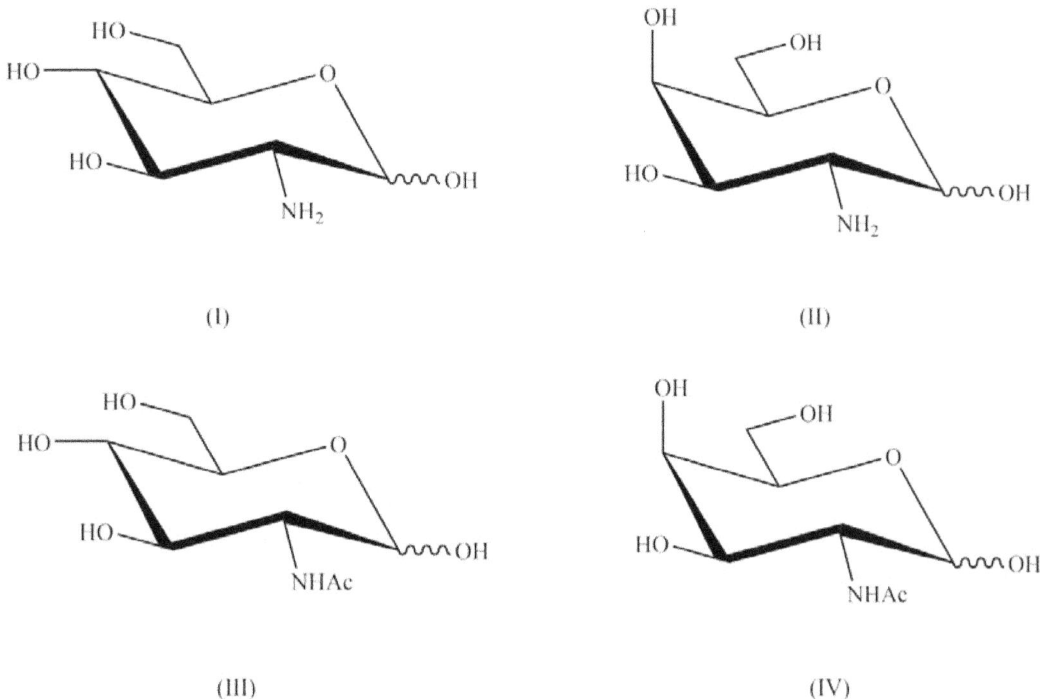

(I)   (II)

(III)   (IV)

Out of more than 60 types of known amino sugars, N-Acetyl-d-glucosamine the most abundant (main component of chitin). It is also worthy to note that the derivatives of many amine-containing sugars like sialic acid or N-acetylglucosamine are also considered amino sugars, despite the fact that nitrogens in them are part of more complex functional groups rather than formal amines. Furthermore, aminoglycosides (conjugates of amino sugars and aminocyclitols) form a group of antimicrobial substances that hinder the synthesis of bacterial proteins.

❖ Branch Chain Sugars

Branched-chain sugars may simply be defined as the sugars in which at least one hydrogen atom or hydroxyl group is replaced by a carbon group.

The common types of branched-chain sugars (both kinds) are found in biomolecules. Some typical examples are given below.

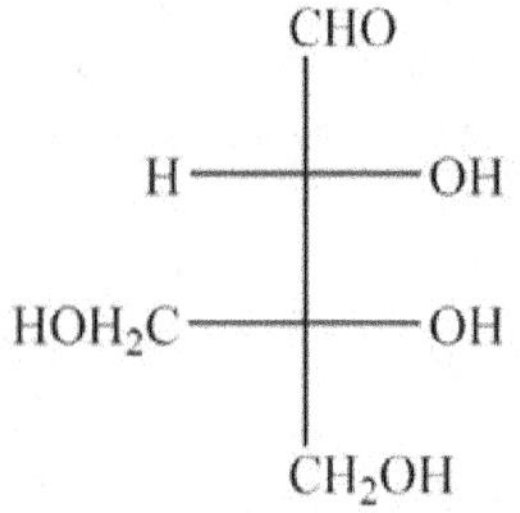

(I)

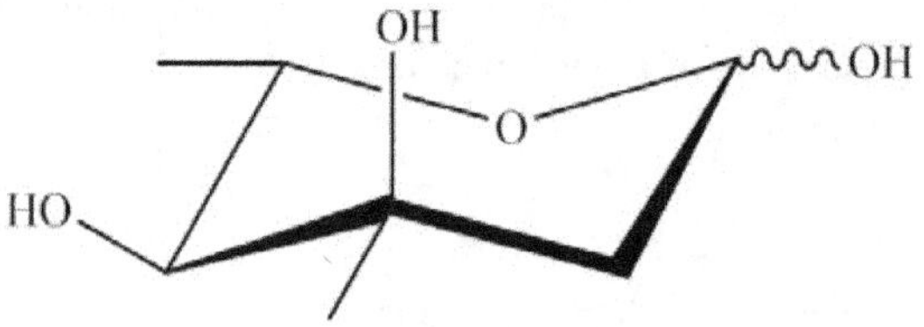

(II)

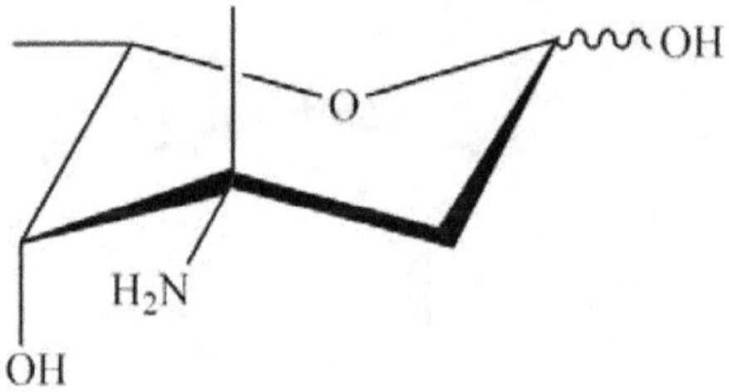

(III)

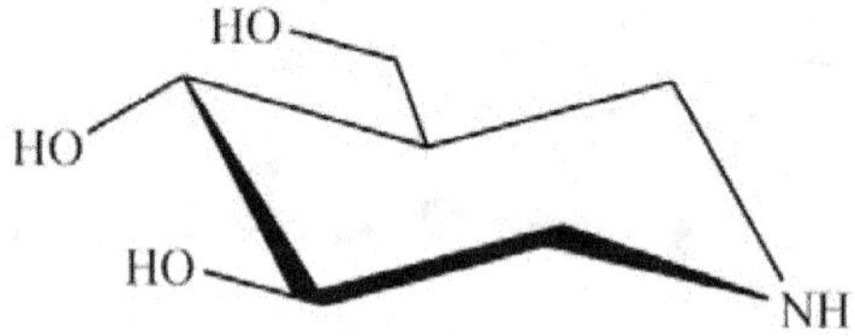

(IV)

It is also worthy to note that branched-chain sugars were considered as "rare" until 1960; however, after the confirmation of their presence in many antibiotics (in form of their glycosidic component), many synthetic chemists put effort to widen the availability domain. Branched-chain sugars are generally obtained from uloses or by the opening of epoxides.

❖ General Methods of Determination of Structure and Ring Size of Sugars with Particular Reference to Maltose, Lactose, Sucrose, Starch and Cellulose

Almost all of the oligo- or polysaccharides are white powdery substances that cannot be dissolved in a typical organic solvent but show good solubility in water; so, we cannot tell how the atoms are connected with each other just by looking at it or by other simple routes. Also, owing to the presence of a large number of atoms (especially in the case of polysaccharides), the routine procedure for the "structure elucidation of organic compounds" cannot be employed. Therefore, instead of studying the entire molecule via mass spectrometry or NMR spectroscopy at once, a very sophisticated route has been developed by the researchers over the years in which building blocks of the same carbohydrate are studied. In this section, we will study the general route for the structure determination of oligo- and polysaccharides and then we will apply the same to evaluate the structure of maltose, lactose, sucrose, starch, and cellulose.

➤ *General Route to Find the Structure of Oligo- or Polysaccharides*

The structure elucidation of complex carbohydrates is based on the principle that an oligo- or polysaccharides can be disconnected into monosaccharide units, which in turn can be studied, and finally can be recombined to produce intact molecule mentally. The general route for the structure determination of oligo- and poly-saccharides involves the following steps.

1. Monomeric analysis: This is the first step which involves the disconnection of a given oligo- or poly-saccharide (typically at glycosidic linkages) into its monosaccharide components. This is typically achieved by the acid hydrolysis of glycosidic bonds. The rate of acid hydrolysis of glycosidic joints differs for the size of the cycle, nature of the bond, and the corresponding configuration also.

2. Study of the monosaccharide units: Once the monosaccharides are obtained, we need to study those building blocks (such as chain length or ring size) by conventional modern spectroscopic techniques like NMR or X-ray analysis. The necessary and optional subtypes of this step are given below.

i) Identification: The identification of methylated sugars or monosaccharides means that we try to identify the building block by search-match its experimental parameters (like the melting point or specific rotation) to previously reported literature i.e., handbooks.

ii) Historical Method: The earliest work to determine the configuration at asymmetric carbon in E. Fisher. He primarily used two routes to study typical monosaccharides; cyanohydrin synthesis and oxidation using nitric acid. In the later period, he used the same approach to get the relative configuration of many other pentoses and hexoses.

iii) Mass spectrometry: The step employs mass spectrometry to find out the structural data of the monosaccharide but gives no information about the stereochemical notation.

iv) NMR analysis: Once the structure of the monosaccharide is known, NMR spectroscopy is used to determine the configuration at the asymmetric center.

3. Finding the nature of linkage: The monomeric analysis and study of monosaccharide units tell us the nature of the cyclic form of monosaccharide units, the bonding poisons, and whether the polysaccharide is branched or unbranched. Therefore, the complete structure of the polysaccharide can be known only after the configuration of glycosidic bonds and the absolute sequence of monosaccharide units in the complete chain. This can be achieved by cleavage selectivity of glycosidic bonds and theoretical mono- or disaccharide yield. In other words, acid hydrolysis's selectivity for most of the polysaccharides is quite unique and can successfully be employed to find out the nature of the linkage.

> ➤ *Structure Determination of Maltose, Lactose, Sucrose, Starch and Cellulose*

The general route of structure determination of some typical oligosaccharides and polysaccharides is given below.

1. Structure determination of maltose: Maltose is a natural sugar with formula $C_{12}H_{22}O_{11}$ that reduces Tollens' reagent and Fehling solution which indicates its reducing character. The structure of maltose is obtained as given below.

i) Monomeric analysis: The hydrolysis of maltose with maltase or mineral acids yields two molecules of D-(+)-glucose. Furthermore, the oxidation of maltose with bromine water results in the maltobionic acid inferring that one of the two glucose units has reactive hemiacetal form at aldehydic carbon.

$$(C_{11}H_{21}O_{11})CHO \xrightarrow{Br_2/H_2O} (C_{11}H_{21}O_{11})COOH \tag{1}$$

ii) Study of the monosaccharide units: The full methylation of maltobionic acid with sodium hydroxide and $(CH_3)_2SO_4$, and then carrying out acid hydrolysis results in a mixture of 2, 3, 4, 6-tetra-O-methyl-D-glucose and 2, 3, 5, 6-tetra-O-methyl-D-gluconic acid.

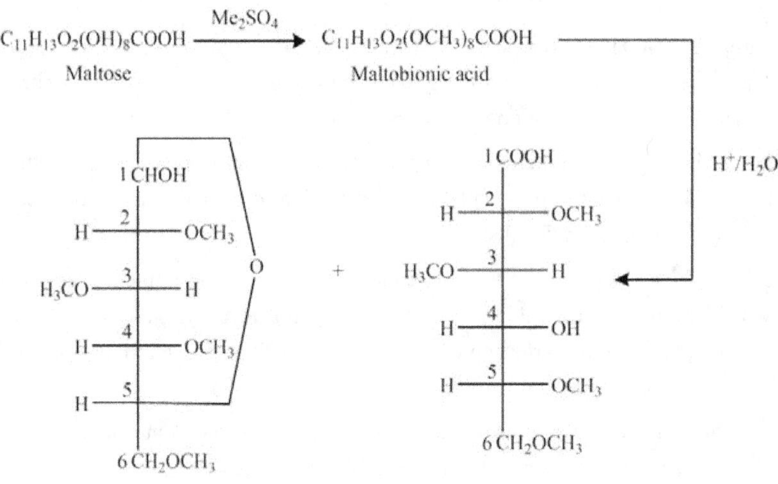

2, 3, 4, 6-tetra-O-methyl-D-glucose   2, 3, 5, 6-tetra-O-methyl-D-gluconic acid

The free –OH present at C_4 in 2, 3, 5, 6-tetra-O-methyl-D-gluconic acid infers that methylation was not possible in maltobionic acid, which in turn proves that C_4–OH must be engaged in glycosidic bonding in maltobionic acid, and therefore in maltose too. Now we are left with the possibility of C_5–OH participating in the formation 6-membered ring structure (pyranose form) in reducing glucose unit. On the other hand, the free –OH present at C_5 in 2, 3, 5, 6-tetra-O-methyl-D-glucose infers that methylation was not possible in maltobionic acid, which in turn proves that C_5–OH must be engaged in glycosidic bonding in maltobionic acid, and therefore in maltose too. Now we are left with the possibility of C_5–OH participating in the formation 6-membered ring structure (pyranose form) in a non-reducing glucose unit.

iii) Nature of linkage: Since we know that the maltose's hydrolysis by yeast (maltase enzyme) has specificity for the hydrolysis of α-glucosidic linkages, α-C_4–OH of reducing glucose unit must be connected to the α-C_1–OH of non-reducing glucose unit. The maltose is made up of two glucose units, one reducing and one none reducing, with are joined together by the glycosidic linkage.

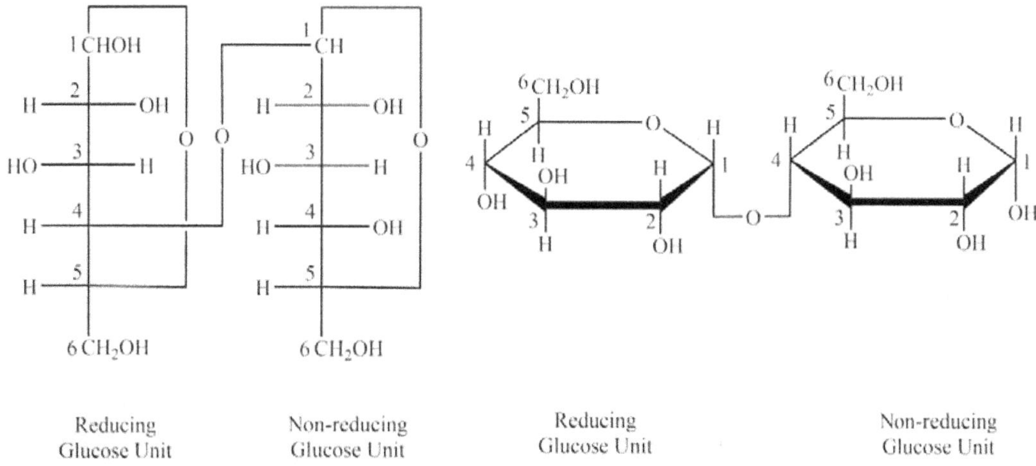

Furthermore, we can also conclude that the α-form of (+)-maltose differs from β-form w.r.t the configuration at anomeric carbon. In other words, the OH group at C_1 in β-form is above the plane whereas it lies below the plane in α-form.

2. Structure determination of lactose: Lactose is a natural sugar with formula $C_{12}H_{22}O_{11}$ that reduces Tollens' reagent and Fehling solution which indicates its reducing character. The structure of lactose is obtained as given below.

i) Monomeric analysis: The hydrolysis of lactose with lactase or mineral acids yields an equimolar mixture of D-(+)-glucose and D-(+)-galactose. Furthermore, the oxidation of lactose with bromine water results in lactobionic acid. The hydrolysis of lactobionic acid results in D-(+)-gluconic acid and D-(+)-galactose inferring that D-(+)-galactose is non-reducing while D-(+)-glucose must be reducing in nature.

ii) Study of the monosaccharide units: The full methylation of lactobionic acid with sodium hydroxide and $(CH_3)_2SO_4$, and then carrying out acid hydrolysis results in a mixture of 2, 3, 4, 6-tetra-O-methyl-D-galactose and 2, 3, 5, 6-tetra-O-methyl-D-gluconic acid

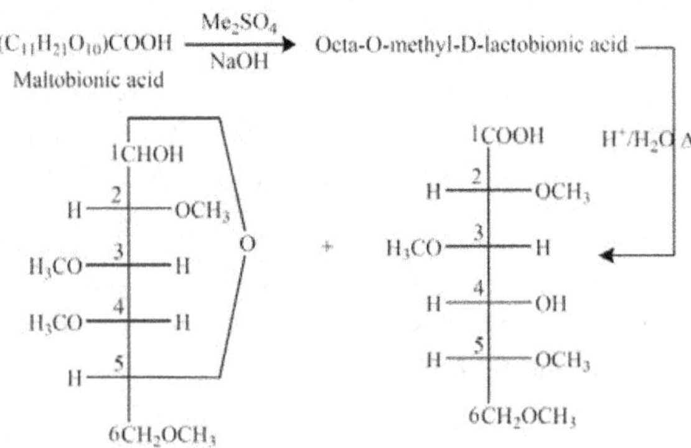

2, 3, 4, 6-tetra-O-methyl-D-galatcose   2, 3, 5, 6-tetra-O-methyl-D-gluconic acid

The free –OH present at C_4 in 2, 3, 5, 6-tetra-O-methyl-D-gluconic acid infers that methylation was not possible in lactobionic acid, which in turn proves that C_4–OH must be engaged in glycosidic bonding in lactobionic acid, and therefore in lactose too. Now we are left with the possibility of C_5–OH participating in the formation 6-membered ring structure (pyranose form) in reducing glucose unit. On the other hand, the free –OH present at C_5 in 2, 3, 4, 6-tetra-O-methyl-D-galactose infers that methylation was not possible in lactobionic acid, which in turn proves that C_5–OH must be engaged in glycosidic bonding in latobionic acid, and therefore in maltose too. Now we are left with the possibility of C_5–OH participating in the formation 6-membered ring structure (pyranose form) in a non-reducing galactose unit.

iii) Nature of linkage: Since we know that the lactose's hydrolysis by yeast (lactase enzyme) has specificity for the hydrolysis of β-glycosidic linkages, β-C_1–OH of non-reducing galactose unit must be connected to the C_4–OH of reducing glucose unit.

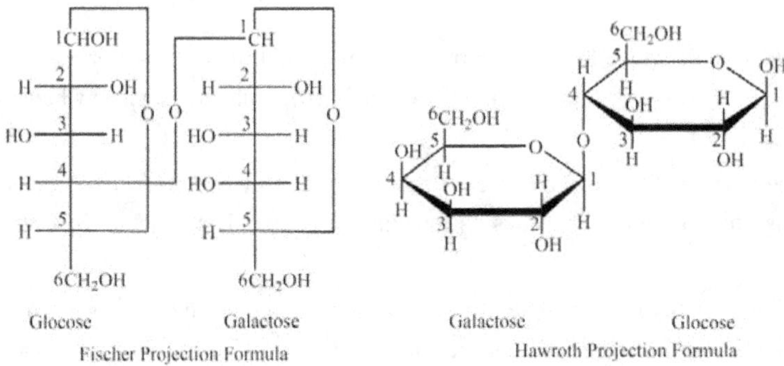

Glocose   Galactose   Galactose   Glocose

Fischer Projection Formula   Hawroth Projection Formula

Furthermore, we can also conclude that the α-form of (+)-lactose differs from β-form w.r.t the configuration at anomeric carbon (OH group at C_1 in β-form is above the plane whereas it lies below the plane in α-form).

3. Structure determination of sucrose: Sucrose is a natural sugar with formula $C_{12}H_{22}O_{11}$ that does not reduce Tollens' reagent and Fehling solution which indicates its non-reducing character. The structure of sucrose is obtained as given below.

i) Monomeric analysis: The hydrolysis of sucrose with invertase or mineral acids yields an equimolar mixture of D-(+)-glucose and D-(+)-fructose.

$$C_{12}H_{22}O_{11} \xrightarrow{\text{H}^+/\text{Invertase}} \underset{\text{Glucose}}{C_6H_{11}O_6} + \underset{\text{Fructose}}{C_6H_{11}O_6} \qquad (3)$$

ii) Study of the monosaccharide units: The full methylation of sucrose with sodium hydroxide and $(CH_3)_2SO_4$, and then carrying out acid hydrolysis results in a mixture of 2, 3, 4, 6-tetra-O-methyl-D-glucose and 1, 3, 4, 6-tetra-O-methyl-D-fructose.

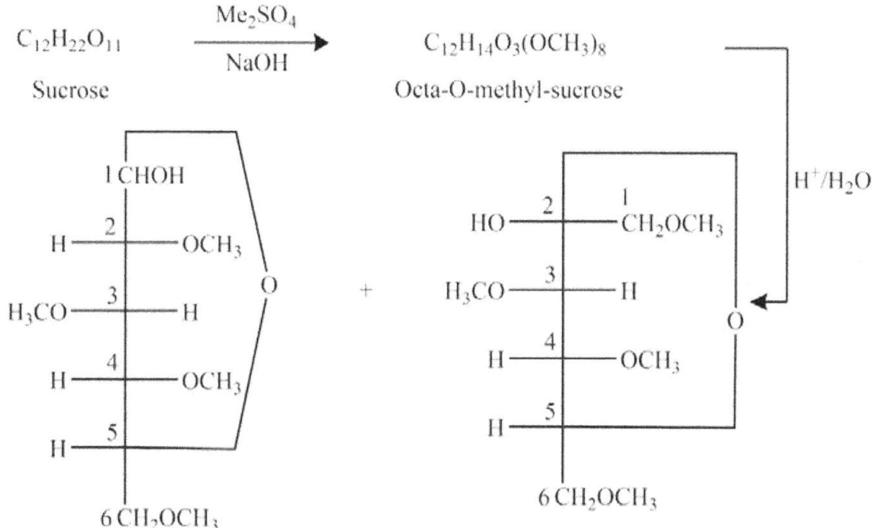

2, 3, 4, 6-tetra-O-methyl-D-glucose   1, 3, 4, 6-tetra-O-methyl-D-fructose

The NMR analysis and Fisher method showed that C_1–OH and C_5–OH in 2, 3, 4, 6-tetra-O-methyl-D-glucose are involved in hemiacetal formation unveiling a six-membered ring structure (pyranose form). On the other hand, C_2–OH and C_5–OH in 1, 3, 4, 6-tetra-O-methyl-D-fructose are involved in hemiacetal formation unveiling a five-membered ring structure (furanose form).

iii) Nature of linkage: Owing to the non-reducing nature of sucrose, we can conclude that glucose and fructose units must be connected via corresponding glycosidic or reducing sites. Now, the overall structure of sucrose can be visualized by considering two important experimental results. First is that maltase hydrolyses α-glucopyranosides and sucrose showing that the glucose unit must be in α-form. The second one is that invertase hydrolyses β-fructofuranosides and sucrose showing that the fructose unit must be in β-form.

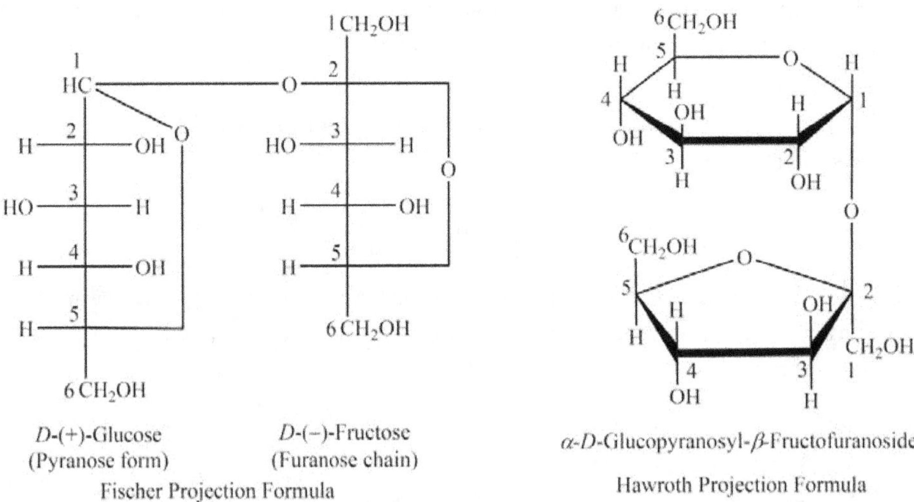

D-(+)-Glucose   D-(–)-Fructose
(Pyranose form)   (Furanose chain)   α-D-Glucopyranosyl-β-Fructofuranoside

Fischer Projection Formula   Hawroth Projection Formula

Hence, we can say that sucrose is made up of one glucose and one fructose unit, which are joined together by the glycosidic linkage.

4. Structure determination of starch: Starch is the most abundant carbohydrates' source for human-kind which is primarily found in seeds, plant roots, fruits, and tubers. Moreover, it is also very important to note that starch is not a single compound but a mixture of two; amylose (20%) and amylopectin (80%). The structure of maltose is obtained as given below.

i) Monomeric analysis: The hydrolysis of both compounds of starch (amylose and amylopectin) with boiling mineral acids yields D-(+) glucose units.

$$(C_6H_{10}O_5)_n \xrightarrow{H^+/\Delta} (C_6H_{10}O_5)_n \xrightarrow{H^+/\Delta} C_{12}H_{22}O_{11} \xrightarrow{H^+/\Delta} (C_6H_{10}O_5)_n \qquad (4)$$

$$\textit{Amylose} \qquad\qquad \textit{Dextrin} \qquad\qquad \textit{D(+)Maltose} \qquad\qquad \textit{D(+)Glucose}$$

$$(C_6H_{10}O_5)_n + nH_2O \xrightarrow{H^+/\Delta} (C_6H_{10}O_5)_n \qquad (5)$$

$$\textit{Amylopectin} \qquad\qquad\qquad \textit{D(+)Glucose}$$

Hence, we can say that amylose, as well as amylopectin, both are made-up of D-(+) glucose units. Furthermore, it is also worthy to note that n has a range of 200–300 and 1000–3000 for amylose and amylopectin, respectively.

ii) Study of the monosaccharide units: The partial hydrolysis of amylose and amylopectin with β-amylase (diastase enzyme) results in the D-(+)-maltose i.e. a single disaccharide.

$$(C_6H_{10}O_5)_n + n(H_2O) \xrightarrow{H^+/\Delta} nC_{12}H_{22}O_{11} \qquad (6)$$

$$\textit{Starch} \qquad\qquad\qquad \textit{D(+)Maltose}$$

iii) Nature of linkage: Since we know that C_4–OH of reducing glucose unit is connected to the α-C_1–OH of non-reducing glucose unit i.e., α-glucosidic linkage, amylose can be considered as a chain of D-(+) glucose units connected by α-glucosidic bonds.

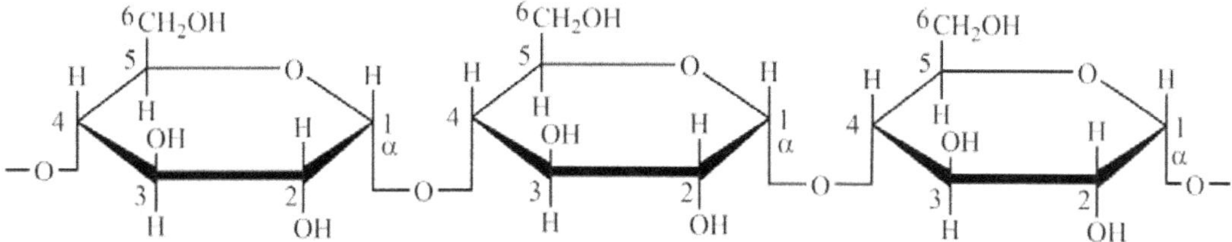

Amylose Structure
(Hawroth Projection Formula)

Just like the case of amylose, the C_4–OH of the reducing glucose unit is connected to the α-C_1–OH of the non-reducing glucose unit (α-glucosidic linkage) to form a chain of D-(+) glucose units. However, unlike amylose, the generation of D-(+)-maltose by hydrolysis of amylopectin with β-amylase (diastase enzyme) is feasible only up to fifty percent. This infers that amylopectin also has some other kind of bonds that are immune to the diastase enzyme. The acid hydrolysis of fully methylated amylopectin yields 2, 3, 6-tri-O-methyl-D-glucose (90%), 2, 3, 4, 6-tetra-O-methyl-D-glucose (5%) and 2, 3-di-O-methyl-D-glucose (5%); which infer α-1, 4-linkages, some non-reducing ends and α-1, 4-linkages, respectively.

Amylose Structure
(Hawroth Projection Formula)

Finally, it is also worthy to note that unlike amylose, a very large extent of branching has been observed in amylopectin in which short-chain (about 25 glucose units) with α-linkage.

5. Structure determination of cellulose: The most abundant organic compound on earth is cellulose which forms all plants' cell walls. It is the primary component of jute, wood (50%), and cotton (95%). The structure of maltose is obtained as given below.

i) Monomeric analysis: The complete hydrolysis of cellulose with dilute sulphuric acid yields D-(+)-glucose as the only product. Therefore, one can conclude that cellulose is made up of D-(+)-glucose units (just like the case of starch).

$$(C_6H_{10}O_5)_n + nH_2O \xrightarrow{\;H_2SO_4\;} nC_6H_{12}O_6 \qquad\qquad (7)$$

ii) Study of the monosaccharide units: The reaction of cellulose with a mixture of sulphuric acid and acetic anhydride results in hydrolysis and acetolysis simultaneously to produce cellobiose (a disaccharide).

$$\underset{Cellulose}{(C_6H_{10}O_5)_n} \xrightarrow{\;(MeCO)_2/H_2SO_4\;} Cellobiose \qquad\qquad (8)$$

iii) Nature of linkage: The cellobiose's structure resembles to the structure of maltose excepting the fact that its hydrolysis is carried out by 'emulsin' instead of β-amylase or maltase. Emulsin enzyme is specific for β-glycosidic linkage whereas β-amylase is specific for α-glycosidic bonds. Hence, we can conclude that D-(+)-glucose units joined together via β-glucosidic linkage to form cellobiose. All this information infers that cellulose can be considered as glucose units' chain via β-1, 4-glucosidic bonds.

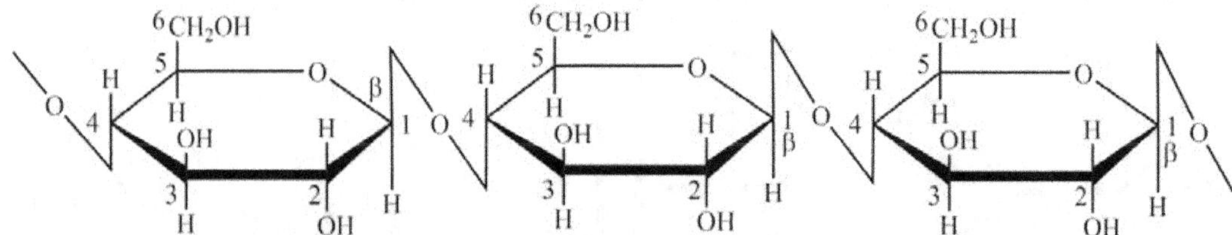

Cellulose Structure
(Hawroth Projection Formula)

One of the glaring differences between amylose and cellulose is that the glucose units in amylose are connected by α-1, 4-glucosidic bonds; whereas in cellulose, glucose units bind together via β-1, 4-glucosidic bonds.

❖ Problems

Q 1. What are natural sugars? Discuss the structure properties of Maltose.

Q 2. What do you mean by deoxy sugars? Explain with suitable examples.

Q 3. Define amino sugars?

Q 4. What do you understand by branched chain sugars? How they are different from normal sugars?

Q 5. Discuss the primary methods of structure determination of disaccharide with special refence to maltose, lactose, sucrose.

Q 6. Discuss the structure determination of starch.

Q 7. What is cellulose? How would you find its structure?

❖ Bibliography

1. R. V. Stick, S. J. Williams, *Carbohydrates: The Essential Molecules of Life*, Elsevier, Oxford, UK, 2009.

2. A. F. Bochkov, G. E. Zaikov, V. A. Afanasev, *Carbohydrates*, Utrecht, Netherlands, 1991.

3. D. Klein, *Organic Chemistry*, John Wiley & Sons, Inc., New Jersey, USA, 2015.

4. J. Clayden, N. Greeves, S. Warren, *Organic Chemistry*, Oxford University Press, Oxford, UK, 2012.

5. R. L. Madan, *Organic Chemistry*, Tata McGraw Hill, New Delhi India, 2013.

CHAPTER 5

Natural and Synthetic Dyes

❖ Various Classes of Synthetic Dyes Including Heterocyclic Dyes

A dye may simply be defined as a colored compound that binds chemically or physically to the substrate to which it is applied; and depending upon the type of source, they are classified as synthetic or natural. In this section, we will further classify the synthetic dyes (azo and non-azo types) based upon their method of application.

➤ Acid Dyes

An acid dye may simply be defined as the sodium salt of azo dyes which contains carboxylic or sulphonic acid.

These types of dyes are applied to fiber from their acid solution and are typically employed to color wool, nylon, polyurethane, and silk. Acid dyes have a greater affinity for nylon which is obviously due to the presence of more free amino groups in polycaprolactam. These types of dyes cannot be applied to cotton due to little to no affection. Some of the typical examples of acid dyes are orange-I, orange-II, congo red, and methyl orange.

Orange I

Orange II

Methyl Orange

Congo Red

It is also worthy to note that besides dying the fabric, some special acid dyes are also used as food colorants, to stain organelles in the medical field.

> ### *Basic Dyes*

A basic dye may simply be defined as the salt of colored bases which contains amino groups as auxochromes.

The amino groups of basic dyes generate water-soluble cations; which in turn, can bind with the anionic sites on the fabric used. These dyes can be used to color modifies nylons and polyesters. Some of the typical examples of basic dyes are aniline yellow, butter yellow, chrysoidine G, and malachite green.

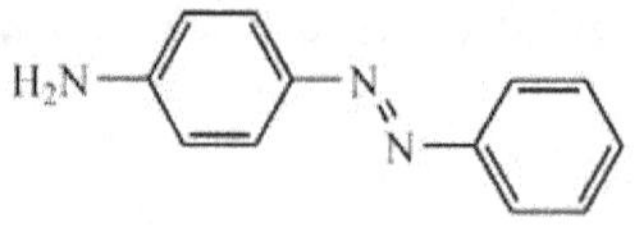

Aniline Yellow

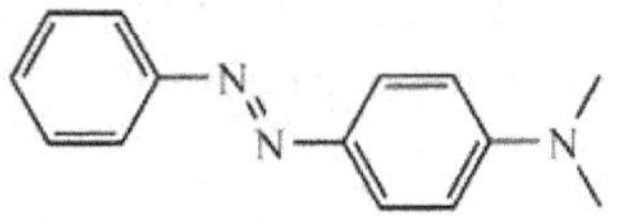

Butter Yellow

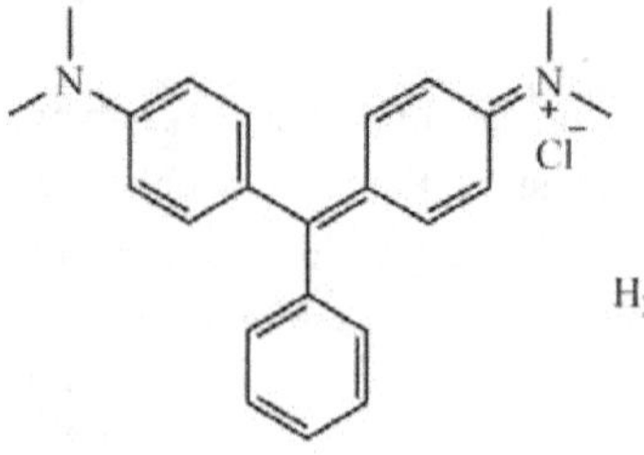

Malachite Green

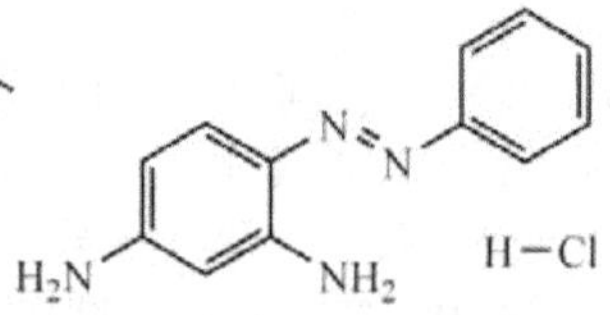

Chrysoidine G

It is also worthy to note that besides dying the fabric, some special basic dyes are also used as paper colorants in the paper industry.

> ### *Direct Dyes*

A direct dye may simply be defined as the dye which can be applied to the fabric directly from its aqueous solution.

These dyes are water-soluble and appropriate for the fabrics that can bind to the dye molecule via hydrogen bonding. Direct dyes are primarily used to color nylon, silk, rayon, cotton, and wool. Some of the typical examples of direct dyes are martius yellow and congo red.

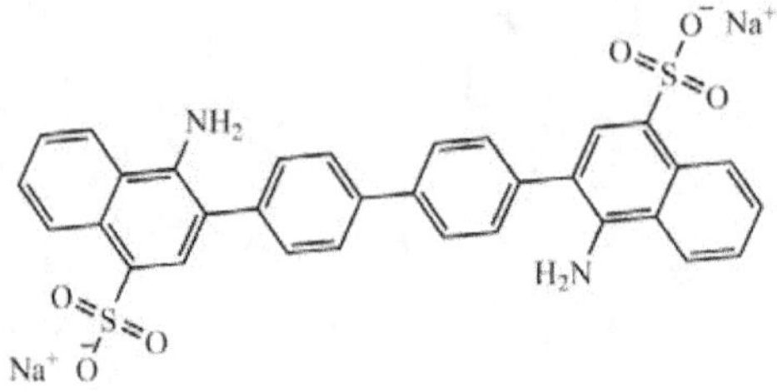

Congo Red

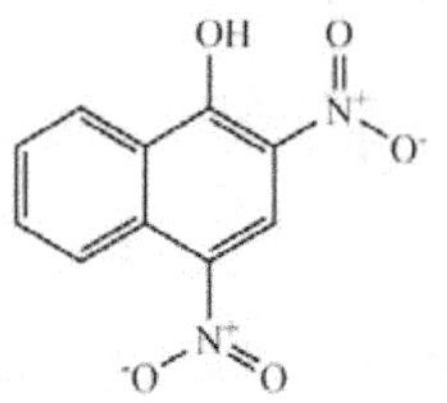

Martius Yellow

> ➤ *Disperse Dyes*

A disperse dye may simply be defined as the dye which can be applied to the fabric in the form of dispersion dye material in a soap solution stabilized by benzoic acid, cresol, or phenol.

These dyes are water-insoluble (correspond to the anthraquinone class) and are appropriate for the synthetic polyamide fibers. Some of the typical examples of disperse dyes are celliton fast pink B and celliton fast blue B.

Celliton Fast Pink B   Celliton Fast Blue B

> ➤ *Reactive Dyes*

A reactive dye may simply be defined as the dye that has reactive groups capable of binding with the amino or hydroxyl groups of the fiber.

These dyes bind very strongly with the fiber which can be attributed to the chemical nature of the bonding interaction. These dyes are primarily used to color wool, silk, and cotton. Some of the typical examples of disperse dyes are reactive blue 4 and 2-[[4-[[4-[bis(2-hydroxyethyl)amino]-6-chloro-1,3,5-triazin-2-yl]amino]phenyl]azo]-p-cresol.

Reactive Blue 4

2-[[4-[[4-[bis(2-hydroxyethyl)amino]-6-chloro-1,3,5-triazin-2-yl]amino]phenyl]azo]-p-cresol

DALAL INSTITUTE

> *Ingrain Dyes*

An ingrain dye may simply be defined as the dye that is produced within the fiber from chemical precursors and binds itself by an irreversible chemical change in such a way that fastness is improved upon the dyeing.

These dyes are created by the coupling of the diazonium salt with naphthols, phenols, aminophenols, or arylamines on the fiber's surface. Ingrain dyes are not considered as fast because the dye molecules bind to the fiber via adsorption. These dyes are primarily used to color nylon, silk, polyester, leather, polypropylene, polyacrylonitriles, polyurethanes, and cellulose. Examples of ingrain dyes are para red and Ingrain blue 1.

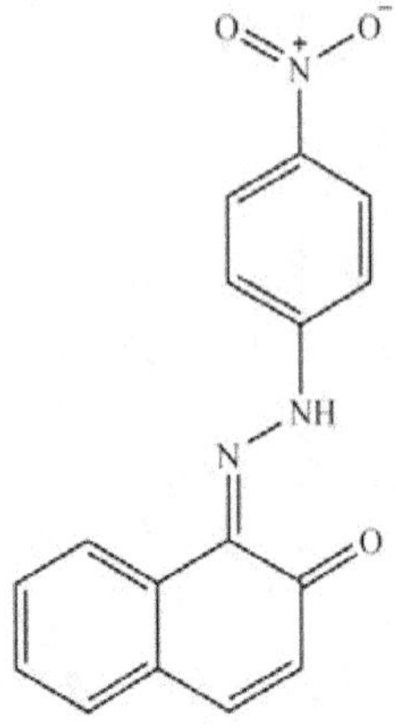

Para Red

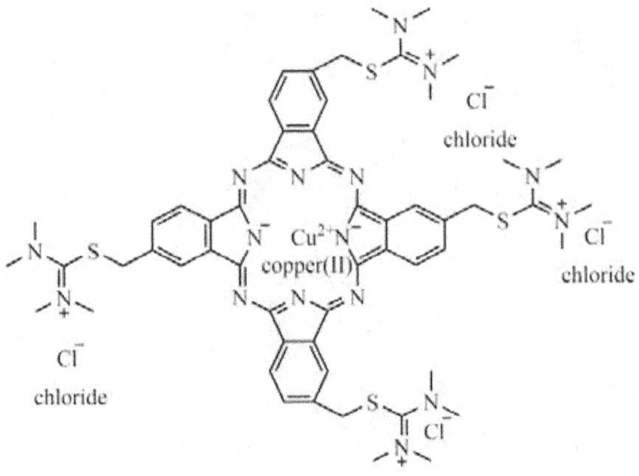

Ingrain Blue 1

> *Vat Dyes*

A vat dye may simply be defined as the dye that is applied in its reduced form to the fabric in a bucket or vat with a reducing agent like sodium hydrosulfite's alkaline solution.

These dyes are water-insoluble and cannot be applied directly to the fiber. However, their reduced form (generally colorless) becomes soluble in water and does have an affinity for cellulose fiber. Owing to this special feature, vat dyes are primarily used to color cotton cloths. Some of the typical examples of vat dyes are indigo (also a naturally occurring dye) and vat orange 3.

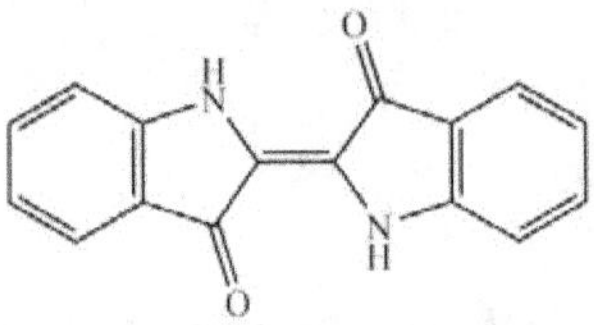

Indigo

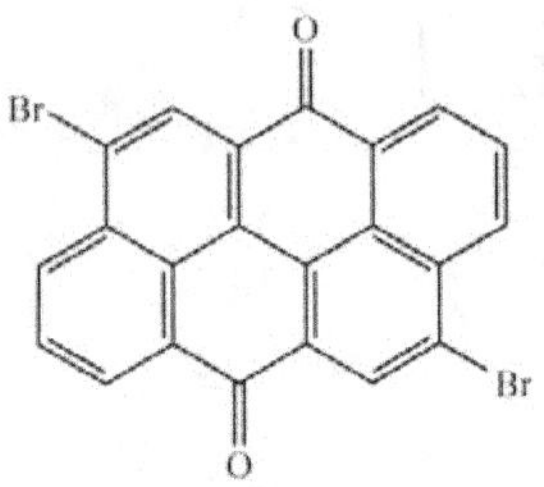

Vat Orange 3

> *Mordant Dyes*

A mordant dye may simply be defined as the dye that is applied to the fabric via some mordant acting as a binding agent between the fiber and dye.

Tannic acid and metal ions are used as mordant basic and acid dyes, respectively. The fabric is first soaked in the mordant provider and then the dye is applied. These dyes are primarily used to color wool. Two of the typical examples of mordant dyes are mordant brown 1 and mordant red 19.

Mordant Brown 1   Mordant Red 19

It is also worthy to note that mordants not only increase the fastness of the dye but can also change the color of both the dye-plus-mordant solution and influence the shade of the overall product.

❖ Interaction Between Dyes and Fibers

In the process of dying, the dye molecules bind with the fiber by some sort of attractive force which can be physical or chemical in nature. The fastness of the dye depends upon the nature and extent of these dye-fiber forces. In this chapter, we will discuss the types of interactions between dye and fiber.

> *Physical Theory*

According to the physical theory of dye-fiber interaction, dye molecules are retained by the fiber via van der Waal forces or hydrogen bonding.

Some of the most important characteristic features of the physical theory of dye-fiber interaction are given below.

1. The fastness of dye in this type of interaction increases with increasing molecular size.

2. Dye's fastness decreases as the solubility increases.

One of the most common examples of dyes showing this type of interaction is the coloring of the cellulosic by solubilized vat, direct, sulfur, and vat dyes.

> ### *Chemical Theory*

According to the chemical theory of dye-fiber interaction, dye molecules are retained by the fiber via the chemical bond.

Some of the most important characteristic features of the chemical theory of dye-fiber interaction are given below.

1. The dye as well as fiber, are required to have reactive groups.

2. Dye's fastness decreases as the number of reactive sites increases.

3. In most cases, the bonding's nature is ionic but can also be covalent in some.

4. An electrolyte is added after the 'half dying process' to exhaust the bath.

5. Sometimes the dying-rate is so high that leveling agents are needed for a more uniform effect.

Some of the most common examples include dying of wool or nylon or silk with acid dye, dying of acrylic or silk or cationic polyester with a basic dye, and coloring of anionic polyester with sulfur, vat, reactive or direct dyes. Furthermore, the coloring of cotton by reactive dye material also results in the formation of chemical bonds (covalent interaction).

> ### *Physio-Chemical Theory*

According to the physio-chemical theory of dye-fiber interaction, dye molecules are retained by the fiber via physical bonds.

Some of the most important characteristic features of the physio-chemical theory of dye-fiber interaction are given below.

1. The fastness of dye in this type of interaction is enhanced by increasing molecular size by the reaction of fiber-retained dye with some other chemical species.

2. One component must be the dye; however, the other component can be the dye or non-dye chemical.

Some of the most common examples of dyes showing this type of interaction are mordant dyeing of wool, mordanting of cotton in basic dyeing, back tanning of dyed protein fiber.

> ### *Fiber-Complex Theory*

According to the fiber-complex theory of dye-fiber interaction, dye molecules are retained by the fiber via the formation of a complex.

Some of the most important characteristic features of the fiber-complex theory of dye-fiber interaction are given below.

1. The dye on its own is not able to enter the fiber's matrix due to lack of affinity or large molecular structure.

2. The reaction of two different compounds under feasible conditions.

One of the most common examples of dyes showing this type of interaction is cotton's coloring with insoluble azoic.

> *Solid Solution Theory*

According to the solid-solution theory of dye-fiber interaction, dye molecules are trapped inside by the fiber material under suitable conditions.

Some of the most important characteristic features of the solid-solution theory of dye-fiber interaction are given below.

1. The dyestuff as well fiber, both are in the solid phase.

2. Dye gets trapped in the hydrophobic fiber, forming a solid solution.

3. The dyeing is carried out at a higher temperature to facilitates the passage of dye molecules into the fiber.

One of the most common examples of dyes showing this type of interaction is dying man-made material with disperse dye.

> *Pigment or Mechanical Theory*

According to the pigment or mechanical theory of dye-fiber interaction, dye molecules are attached to the fiber via a binding agent.

Some of the most important characteristic features of the pigment theory of dye-fiber interaction are given below.

1. The dye has no reactive sites, no affinity for the fiber, and is insoluble in most of the solvents.

2. The fiber becomes more stiff and the dye's fastness depends upon the film's longevity.

Some of the most common examples of dyes showing this type of interaction are the dying of fabric, pigment colors.

❖ Structure Elucidation of Indigo and Alizarin

Two of the most popular examples of natural dyes are indigo and alizarin which have a cultural history all across the globe. In this section, we will discuss the general method of structure elucidation of alizarin and indigo as well.

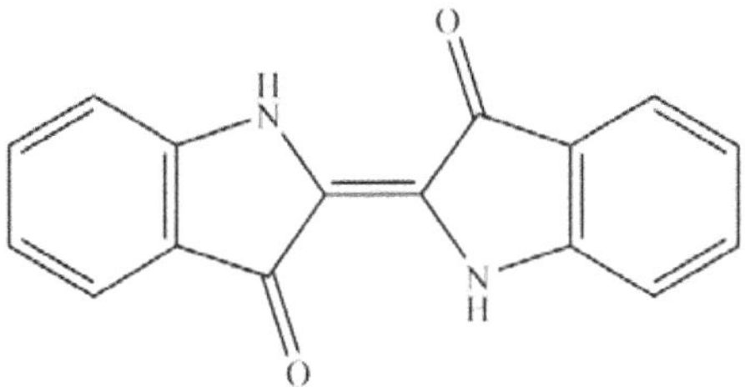

Indigo

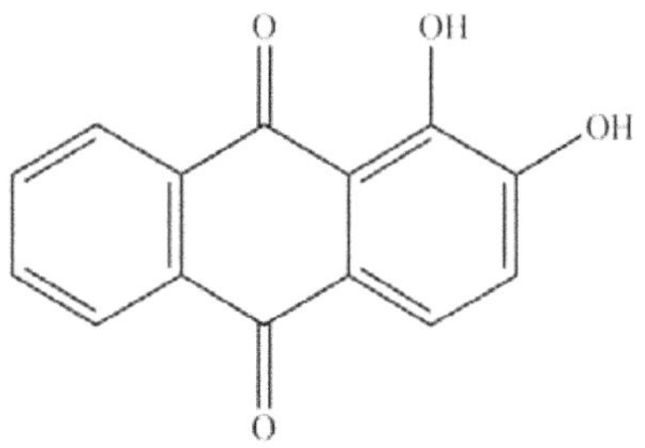

Alizarin

> ➤ *Structure of Indigo*

Indigo is a naturally occurring blue dye whose molecular structure can be elucidated by the general method as given below.

1. The combined study of elemental analysis and mass spectrum suggested that $C_{16}H_{10}O_2N_2$ is the molecular formula of indigo.

2. The double bond equivalent (D.B.E) for indigo was found to be 13 suggesting that there must be more than one benzene ring.

3. Indigo's fusion at low-temperature results in anthranilic acid suggesting that there must be at least one ortho-substituted benzene ring where one position is bind by nitrogen and the other by carbon.

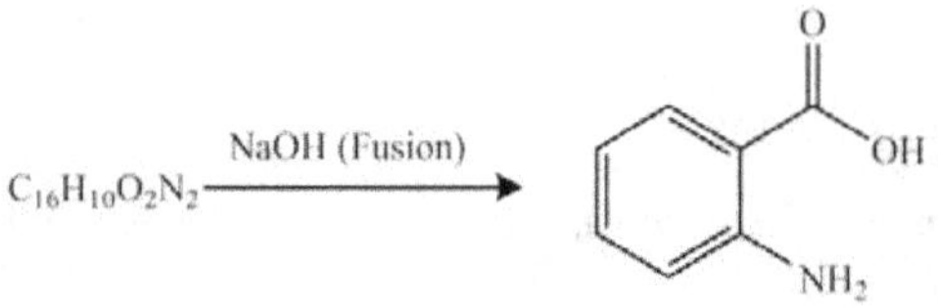

Indigo   Anthranilic Acid

4. Indigo's oxidation with nitric acid results in two moles of isatin indicating two identical units in indigo that are connected with each other such that one molecule of isatin is generated from each unit.

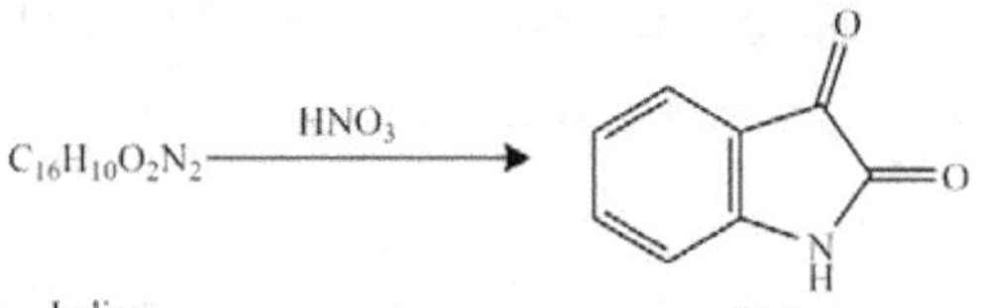

Indigo   Isatin

5. Three possible structures can be concluded from the isatin structure.

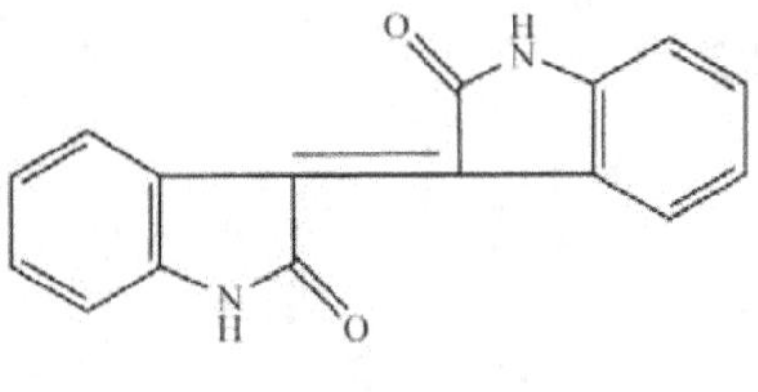

Structure 1 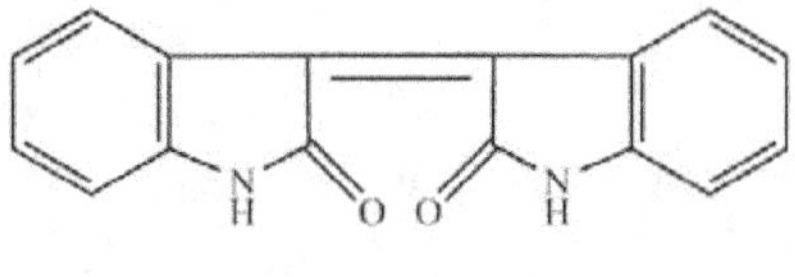   Structure 2

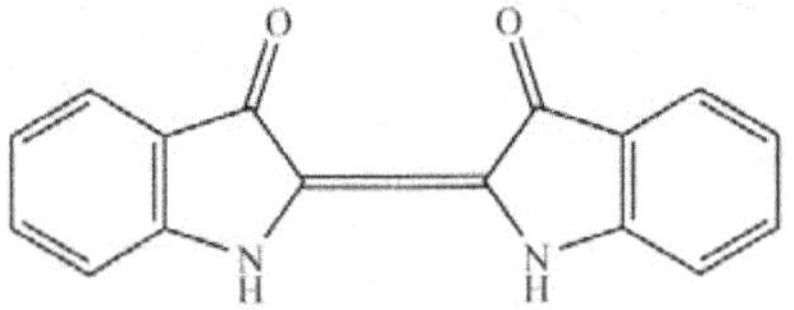

Structure 3

6. The treatment of indigotin with dilute alkali results in the formation of indoxyl-2-aldehyde and anthranilic acid.

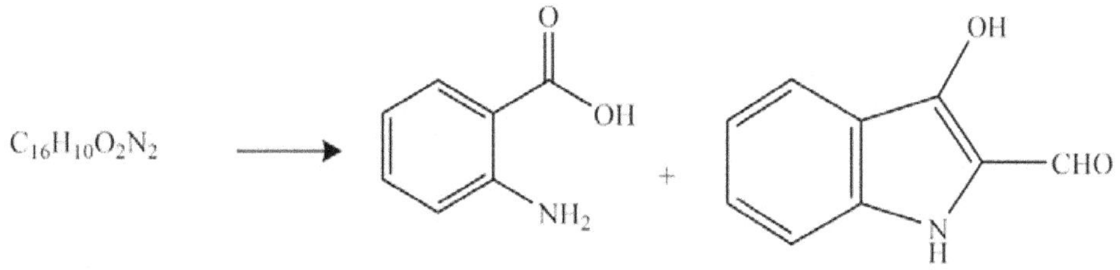

$$C_{16}H_{10}O_2N_2 \longrightarrow$$

Indigo   Anthranilic Acid   Indoxyl-2-aldehyde

7. Finally, we may conclude that two identical units must be joined via the second position of the indoxyl fragment, and an oxygen-containing functional group must be present at position 3. All this suggests following two structures for indigotin.

Stablized by
hydrogen
bonding

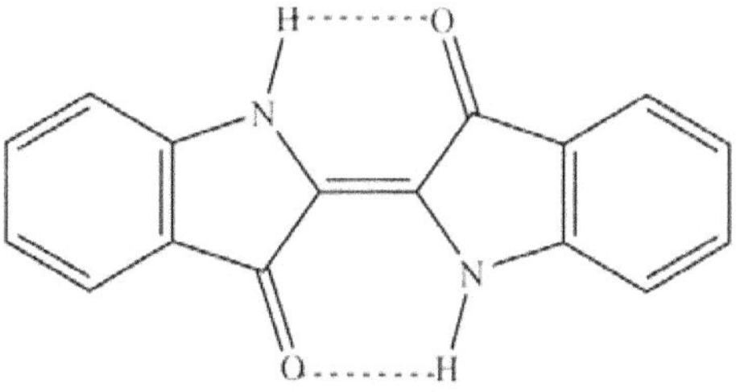

trans-Indigo

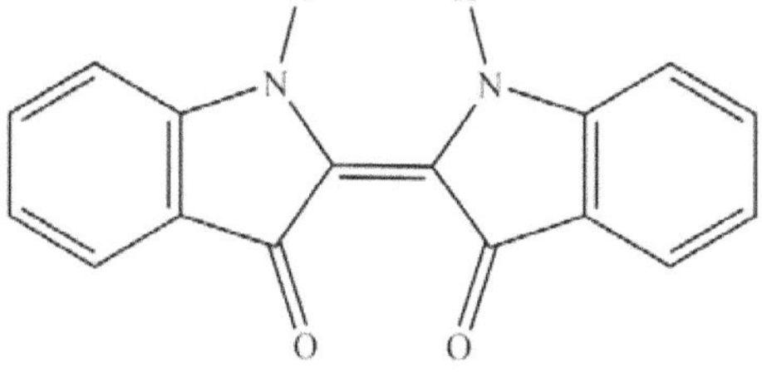

Not stablized by
hydrogen
bonding

cis-Indigo

Now since trance configuration is stabilized by H-bonding, it must be the primary choice. Similar results were obtained by qualitative analysis of mass spectra and NMR data.

> ### *Structure of Alizarin*

Indigo is a naturally occurring blue dye whose molecular structure can be elucidated by the general method as given below.

1. The combined study of elemental analysis and mass spectrum suggested that $C_{14}H_8O_4$ is the molecular formula of alizarin.

2. The double bond equivalent (D.B.E) for alizarin was found to be 11 suggesting that there must be at least one benzene ring.

3. The reduction of alizarin with zinc at 675K results in anthracene indicating that alizarin must a derivative of the anthracene molecule.

4. The reaction of alizarin with acetic anhydride results in diacetate showing two hydroxyl groups are expected in the alizarin molecule.

5. Since the condensation of phthalic anhydride with catechol at 455K results in alizarin, it can be concluded that this dihydroxy derivative of anthraquinone must be having OH groups in the same cycle.

Phthalic Anhydride   Catechol H_2SO_4 Alizarin

6. Therefore, following structures can be proposed for alizarin molecule.

Structure I   Structure II

D **DALAL**
INSTITUTE

7. The nitration of alizarin results in two isomers of mono-nitro derivative, which in turn produce phthalic acid upon oxidation; indicating the presence of the nitro group in the same cycle as the hydroxy group. Now since structure I can give two mono-nitro derivatives but the second structure can produce only one mono-nitro derivative, the correct answer should be structure I.

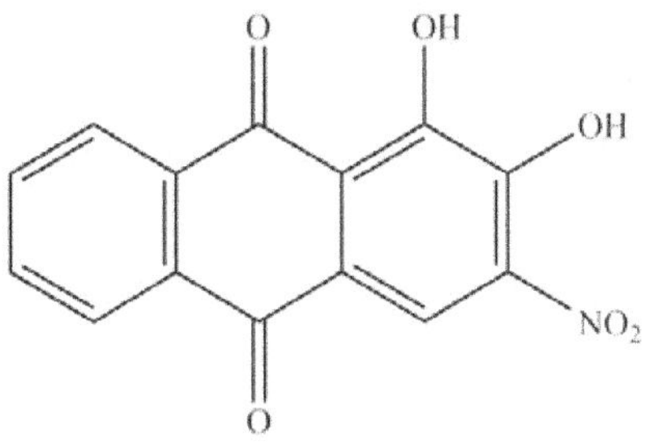

Structure IA

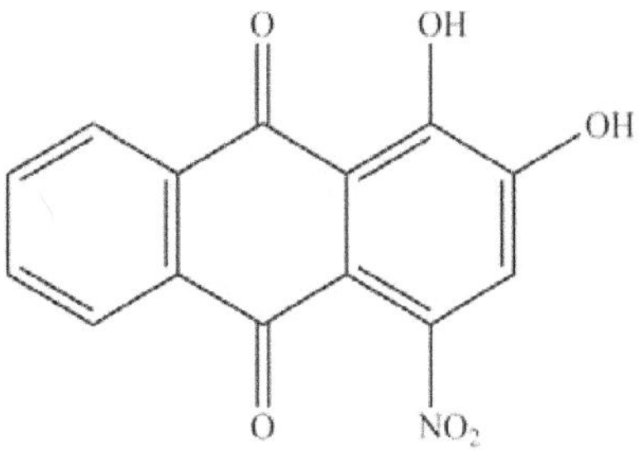

Structure IB

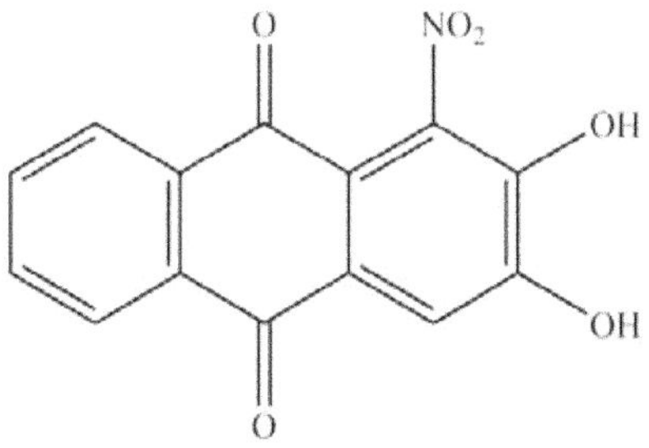

Structure IIA

Now since trance configuration is stabilized by H-bonding, it must be the primary choice. Similar results were obtained by qualitative analysis of mass spectra and NMR data

❖ Problems

Q 1. What is a dye? How would you classify them based on their source?

Q 2. Discuss the classification of various forms of dyes on basis of their method of application.

Q 3. What is the difference between mordant and vat dyes?

Q 4. Give the points of similarity and differences between the physical and chemical theory of dye-fiber interaction.

Q 5. State solid-solution theory of dye-fiber interaction. Also give is salient features.

Q 6. Discuss the general method of structure determination of indigo.

Q 7. What is the chemical formula of alizarin? Also, discuss its structure elucidation.

❖ Bibliography

1. D. Klein, *Organic Chemistry*, John Wiley & Sons, Inc., New Jersey, USA, 2015.

2. J. Clayden, N. Greeves, S. Warren, *Organic Chemistry*, Oxford University Press, Oxford, UK, 2012.

3. R. L. Madan, *Organic Chemistry*, Tata McGraw Hill, New Delhi, India, 2013.

4. L. Knutson, *Synthetic Dyes for Natural Fibers, Interweave Press*, Colorado, USA, 1986.

5. P. S. Vankar, *Natural Dyes for Textiles: Sources, Chemistry and Applications*, Woodhead Publishing, Cambridge, UK, 2017.

CHAPTER 6

Aliphatic Nucleophilic Substitution

❖ The SN₂, SN₁, Mixed SN₁ and SN₂, SN$_i$, SN₁′, SN₂′, SN$_i$′ and SET Mechanisms

Although the number of mechanisms by which the nucleophilic substitutions proceed is very large, certain patterns can still be used to profile them for more systematic and simplistic analysis. Some of the prominent types of aliphatic nucleophilic substitutions are given below.

➤ *SN₂ (Substitution Nucleophilic Bimolecular) Mechanism*

In SN₂ reactions, the "SN" stands for "nucleophilic substitution", and "2" means that the rate-determining step is bimolecular. In other words, a stronger nucleophile displaces a weaker one via the formation of a transition state.

Illustrative reaction: One of the most common examples of the SN₂ reaction is the attack of Br⁻ on ethyl chloride results in ethyl bromide, with chloride ejected as the leaving group.

Mechanism involved: The proposed mechanism for the reaction given above involves a single step which must be discussed before we give the salient features of the same. The process occurs most often at the sp^3 hybridized carbon with a stable electronegative leaving group attached to it (usually halide X⁻).

The breaking of the carbon–halogen bond and the formation of the new covalent bond takes place simultaneously via a transition state in which the carbon under nucleophilic attack is in 5-coordination with probable sp^2 hybridization. The nucleophilic attack at the carbon takes place at 180° w.r.t the leaving group to provides a good overlap between the nucleophile's lone pair and the antibonding orbital (σ*) C–X bond. The leaving group is then detached from the opposite side and the product is formed with inversion of the tetrahedral geometry at the central carbon atom if the substrate is chiral.

Salient Features: The main features of the mechanism involved in nucleophilic substitution bimolecular or SN_2 type reactions are given below.

i) SN_2 reactions follow second-order kinetics with the rate law

$$Rate = k[RX][Nu]$$

Where k is the rate constant. The symbol $[RX]$ and $[Nu]$ represent the molar concentration of the substrate and attacking nucleophiles, respectively.

ii) If the alkyl halide is chiral, then this often leads to an inversion of configuration, called the Walden inversion.

iii) The rate of the substitution becomes independent of the concentration of the attacking reagent if its concentration is extremely high in comparison to the substrate.

iv) The rate of the substitution increases as the steric bulk around the carbon center decreases.

v) The SN_2 reactions are favored in polar aprotic solvents.

> ➤ *SN₁ (Substitution Nucleophilic Unimolecular) Mechanism*

In SN_1 reactions, the word "SN" stands for "nucleophilic substitution", and "1" means that the rate-determining step is unimolecular in nature. In other words, a stronger nucleophile displaces a weaker one via the formation of an intermediate.

Illustrative reaction: The most common example of an SN_1 reaction is the formation of alcohols from alkyl halides as shown below.

Mechanism involved: The proposed mechanism for the reaction given above involves two steps which must be discussed before we give salient features of the same.

i) Formation of intermediate:

The carbocation formed during this step is trigonal planar in geometry and is open for attack from both sides. Now since the carbocations are electron-deficient species and very reactive, The OH^- will attack from either side to give the same product, which will be the second step of the reaction.

ii) Attack by the Nucleophile:

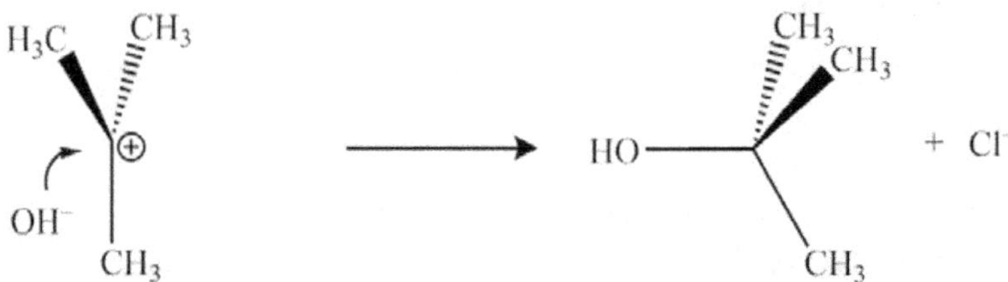

Now since the faces of the carbocations formed are homotopic, the OH⁻ can attack from either side to give the same product.

Salient Features: The main features of the mechanism involved in nucleophilic substitution unimolecular or SN_1 type reactions are given below.

i) SN_1 reactions follow first-order kinetics with the rate law

$$Rate = k[RX]$$

Where k is the rate constant and $[RX]$ represents the molar concentration of the substrate (tert-butyl halide in this case).

ii) If the alkyl halide has one or more asymmetric carbons, two stereoisomers (diastereomers or enantiomers) will be formed.

iii) The rate of the nucleophilic substitution unimolecular is almost independent of the concentration of the attacking reagent.

iv) The rate of the substitution increases as the steric bulk around the carbon center increase.

v) Since an unstable intermediate carbocation is formed in course of the SN_1 reactions (rate-determining step), any factor that can support this will boost up the rate. Normal solvents of choice are both protic (to hydrolyze the leaving group in particular) and polar (to simply stabilize ionic intermediates). Archetypal polar protic solvents include alcohol and water, which are also capable of acting as nucleophiles (i.e. support solvolysis). Therefore, we can conclude that the SN_1 reactions are favored in polar and protic solvents.

vi) Since the intermediate formed is carbocation, the possibility of rearrangement to form more stable carbocation and yielding different products is also there.

vii) The substitution at bridgehead carbon is either absent or takes place very slowly because the carbocation in such cases cannot attain planar geometry.

viii) In asymmetric alkyl halides, racemization does not take place fully all the time because the nucleophile attacks even before the complete detachment of leaving group. This leads to some inversion also causing unequal racemic mixture.

> *Mixed SN₁ and SN₂ Mechanism*

Most of the organic reactions are either SN_1 or SN_2 over a vast range of experimental conditions. However, some reactions show both types of characteristic features under certain conditions indicating that they are neither SN_1 nor SN_2 but a mixture of two. In other words, some nucleophilic substitution reactions proceed via mixed SN_1 and SN_2 mechanisms.

Illustrative reaction: The common depictive example of SN_1-SN_2 mixed-mechanism is shown below.

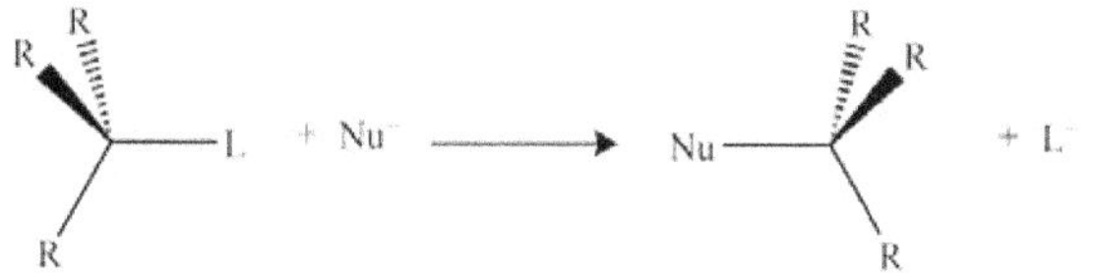

Mechanism involved: There are two theories that are typically used to rationalize the borderline nucleophilic substitution mechanism as given below.

i) Simultaneity of SN₁ and SN₂: As the name suggests, this theory says that the reaction proceeds simultaneously via SN_1 and SN_2 pathways. The pictorial representation of this theory is given below.

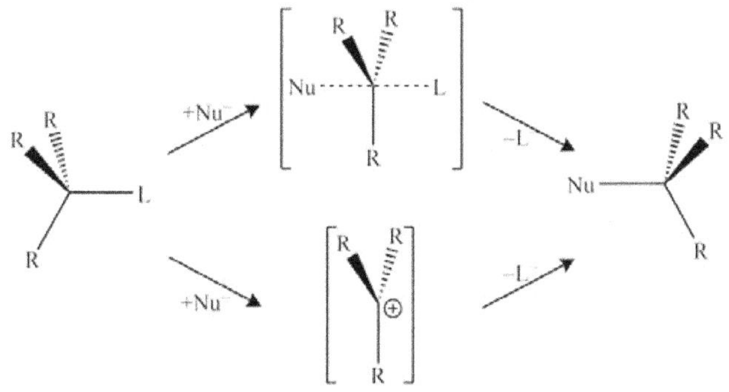

ii) Intermediatory mechanism: This theory assumes that the reaction takes place via the formation of an intermediary ion-pair as shown below.

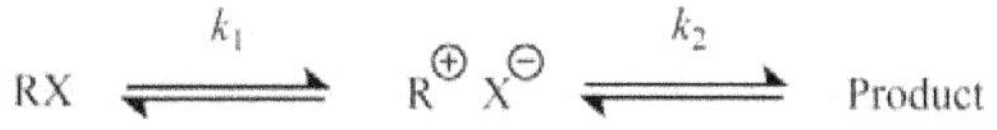

$$RX \underset{}{\overset{k_1}{\rightleftharpoons}} R^{\oplus} X^{\ominus} \underset{}{\overset{k_2}{\rightleftharpoons}} Product$$

When the formation of ion-pair is the rate-determining step, the reaction becomes SN_1; whereas, if the conversion of ion-pair into the product is the rate-determining step, the reaction becomes SN_2; if $k_1 = k_2$, we get a borderline case.

Salient Features: The main features of the mixed SN_1 and SN_2 mechanism are given below.

i) SN_1 pathway competes with the SN_2 route to dominate the products' ratio for asymmetric reactants.

ii) The ion-pair theory can be applied to both SN_1 and SN_2 as well.

> ## *SN_i (Substitution Nucleophilic Internal) Mechanism*

In SN$_i$ reactions, the "SN" stands for "nucleophilic substitution", and the "i" means that the substitution takes place internally in the molecule.

Illustrative reaction: One of the most common examples of the SN$_i$ reaction is the displacement of OH$^-$ of alcohols by Cl$^-$ in the presence of SOCl$_2$.

Mechanism involved: The SOCl$_2$ first reacts with the alcohol to give rise to an alkyl chloro sulfite (i.e. intimate ion pair). The next step is concerted and involves the loss of an SO$_2$ molecule and its displacement by its own chloride group.

The major point of difference between SN$_i$ and SN$_1$ is actually that the ion pair is not completely separate, and therefore, no actual carbocation is generated (otherwise we would have got racemized product).

Salient Features: The main features of the mechanism involved in nucleophilic substitution internal or SN$_i$ type reactions are given below.

i) SN$_i$ reactions follow second-order kinetics with the rate law

$$Rate = k[ROH][SOCl_2]$$

Where k is the rate constant. The symbol $[ROH]$ and $[SOCl_2]$ represent the molar concentration of the substrate and species with attacking nucleophiles, respectively.

ii) If the alcohol is chiral, then this leads to the retention of configuration.

> ➤ *SN₁' (Substitution Nucleophilic Unimolecular Prime) Mechanism*

In SN_1' reactions, the word "SN" stands for "nucleophilic substitution", "1" means that the rate-determining step is unimolecular in nature, and prime indicates that there is a double bond in the vicinity of leaving group. In other words, a stronger nucleophile displaces a weaker one via the formation of an intermediate that has a delocalization of π-electron density.

Illustrative reaction: The most common example of SN_1' reaction is the formation of but-2-ene-1-ol from 3-bromobuta-1-ene as shown below.

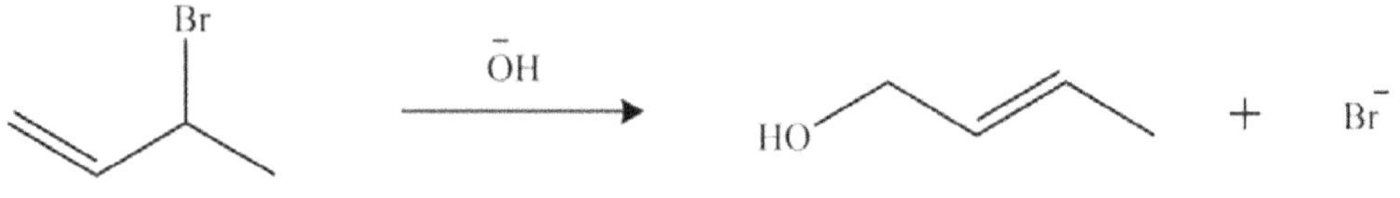

3-bromobuta-1-ene (E)-but-2-en-1-ol

Mechanism involved: The proposed mechanism for the reaction given above involves two steps which must be discussed before we give salient features of the same. At first, the detachment of bromide ion gives rise to an allylic carbocation system in which the positive charge is distributed at 1st and 3rd carbon atoms. Now since the terminal carbon is primary but 3rd carbon is secondary, the first carbon is more electron deficient, and therefore, will become the first choice for attacking nucleophile to yield our product.

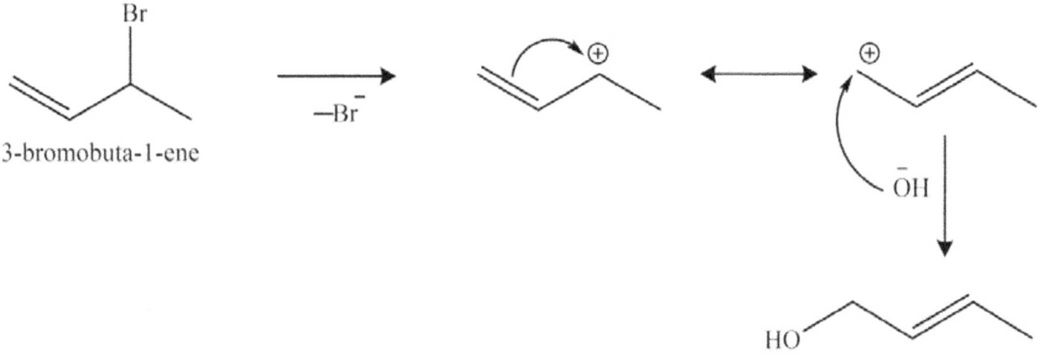

3-bromobuta-1-ene

(E)-but-2-en-1-ol

One more reason that why the nucleophile did not attack at the 3rd carbon to give normal SN_1 is that there is more steric hindrance at the 3rd carbon than what it is at 1st.

Salient Features: Almost all of the features of SN_1' prime are similar to the SN_1 mechanism with some exceptions as given below.

i) The carbonation formed in SN_1 was simple but rearrangeable in the case of allylic systems.

ii) The nucleophile attack on γ-carbon rather than the α- one.

> ### *SN₂′ (Substitution Nucleophilic Bimolecular Prime) Mechanism*

In SN_2' reactions, the "SN" stands for "nucleophilic substitution", "2" means that the rate-determining step is bimolecular, and prime indicates that there is a double bond in the vicinity of leaving group. In other words, a stronger nucleophile displaces a weaker one via the formation of a transition state; though the attachment and detachment are at different carbons.

Illustrative reaction: One of the most common examples of the SN_2' reaction is the conversion of 3-bromo-3-methylcyclohex-1-ene into 3-methylcyclohex-2-en-1-ol, with bromide ejected as the leaving group.

3-bromo-3-methylcyclohex-1-ene   3-methylcyclohex-2-en-1-ol

Mechanism involved: The proposed mechanism for the reaction given above involves the use of a double bond as a relay system of electron density. Instead of attacking at the 3rd carbon in the cycle (would have yield normal SN_2 product), the incoming nucleophile attacks at 1st carbon due to its greater electron deficiency than the 3rd one which is obviously caused by electrons' relay from first carbon to bromine.

3-bromo-3-methylcyclohex-1-ene   3-methylcyclohex-2-en-1-ol

One more reason that why the methoxide ion did not attack at the 3rd carbon to give normal SN_2 is that there is more steric hindrance at the 3rd carbon than what it is at 1st. In other words, the greater electron deficiency and a less steric hindrance at first carbon make it a better site for nucleophilic attack.

Salient Features: Almost all of the features of SN_2' prime are similar to the SN_2 mechanism with some exceptions as given below.

i) The nucleophilic attack and the detachment of leaving group takes place at different carbon atoms.

ii) The double bond is used as an electrons' relay system.

> ### *SNᵢ' (Substitution Nucleophilic Internal Prime) Mechanism*

In SN_i' reactions, the "SN" stands for "nucleophilic substitution", the "*i*" means that the substitution takes place internally in the molecule, and prime indicates that there is a double bond in the vicinity of leaving group.

Illustrative reaction: One of the most common examples of the SN_i' reaction is the displacement of OH of in but-3-en-2-ol by Cl in the presence of $SOCl_2$.

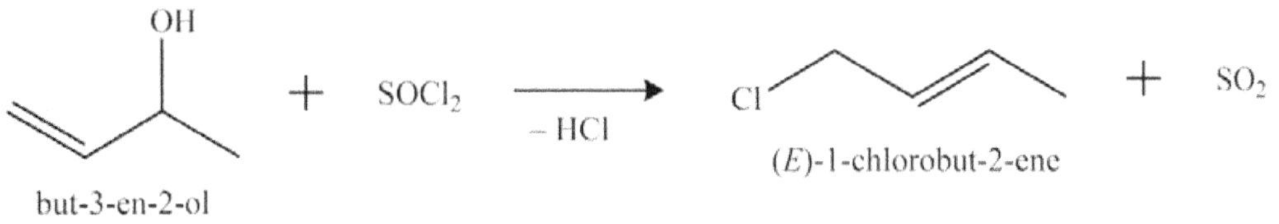

OH

but-3-en-2-ol + $SOCl_2$ → (− HCl) Cl (*E*)-1-chlorobut-2-ene + SO_2

Mechanism involved: The proposed mechanism for the reaction given above initially proceed normally like SN_i; however, the detachment of sulfurochloridite ion gives rise to an allylic carbocation system in which the positive charge is distributed at 1st and 3rd carbon atoms. Now since the terminal carbon is primary but 3rd carbon is secondary, the first carbon is more electron deficient, and therefore, will become the first choice for attacking nucleophile to yield our product.

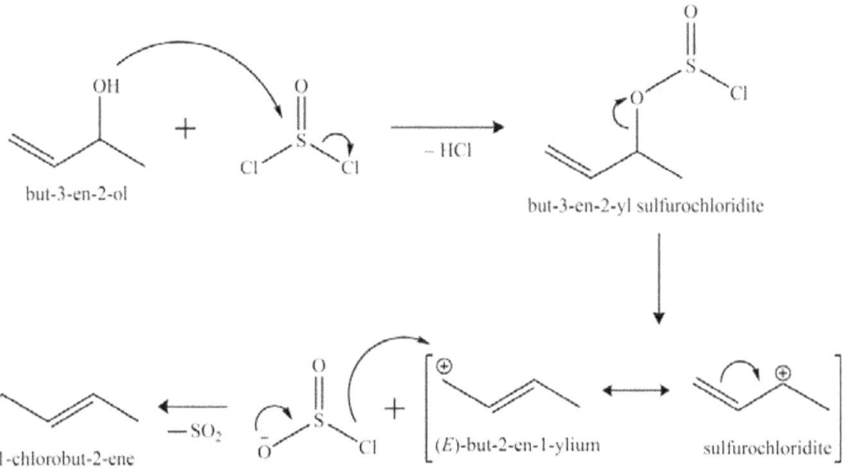

but-3-en-2-ol + (− HCl) but-3-en-2-yl sulfurochloridite

(*E*)-1-chlorobut-2-ene ← (−SO_2) + [(*E*)-but-2-en-1-ylium ↔ sulfurochloridite]

One more reason that why the nucleophile did not attack at the 2nd carbon to give normal SN_i is that there is more steric hindrance at the 2nd carbon than what it is at 4th.

Salient Features: Almost all of the features of SN_i' prime are similar to the SN_i mechanism with some exceptions as given below.

i) The carbonation formed in SN_i was simple but rearrangeable in the case of allylic systems.

ii) The nucleophile attack on the 1st carbon rather than the 3rd one.

> ➤ *SET (Single-Electron Transfer) Mechanism*

SET (single electron transfer) reactions may simply be defined as the organic reaction mechanism in which an electron-rich molecule gives away one of its electrons to an electron-poor molecule to form radical cation and radical anion, respectively. Furthermore, these radical anions and cations can bind to give new bonds or may react in some other way to yield strange products.

Illustrative reaction: One of the most common examples of the SET reactions is the transformation of benzophenone into 1,1-diphenylmethanol in the presence of metallic sodium.

benzophenone   diphenylmethanol

Mechanism involved: The proposed mechanism for the reaction given above involves four steps which must be discussed before we give salient features of the same. The process occurs most often via the formation of benzophenone anion and dianion as given below.

benzophenone

diphenylmethanol

At this stage, some protons are added in the form of very weak (NH_4Cl) or strong acid (HCl). The protonation of benzophenone dianion would yield 1,1-diphenylmethanol.

Salient Features: The main features of the mechanism involved in simple electric transfer or SET type reactions are given below.

i) The electron transfer results in radical cation and radical cation.

ii) The SET mechanisms can be distinguished from polar mechanisms by careful analysis of end products.

❖ The Neighbouring Group Mechanisms

There are many nucleophilic substitution reactions that give rise to the same configuration (i.e., retention) instead of inversion or racemization. Also, the rate of reaction for such reactions is so high that we cannot rationalize them by simple nucleophilic substitutions. However, it has been observed that one feature that is common in these reactions is a group or atom at β-position to the leaving group. The mechanism responsible for such transformations is labeled as neighboring group participation and can be parted into two normal SN_2 consecutive steps. Now since the first SN_2 reaction gives inversion (neighboring group as the nucleophile), the subsequent SN_2 changes will revert the configuration to the original (neighboring group as leaving group).

The faster rate of reaction can be rationalized in terms of the ready availability of nucleophilic attack in the first step (rate-determining step). Furthermore, it should also be noted that the generation of the cyclic intermediate is a characteristic feature of the neighboring group participation. Some typical cases of neighboring group participation are discussed below.

➤ *Reactions Involving Oxygen as Neighbouring Group*

One of the most common examples of this type of neighboring group mechanism is the reaction of 2-bromopropanoic acid with a dilute solution of NaOH.

It is obvious from the above route that the configuration has remained the same (R)-lactate anion), unlike SN_2 where we would have obtained the (S)-lactate anion.

➤ *Reactions Involving Nitrogen as Neighbouring Group*

One of the most common examples of this type of neighboring group mechanism is the reaction of 2-chloro-N,N-diethylpropan-1-amine with a dilute solution of NaOH.

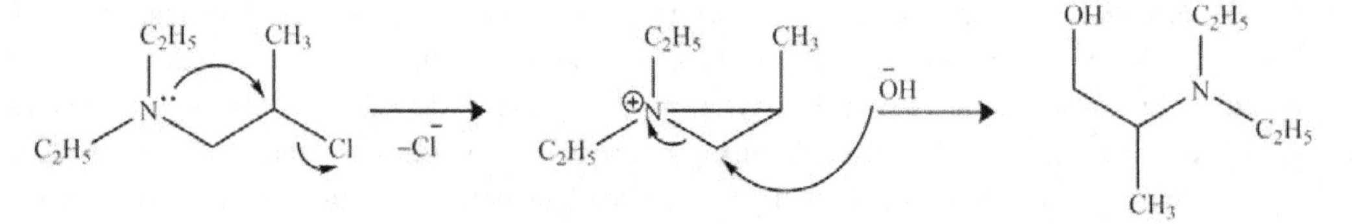

2-chloro-*N,N*-diethylpropan-1-amine   1,1-diethyl-2-methylaziridin-1-ium   2-(diethylamino)propan-1-ol

It is obvious from the above route that the configuration has remained the same, unlike SN_2 where we would have obtained the 1-(diethylamino) propan-2-ol.

➤ *Reactions Involving Halogen as Neighbouring Group*

One of the most common examples of this type of neighboring group mechanism is the reaction involving the acetolysis of trans-2-iodocyclohexyl brosylate in which the configuration remains the same at the asymmetric center.

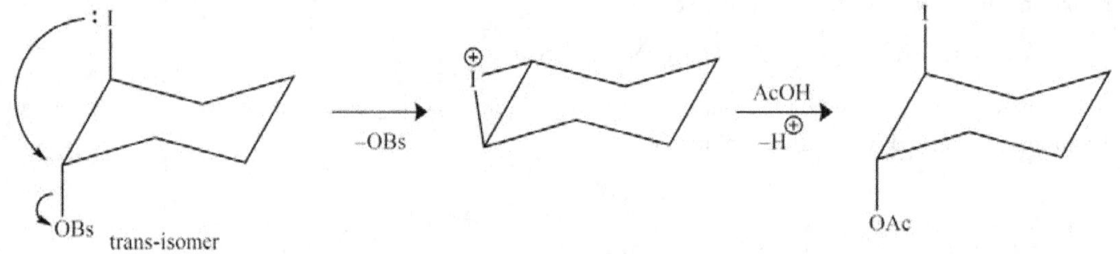

trans-isomer

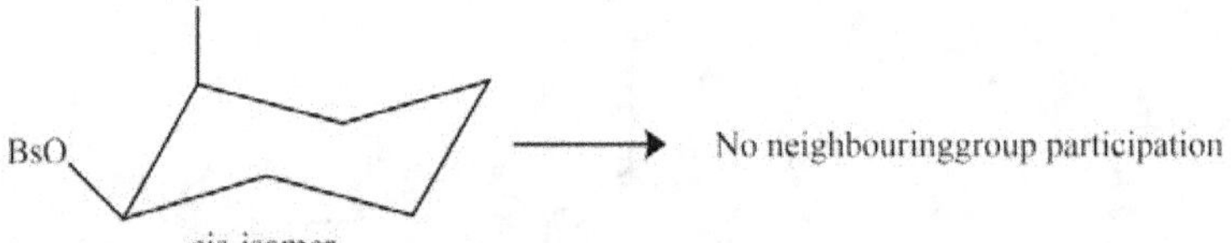

cis-isomer   No neighbouringgroup participation

The is obvious from the above route that the configuration would have changed if the reaction had taken place via normal SN_2, which is observed for cis isomer.

> *Reactions Involving Sulphur as Neighbouring Group*

One of the most common examples of this type of neighboring group mechanism is the reaction involving the hydrolysis of bis(2-chloroethyl) sulfane in which the configuration remains the same at the asymmetric center.

bis(2-chloroethyl)sulfane

(mustard gas)

$-Cl^-$

1-(2-chloroethyl)thiiran-1-ium

(sulphonium ion)

It is obvious from the above route that the configuration would have changed if the reaction had taken place via the normal SN_2 route.

❖ Neighbouring Group Participation by π and σ Bonds

Besides oxygen, nitrogen, sulfur, and halogen; π- and σ-bonds can also act as a nucleophile in 'neighboring group mechanism' causing the retention of the original configuration. In this section, we will study the neighboring group participation by π and σ bonds with illustrative examples.

> *π Bond as Neighbouring Group*

This type of neighboring group participation can primarily be classified into two categories as discussed below.

1. Neighbouring group participation by an alkene: The π orbitals of an alkene can stabilize a transition state by helping to delocalize the positive charge of the carbocation. For instance, the saturated tosylate will react very slowly (10^{11} times slower in solvolysis) with a nucleophile than the unsaturated tosylate.

The positively charged intermediate will be stabilized by the phenomenon of resonance where the positive charge is spread over many atoms as shown below.

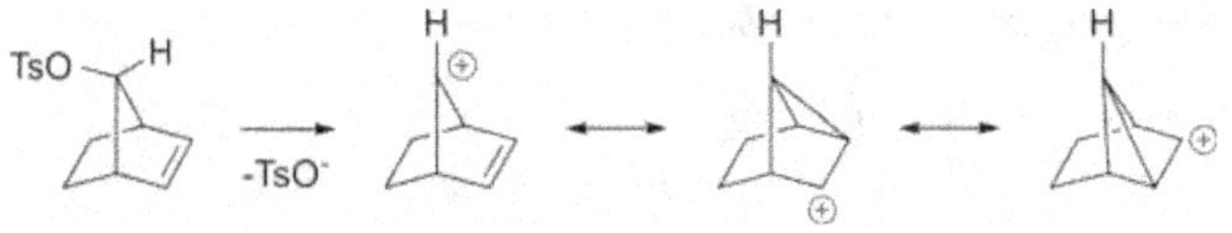

A different view of the same intermediate is also given below.

Even if the alkene is more distant from the reacting center, it can still act in this way. For instance, in the following alkyl benzenesulfonate, the alkene can delocalize the carbocation's positive charge.

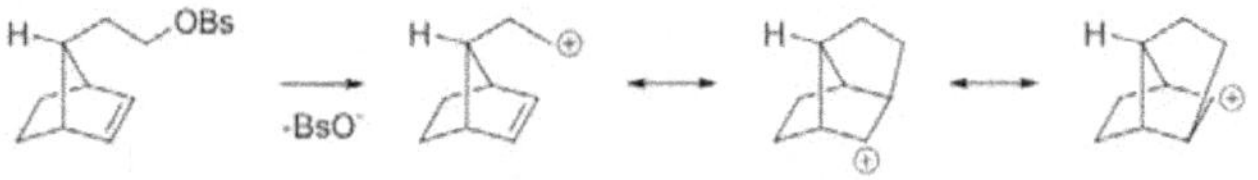

Furthermore, the raise in the SN_2 reaction rate of allyl bromide with a nucleophile relative to the treatment with n-propyl bromide is due to the π-bond's orbitals-overlap with transition state's counterparts. Therefore, we can say that the alkene orbitals overlap with the SN_2-transition-state's orbitals in the allyl systems.

2. Neighbouring group participation by an aromatic ring: Just like allyl systems, higher reactivity for benzyl halide is observed because the SN_2 transition state benefits from a similar overlapping effect. Similarly, many aromatic rings support the formation of an carbocation by delocalizing the positive charge density.

If the tosylate given below reacts with acetic acid via solvolysis instead of the normal SN_2 pathway forming B, a 48:48:4 mixture of A, B (i.e., enantiomers), and C+D was formed.

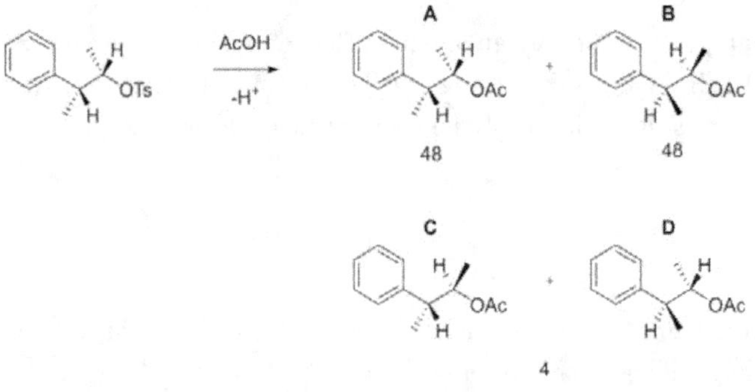

The mechanism which forms A and B is shown below.

It is obvious from the above routes that the configuration would have changed if the reaction had taken place via the normal SN$_2$ route.

> ➤ ***σ Bond as Neighbouring Group***

This type of neighboring group participation can primarily be classified into two categories as discussed below.

1. Neighbouring group participation by cyclopropylmethyl, cyclobutyl, or a homoallyl group: The treatment of cyclopropylmethyl chloride with dilute ethyl alcohol yields a mixture of 5% homoallyl alcohol, 47% cyclobutanol, and 48% cyclopropylmethyl alcohol. Similar results were obtained if we use cyclobutyl chloride or homoallyl chloride instead of cyclopropylmethyl chloride.

(chloromethyl)cyclopropane   or   chlorocyclobutane   or   4-chlorobut-1-ene

cyclopropylmethyl alcohol   +   cyclobutanol   +   homoallyl alcohol

All this suggests that the carbocationic intermediate present in all three reactions must be the same, which in turn, is responsible for the same resulting products.

2. Neighbouring group participation by aliphatic C-C or C-H bonds: The aliphatic C–H or C–C bonds can also give rise to delocalization of charge if these bonds are close enough and antiperiplanar to the leaving group. The intermediates corresponding to these mechanisms are nonclassical in nature; and 2-norbornyl system is the most popular of such type. More precisely, the acetolysis of exo-2-norbonyl brosylate yields a racemic mixture of exo-acetates only and no endo product; suggesting neighboring group participation from σ-bond. Furthermore, a very slow rate is observed if we use endo-2-norbonyl brosylate confirming our guess.

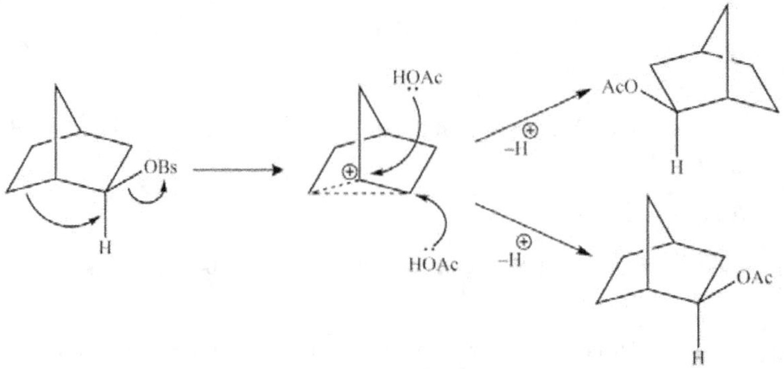

Another example of such neighboring petrification by aliphatic bonds is the methyl system where C–H gives rise to the original configuration as depicted below.

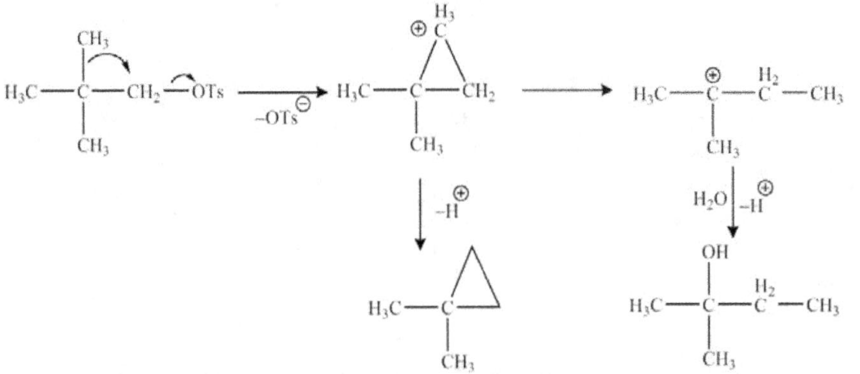

Another example of such neighboring group participation by aliphatic bonds is the methyl system where C–C gives rise to the original configuration as depicted below.

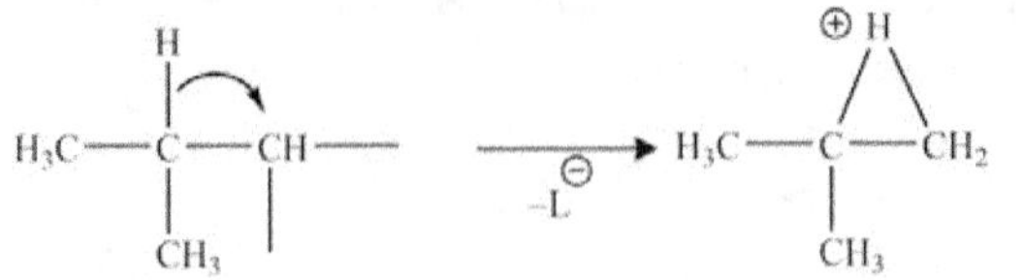

❖ Anchimeric Assistance

The anchimeric assistance may simply be defined as the increase in reaction-rate due to the presence of a neighboring group β- to the leaving group.

It is a well-known fact that many nucleophilic substitution reactions give rise to the same configuration (i.e., retention) instead of inversion or racemization due to neighboring group participation. Also, the rate of reaction for such reactions is very high because the group or atom at β-position to the leaving group is readily available for the nucleophilic attack in the first step (rate-determining step). The schematic representation of the whole process is shown below.

Since the first step is the rate-determining step, the reaction kinetics of the neighboring group mechanism is of the first order, which can be formulated as:

$$Rate = k[RX]$$

Where k is the rate constant and symbol $[RX]$ represents the molar concentration of the substrate.

> ### Anchimeric Assistance by Heteroatom with Lone Pair

The typical example of this type of anchimeric assistance that arises via neighboring group mechanisms is the reaction involving the acetolysis of trans-2-iodocyclohexyl brosylate in which the rate is 1.75×10^6 times greater than what is in acetolysis of cis-isomer.

Also, it is obvious from the above route that the configuration would have changed if the reaction had taken place via the normal SN₂ route.

> ### *Anchimeric Assistance by Alkene*

The typical example of this type of anchimeric assistance that arises via neighboring group mechanism is the reaction involving the solvolysis of unsaturated tosylate in which the rate is 10^{11} times greater than what is in solvolysis of saturated tosylate.

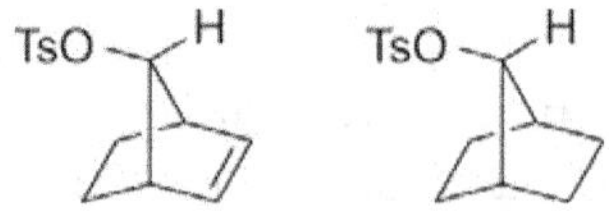

The carbocationic intermediate will be stabilized by resonance where the positive charge is spread over several atoms. In the diagram below this is shown.

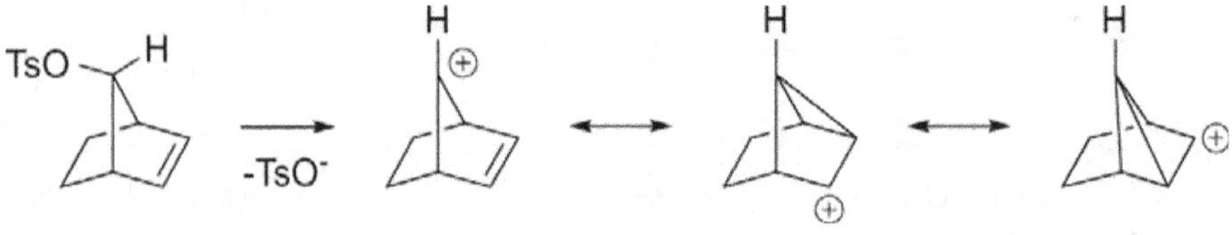

This effect is observed even if the alkene is more remote from the reacting center the alkene can still act in this way.

> ### *Anchimeric Assistance by Aromatic Ring*

The typical example of this type of anchimeric assistance that arises via neighboring group mechanism is the reaction tosylate with acetic acid in solvolysis in which the rate is much greater than what is in solvolysis via aliphatic counterpart.

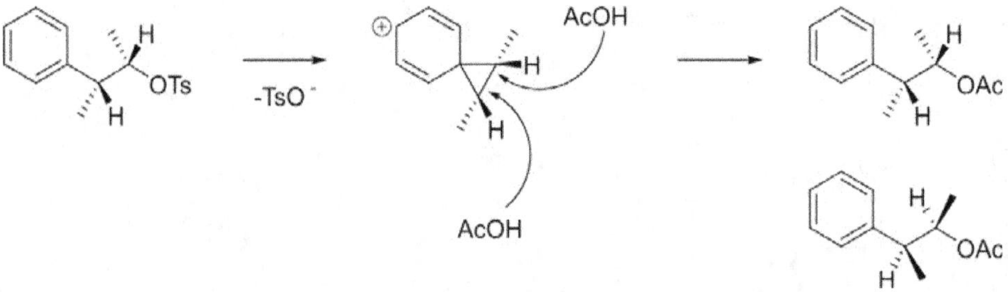

It is obvious from the above routes that the configuration would have changed if the reaction had taken place via the normal SN₂ route. An aromatic ring assists in the formation of a carbocationic intermediate called a phenonium ion by delocalizing the positive charge.

➢ *Anchimeric Assistance by Cyclopropylmethyl System*

The typical example of this type of anchimeric assistance that arises via neighboring group mechanism is the reaction of cyclopropylmethyl chloride with dilute ethyl alcohol in which the rate is much greater than what is expected in the normal SN₂ route. Similar results were obtained if we use cyclobutyl chloride or homoallyl chloride instead of cyclopropylmethyl chloride.

(chloromethyl)cyclopropane   or   chlorocyclobutane   or   4-chlorobut-1-ene

cyclopropylmethyl alcohol   cyclobutanol   homoallyl alcohol

All this suggests that the carbocationic intermediate present in all three reactions must be the same, which in turn, is responsible for the same resulting products.

➢ *Anchimeric Assistance by Aliphatic Bonds*

The typical example of this type of anchimeric assistance that arises via neighboring group mechanism is the reaction of the acetolysis of exo-2-norbonyl brosylate yielding an racemic mixture of exo-acetates in which the rate is much greater than what is for endo-2-norbonyl brosylate.

Therefore, we conclude that the enhanced rate can be due to the anchimeric assistance by the aliphatic carbon-carbon single bond.

❖ Classical and Nonclassical Carbocations

It is a well-known fact that a large number of chemical reactions proceed via the formation of certain chemical species called carbocations in which one the carbon carries a positive charge with only six valence electrons. In this section, we will study the generation, structure, stability, and reactivity of two main types of carbocations called classical and non-classical carbocations one by one.

➢ *Classical Carbocations*

Classical carbocations in the organic chemistry may simply be defined as the carbocations that are stabilized by the delocalization of π-bond, or σ-bond, or lone pair of at conjugated site to the carbon bearing positive charge.

Since the carbon in classical carbocations has only six electrons, it is electron deficient; and therefore, acts as an electrophile in chemical reactions.

1. Generation of classical carbocations: The heterolytic cleavage of the covalent bond is responsible for the generation of most of the classical carbocation species. Some reactions involving the production of carbocations are given.

i) Ionization of alkyl halides in polar solvents:

$$H_3C\!-\!\underset{\underset{CH_3}{|}}{\overset{\overset{CH_3}{|}}{C}}\!-\!Cl \longrightarrow H_3C\!-\!\underset{\underset{CH_3}{|}}{\overset{\overset{CH_3}{|}}{C}}{}^{\oplus} + Cl^-$$

ii) Protonation of alcohols followed by dehydration:

$$R\!-\!\ddot{O}H + H^+ \longrightarrow R\!-\!\overset{\oplus}{\underset{\cdot\cdot}{O}}H_2 \longrightarrow R^+ + H_2O$$

iii) Protonation of unsaturated systems:

$$R\!-\!\underset{H}{\overset{}{C}}\!=\!CH_2 + H^+ \longrightarrow R\!-\!\underset{\oplus}{\overset{H}{C}}\!-\!CH_3$$

iv) Action of superacids on alkyl fluorides:

$$R\!-\!F + SbF_5 \xrightarrow{\text{FSO}_3\text{H}} R^{\oplus} + SbF_6^{\ominus}$$

v) Deamination of primary aliphatic amines by nitrous acid:

$$R\!-\!NH_2 + HNO_2 \xrightarrow[-2H_2O]{} R\!-\!\overset{\oplus}{N}\!\equiv\!N \longrightarrow R^{\oplus} + N_2$$

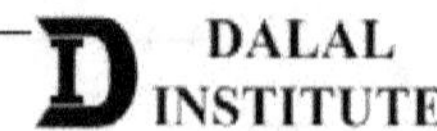

2. Orbital structure of classical carbocations: It has been experimentally found that the classical carbocations are trigonal planar around the carbon bearing positive charge. Now valence bond theory, as well as molecular orbital theory, easily accounted for such structure, it is more comfortable to discuss the valence bond approach. The carbon with the positive charge is in sp^2 hybridization with three hybrid orbitals oriented at 120° in a plane with empty p_z-orbital at a perpendicular.

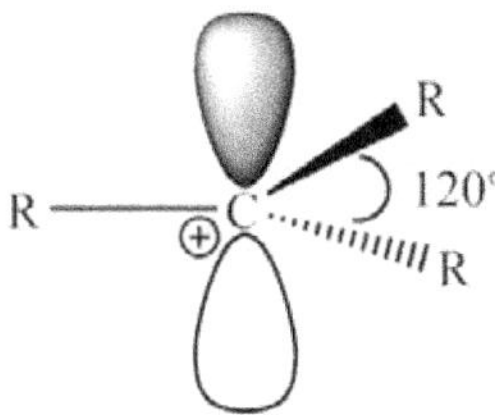

Figure 1. Orbital structure of a classical carbocation.

3. Stability of classical carbocations: Before we discuss the stability of classical carbocations, we need to classify them based on saturation. The first case is alkyl carbocations which are given below.

Methyl carbocation   Ethyl carbocation (1°)   Isopropyl carbocation (2°)   tert-butyl carbocation (3°)

The second case is of unsaturated classical carbocations where the carbon bearing positive charge is directly connected to a carbon participating in multiple bonds i.e.

Allyl carbocation   Benzyl carbocation   Diphenylmethyl carbocation   Triphenylmethyl carbocation

Since the carbon in carbocations has only six electrons, it is electron deficient; and therefore, any effect that can compensate for the deficiency will stabilize the carbocation.

i) Stability of alky carbocations on the basis of inductive effect:

Since the alkyl group has an electron-donating effect (+I), the stability of the classical carbocation will increase as the number of attached donating groups increases. The stability order of alky carbocations based on inductive effect is given below.

$$\overset{\oplus}{CH_3} \quad < \quad H_3C\overset{\oplus}{-CH_2} \quad < \quad H_3C-\underset{CH_3}{\overset{\oplus}{CH}} \quad < \quad H_3C-\underset{CH_3}{\overset{CH_3}{C\oplus}}$$

Methyl carbocation   Ethyl carbocation (1°)   Isopropyl carbocation (2°)   tert-butyl carbocation (3°)

ii) Stability of alky carbocations on the basis of hyperconjugation:

The existence of the hyperconjugation effect can be used to rationalize the relative stability of different carbocations as shown below.

tert-butyl carbocation $\left(\begin{array}{c}\text{3 hyperconjugative structures for one methyl group}\\ \text{Total hyperconjugative structures} = 9\end{array}\right)$

iso-propyl carbocation $\left(\begin{array}{c}\text{3 hyperconjugative structures for one methyl group}\\ \text{Total hyperconjugative structures} = 6\end{array}\right)$

ethyl carbocation $\left(\begin{array}{c}\text{3 hyperconjugative structures for one methyl group}\\ \text{Total hyperconjugative structures} = 3\end{array}\right)$

Hence, as far as the number of possible hyper-conjugative structures possible is concerned, tertiary carbocation should be more stable than secondary, which in turn should be more stable than primary.

iii) Stability of alky carbocations on the basis of steric effect:

      Since the alkyl carbocations are primarily obtained from alkyl halides with tetrahedral geometry, a link between the steric relief and carbocation formed can be established. During the formation of carbocations in such cases, the carbon-carbon bond angles change from 109°28' to 120°. Therefore, the carbon with bulky groups around is expected to get more relief from this carbocationic conversion. The stability order of alky carbocations on the basis of steric effect is given below.

iv) Stability of ally and benzyl carbocations:

      The stability of the carbocations in which the carbon bearing positive charge is adjacent to the double or triple bond can be rationalized in terms of resonance effect. First of all, let us draw the resonance structures allyl and benzyl carbocations.

Resonance stablized allyl cation

Resonance stablized benzyl cation

Now, as the number of phenyl groups attached to carbon bearing positive charge increases, the number of resonating structures will also increase, and hence the stability.

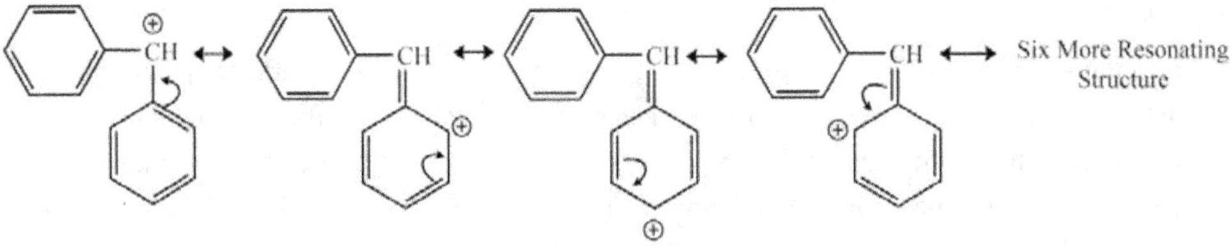

Resonance stablized diphenylmethyl carbocation

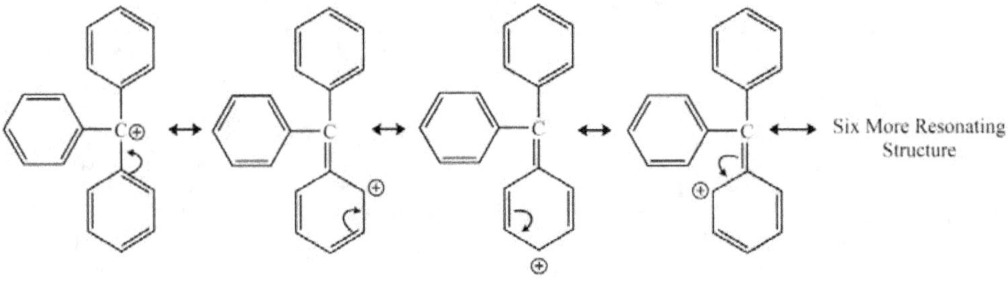

Resonance stablized triphenylmethyl carbocation

Therefore, the expected order of the stability of unsaturated systems with carbon bearing positive charge should be as given below.

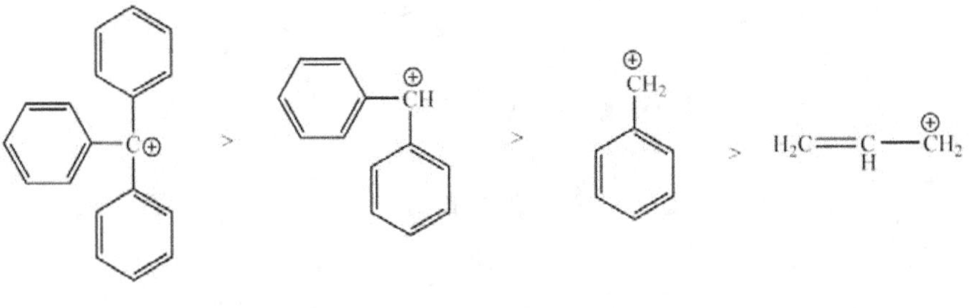

Triphenylmethyl
carbocation

Diphenylmethyl
carbocation

Benzyl
carbocation

Allyl
carbocation

Similarly, the order of stability in phenylcyclopropenyl, diphenylcyclopropenyl and triphenylcyclopropenyl should follow the following order.

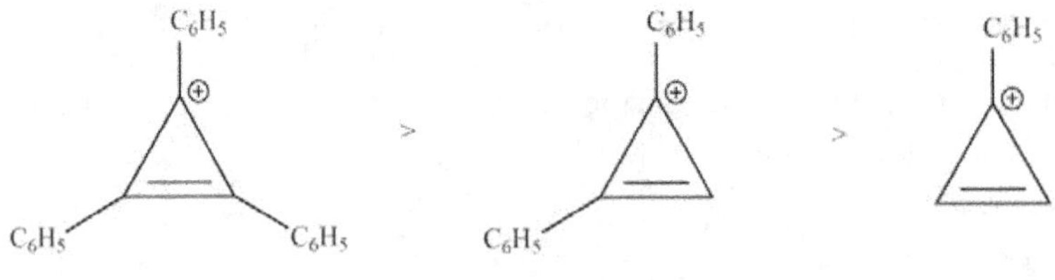

v) Stability of substituted benzyl carbocations: Since the carbon is electron bearing positive charge is electron deficient in nature, any group with +R effect will stabilize the system and vice-versa. The order of stability of some typically substituted carbocations is given below.

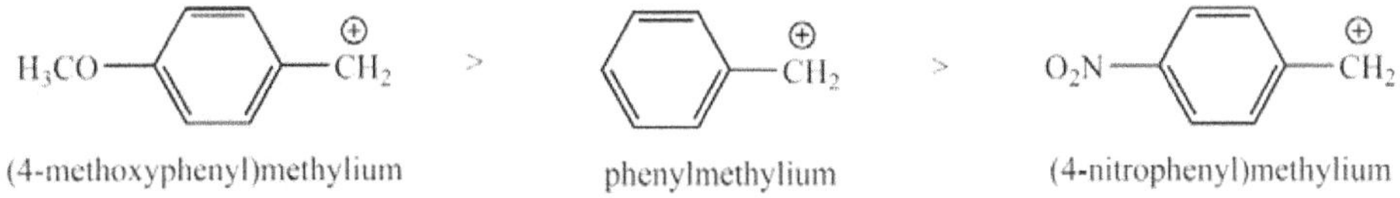

(4-methoxyphenyl)methylium   phenylmethylium   (4-nitrophenyl)methylium

vi) Stability of tropylium ion: The cycloheptatrienyl cation or tropylium ion is exceptionally stable due to its aromatic character (planar and $4n+2$ π electrons). According to molecular orbital theory, its delocalization energy is significantly greater than the delocalization energy of its acyclic counterpart. Similarly, the valence bond theory can also explain its exceptional stability of the basis of resonance as given below.

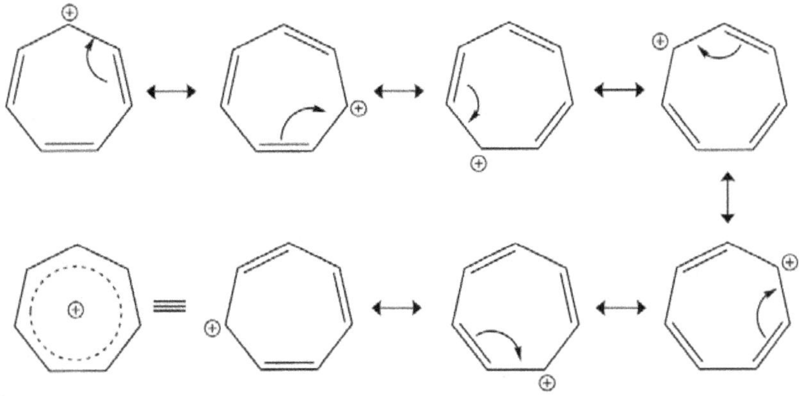

vii) Instability of cyclopentadienyl cation: The cyclopentadienyl cation is very unstable due to its antiaromatic character (planar and $4n$ π electrons). According to molecular orbital theory, its delocalization energy is significantly less than the delocalization energy of its acyclic counterpart.

$4n+2$ π electrons
(antiaromatic)

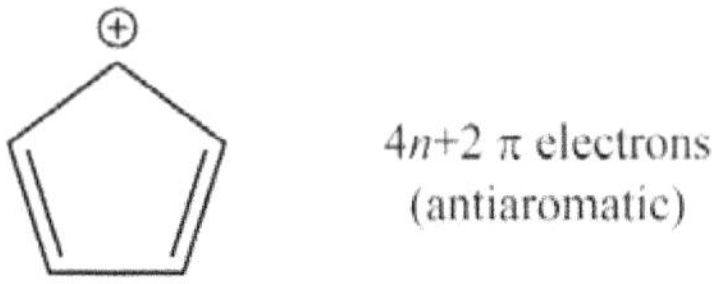

viii) Stability of alkoxyalkyl cation: if the positive charge bearing carbon in the carbocationic species is connected to a hetero atom with lone pair of electrons, the resonance will get it stabilized.

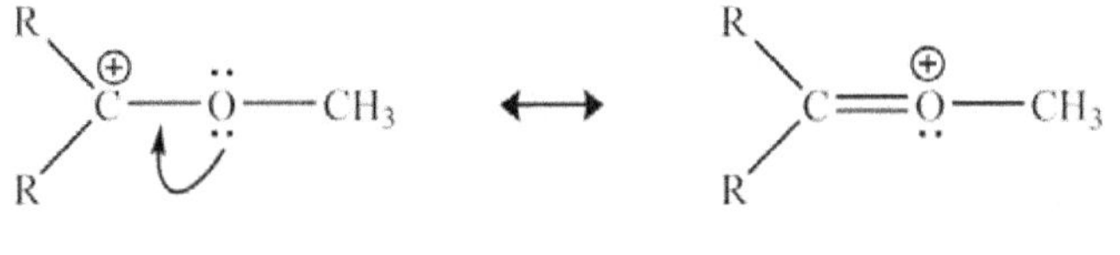

ix) Stability of acyl cation: Just like alkoxyalkyl cation, the resonance will also stabilize the acyl cation as shown below.

$$R-C\equiv\overset{\oplus}{O}\qquad\longleftrightarrow\qquad R-\overset{\oplus}{C}=\ddot{O}$$

x) Instability of phenyl and vinyl cation: If the positive charge is on the double-bonded carbon atom, the system cannot be stabilized because the *sp²* orbital carrying positive charge will be perpendicular to the orbital of the double bond.

$$H_2C=\overset{\oplus}{C}H$$

4. Reactivity of classical carbocations: The principal routes by which the carbocations can react to give rise to stable products are given below.

i) Nucleophilic attack: In these types of reactions, a carbocation may combine with a species by accepting an electron pair. Furthermore, it should also be noted that if all the three groups on the carbocation are different, a racemic mixture will be obtained.

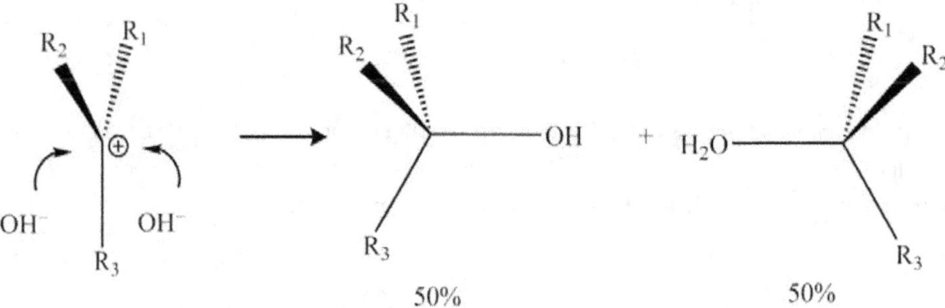

ii) Proton removal: In these types of reactions, a carbocation may result in the removal of a proton from the adjacent atom forming a double bond.

iii) Rearrangement reaction: 1-2 methyl shit or 1-2 hydride shifts are very common in carbocation chemistry to attain a more stable counterpart. For instance, a primary and secondary carbocation will prefer to rearrange themselves into a more stable tertiary carbocation via methyl shifts as given below.

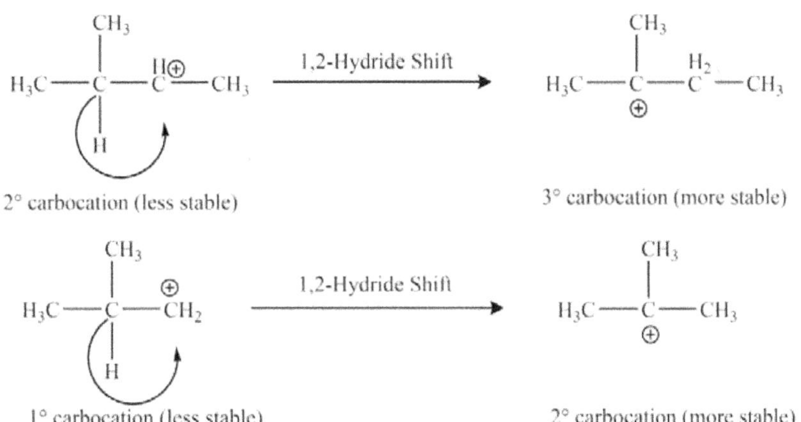

Similarly, primary and secondary carbocation can also rearrange themselves into a more stable tertiary carbocation via hydride shift.

iv) Addition reactions: A carbocation may attack at the triangular face of a double bond to create a new positively charged center as shown below.

> ➤ *Non-Classical Carbocations*

Non-classical carbocations in the organic chemistry may simply be defined as the carbocations in which the π-bond is not conjugation with the carbon bearing positive charge, and gets stabilized by neighboring group participation by π- or σ-bond.

Like classical carbocations, the carbon in a non-classical carbocation has only six electrons, it is also electron deficient; and therefore, acts as an electrophile in chemical reactions. Now, whether the neighboring participant is σ-bond or π-bond (not conjugated), non-classical carbocations can be divided into two categories.

1. Generation of non-classical carbocations: The generation of non-classical carbocations takes place mainly in neighboring group participation via π- or σ-bond. Some reactions involving the production of non-carbocations are given.

i) Neighbouring group participation via π-bond:

In the case of a benzyl halide, the reactivity is higher because the SN₂ transition state enjoys a similar overlap effect to that in the allyl system. An aromatic ring can assist in the formation of a carbocationic intermediate called a phenonium ion by delocalizing the positive charge.

ii) Neighbouring group participation via σ-bond:

The aliphatic C–H or C–C bonds can also give rise to delocalization of charge if these bonds are close enough and antiperiplanar to the leaving group. The intermediates corresponding to these mechanisms are nonclassical in nature; and the 2-norbornyl system is the most popular of such type. More precisely, the acetolysis of exo-2-norbonyl brosylate yields a racemic mixture of exo-acetates only and no endo product; suggesting neighboring group participation from σ-bond. Furthermore, a very slow rate is observed if we use endo-2-norbonyl brosylate confirming our guess.

2. Orbital structure of non-classical carbocations: One of the first (and popular) examples of non-classical ions was the 2-norbornyl cation. Non-classical carbocations are simply the organic cations where the electron density of a filled bonding molecular orbital is distributed over three or more atomic centers and have some sigma-bond property. Satisfying all the requirements, the 2-norbornyl cation is considered an archetypal case.

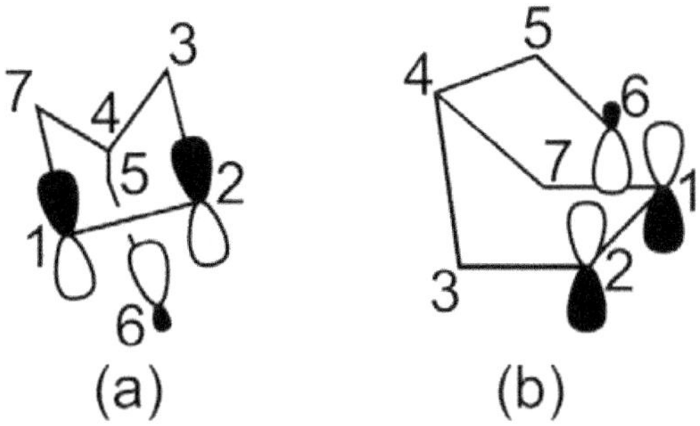

Figure 2. Orbital structure of a non-classical carbocation (2-norbornyl).

The most widely accepted molecular orbital structure of the 2-norbornyl cation is having two p-type orbitals (on carbons 1 and 2) interacting with an sp^3-hybridized orbital on carbon 6 to form the hypervalent bond. Extended Hückel Theory calculations for the 2-norbornyl cation suggest that the orbital on carbon 6 could instead be sp^2-hybridized, though this only affects the geometry of the geminal hydrogens.

3. Stability of non-classical carbocations: The stability profile of nonclassical carbocations can be rationalized by the delocalization of σ- or π-bonds as discussed below.

i) Stability gain via the delocalization of π-bond: The typical example of this type of stabilization is 2-norbornyl cation which is shown below.

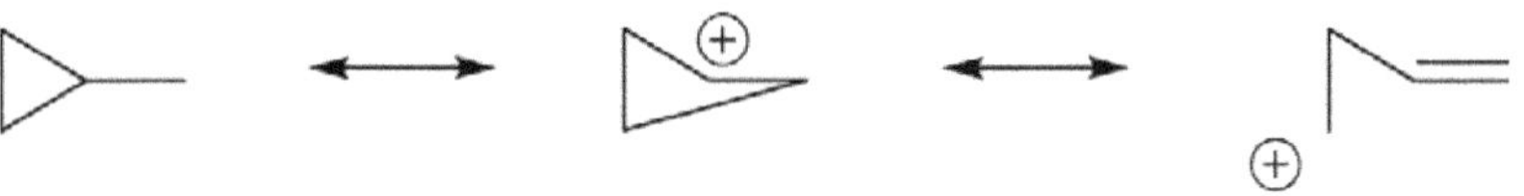

ii) Stability gain via the delocalization of σ-bond: The treatment of cyclopropylmethyl (or cyclobutyl chloride or homoallyl chloride) chloride with dilute ethyl alcohol yields a mixture of 5% homoallyl alcohol, 47% cyclobutanol, and 48% cyclopropylmethyl alcohol. All this suggests that the carbocationic intermediate present in all three reactions must be the same, which in turn, is responsible for the same resulting products.

4. Reactivity of non-classical carbocations: The principal routes by which the non-classical carbocations can react to give rise to stable products are given below.

i) Nucleophilic attack: In these types of reactions, a non-classical carbocation may combine with a species by accepting an electron pair. Furthermore, it should also be noted that if all the three groups on the carbocation are different, a racemic mixture will be obtained.

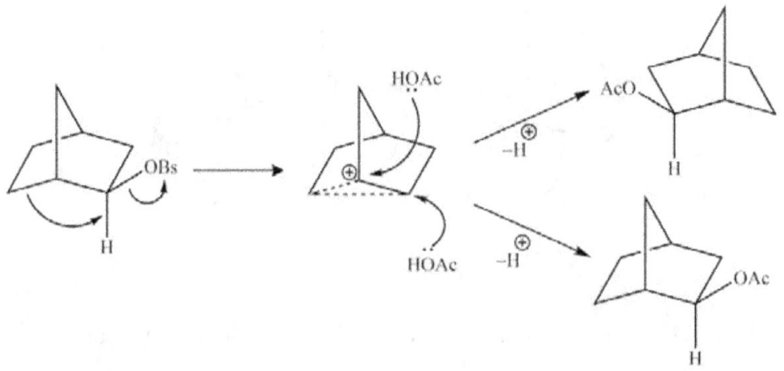

ii) Rearrangement reactions: The treatment of cyclopropylmethyl (or cyclobutyl chloride or homoallyl chloride) chloride with dilute ethyl alcohol yields a mixture of 5% homoallyl alcohol, 47% cyclobutanol and 48% cyclopropylmethyl alcohol.

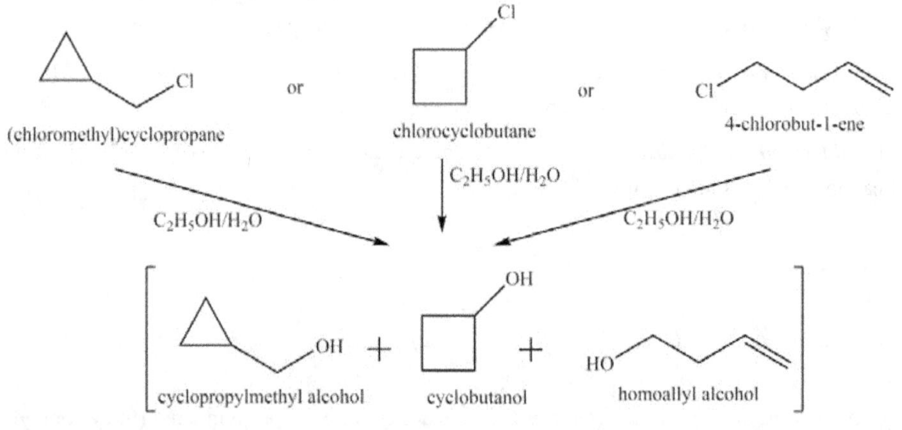

All this suggests that the carbocationic intermediate present in all three reactions must be the same, which in turn, is responsible for the same resulting products.

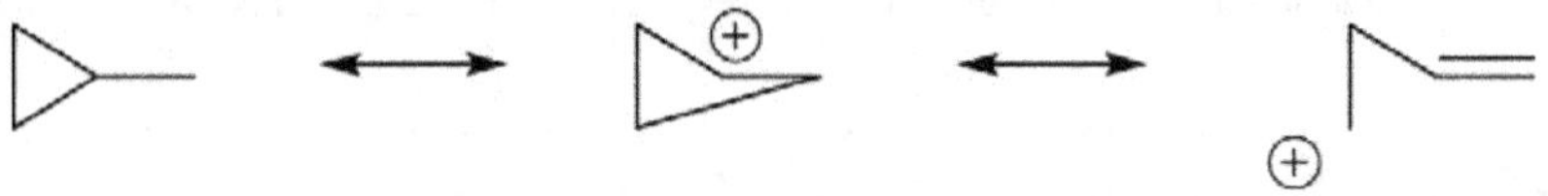

❖ Phenonium Ions

Phenonium ions may simply be defined as cyclohexadienyl cations which are spiro-annulated with a cyclopropane unit.

These ions form an arenium-ions' subclass and greatly affect the reactivity. An aromatic ring can give great support in the formation of a carbocationic intermediate by delocalizing the positive charge. In other words, the aryl group participates in the neighboring mechanism via the formation of phenonium ion yielding retention of the original configuration.

Furthermore, we can have many types of phenonium ions depending upon the degree and nature of substitution affecting its overall reactivity. In addition to the typical phenonium ion given above, two of the most commonly studied disubstituted phenonium ions are meso- and C_2-symmetric phenonium ions. The neighboring group mechanism when we use diastereomeric single enantiomer substrates is shown below.

It is obvious from the above routes that the double inversion in meso-phenonium ion created a racemic mixture whereas double inversion in C_2-symmetric phenonium ion has created a single enantiomeric product.

D DALAL INSTITUTE

❖ Carbocation Rearrangements

Carbocations may undergo rearrangements to yield their more stable counterparts, and this phenomenon is typically labeled as carbocationic rearrangement. In this section, we will discuss the carbocation rearrangement for classical and non-classical carbocations.

➢ *Rearrangement Reactions in Classical Carbocation:*

The 1-2 methyl shit or 1-2 hydride shifts are very common in carbocation chemistry to attain a more stable counterpart. For instance, a primary carbocation will prefer to rearrange itself into a more stable tertiary carbocation.

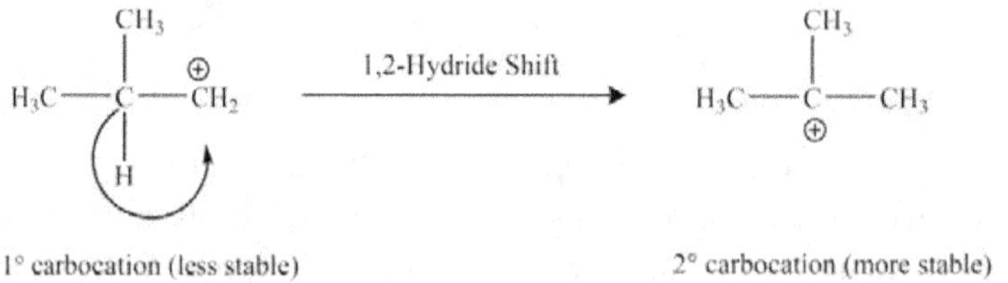

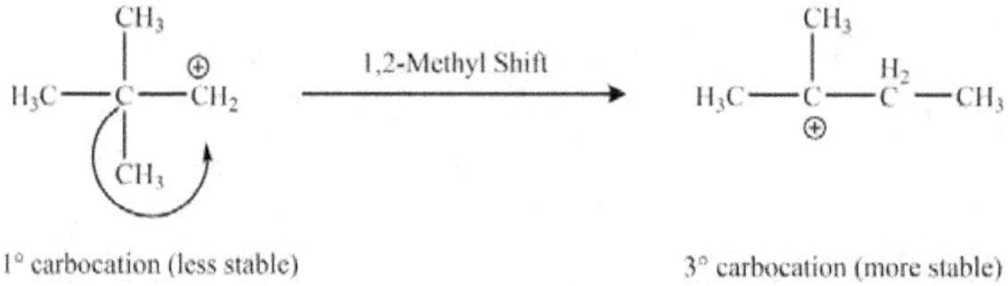

$$1,2\text{-Hydride Shift}$$

1° carbocation (less stable)   2° carbocation (more stable)

$$1,2\text{-Methyl Shift}$$

1° carbocation (less stable)   3° carbocation (more stable)

Some typical examples of methyl or hydride shift are given below.

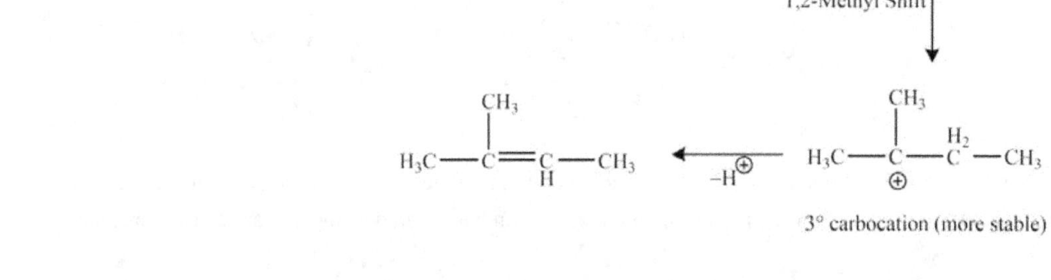

$$H_2SO_4 \ (conc.)$$

$$-H_2O$$

1,2-Methyl Shift

3° carbocation (more stable)

> ### *Rearrangement Reactions in Non-Classical Carbocation:*

The most common example of this type of rearrangement is the cyclopropylmethyl system in with three rearranged forms of the carbocation exist in equilibrium.

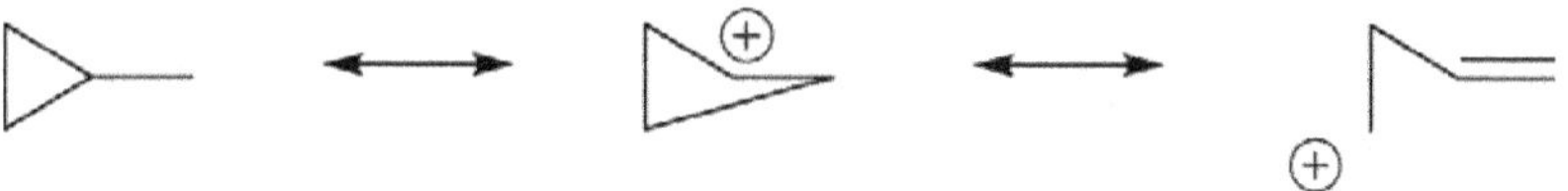

The treatment of cyclopropylmethyl chloride with dilute ethyl alcohol yields a mixture of 5% homoallyl alcohol, 47% cyclobutanol, and 48% cyclopropylmethyl alcohol. Similar results were obtained if we use cyclobutyl chloride or homoallyl chloride instead of cyclopropylmethyl chloride.

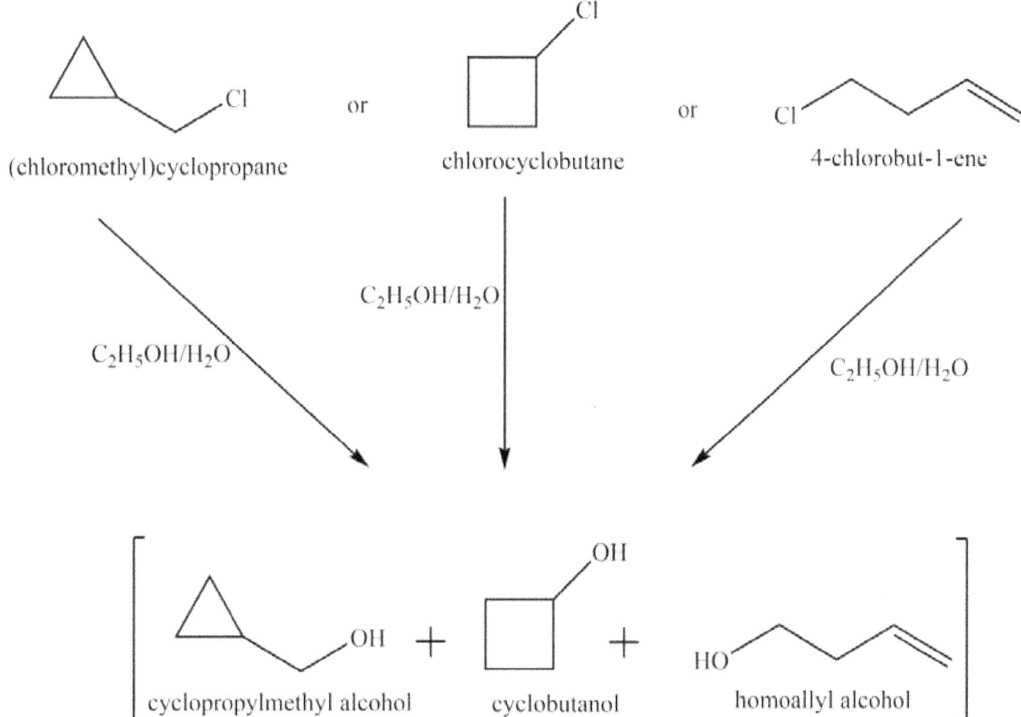

All this suggests that the carbocationic intermediate present in all three reactions must be the same, which in turn, is responsible for the same resulting products.

Furthermore, it is also worthy to note that the participation of σ-bond in the neighboring group mechanism to yield no-classical carbocations results in increased rates of a reaction than what we would have observed in normal SN_2; which is obviously due to anchimeric assistance.

❖ Applications of NMR Spectroscopy in the Detection of Carbocations

Carbocations are the typical example of how the study of reaction intermediates is carried out. Two of the most popular and efficient experimental techniques to analyze the properties and structure of carbocationic species are ^{13}C and PMR (proton magnetic resonance) spectroscopy. The notable work in carbocation chemistry was carried out by George A. Olah, an American chemist who was also awarded the Nobel prize (1994) for the same.

Olah discovered that the NMR spectra of organic precursors in super-acid solutions were indicating relatively stable carbocations. He found that the chemical shifts (^{13}C NMR) of carbocations are much downfield than their parent compounds. For instance, the chemical shift for tertiary carbon in tert-butyl carbocation is at 330 ppm whereas the corresponding carbon in isobutane absorbs at 25.2 only. This can be explained in terms of reduced electron density at the carbon center in the carbonium ion.

$$\delta = 25.2 \text{ ppm} \qquad\qquad \delta = 330 \text{ ppm}$$

^{13}C NMR in SO_2 ClF-SbF_5 Solution

The argument is also supported by other studies such as the NMR spectrum of substituted benzylic carbocations. More precisely, as the electron-withdrawing group becomes stronger at *p*-position in benzylic carbocation, the NMR peaks of cationic carbon shift towards the more down-field region.

Table 1. Benzylic carbocations' peaks in ^{13}C NMR spectra.

Substituents	δ(ppm)
p-OCH₃	219
p-CH₃	243
p-H	255
p-CF3	269

Now from Table 1, we might conclude that all electron-donating groups (like alky substituents) at carbonium center should behave in the opposite manner and should decrease the chemical shifts. Nevertheless, it is found that though the electron-donating alkyl groups directly attached to carbonium centers raise their stability, but little to no effect was found upon the chemical shift.

For instance, when CS_2 was taken as the reference standard, the chemical shift (^{13}C NMR) of secondary carbon in isopropyl carbonium ion appears at -125 δ whereas the tertiary carbon in tert-butyl carbocation shows the peak at -125 δ. Al this suggests that the replacement of hydrogen by methyl has actually supported the withdraw rather than the donation.

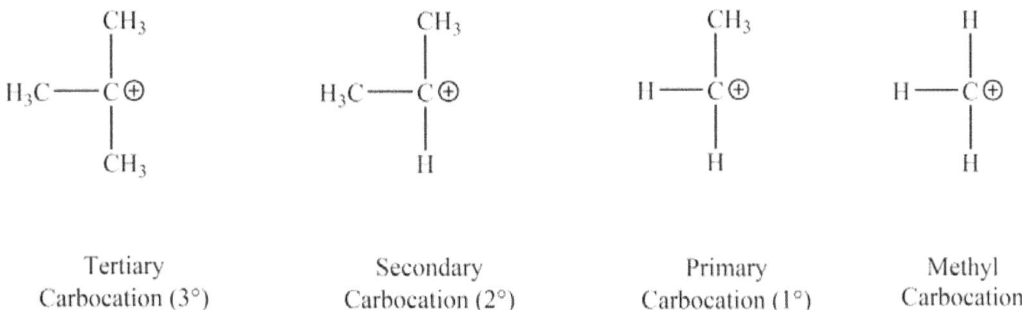

Tertiary   Secondary   Primary   Methyl

Carbocation (3°)   Carbocation (2°)   Carbocation (1°)   Carbocation

Figure 3. Some typical carbocations.

Furthermore, the molecular orbital theory also proves theoretically that the charges at 2° in isopropyl cation and 3° carbons in tert-butyl carbocations are +0.611 and +0.692, respectively.

Furthermore, the simplest arenium ion produced by the protonation of benzene ring strongly acidic solution can also be detected and studied by employing NMR spectroscopy. The chemical shifts for the o- and p- carbons (w.r.t protonation's site) in the ^{13}C NMR of arenium ion show very strong downfield movement which can be rationalized in terms of reduced electron density at the same.

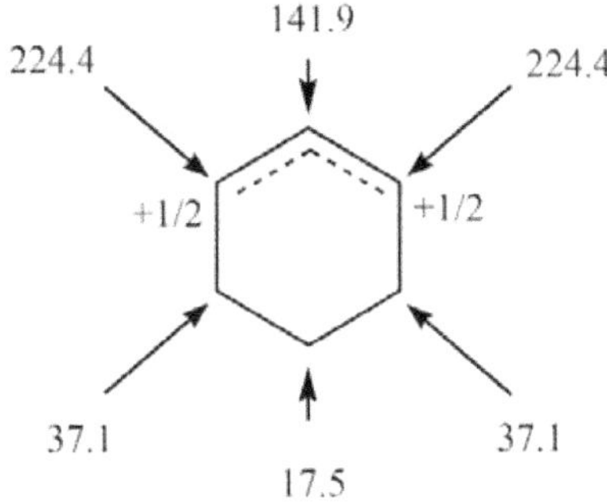

A typical allylic cation with ^{13}CNMR spectrum

It is also worthy to note down from the Figure given above that the similar magnitudes of chemical shifts on the right and left side of carbocation suggest a plane of symmetry in the same.

❖ Reactivity – Effects of Substrate Structure, Attacking Nucleophile, Leaving Group and Reaction Medium

The reactivity of aliphatic nucleophilic substitution reactions is affected by many factors which can be better understood via experimental data and theoretical treatment combined. In this section, we will discuss some major factors that greatly influence the nucleophilic substitution's rate in aliphatic compounds like substrate structure, attacking nucleophile, leaving group, and reaction medium.

> ### Effect of Substrate Structure on the Reactivity of Aliphatic Nucleophilic Substitution

Since the rates in SN_1 and SN_1' reactions are determined by the 1st step (formation of carbocation), the more stable the carbocation is, readily it will be formed, and the faster the rate will be. Also, the neighboring group participation takes place via first-order kinetics; therefore, the enhanced rate (anchimeric assistance) must be analyzed with special reference to the substrate. The effect of substrate structure on the reactivity of aliphatic nucleophilic substitution reactions can be divided into the following categories.

1. Unsaturation at α-carbon: It is quite a well-known fact that aryl halide, vinyl halide, and acetylenic halides show very small reactivity towards nucleophilic substitution reactions because the carbon-halogen bond gains some double bond nature due to resonance. All this makes the breaking of C−X bond more difficult resulting in a lower tendency to undergo nucleophilic substitution.

2. Unsaturation at β-carbon: Unlike the unsaturation at α-carbon, the presence of multiple bonds at β-carbon enhances the favourability of nucleophilic substitution because of the generation of resonance stabilized benzyl or allyl carbocation. This is true for SN_1 as well as for SN_2 pathways. For instance, consider the unimolecular nucleophilic substitution in allyl- and benzyl halides.

Resonance stablized allyl cation

Resonance stablized benzyl cation

Similarly, the transition state formed during the course of the bimolecular nucleophilic substation is also stabilized by the phenomenon of resonance.

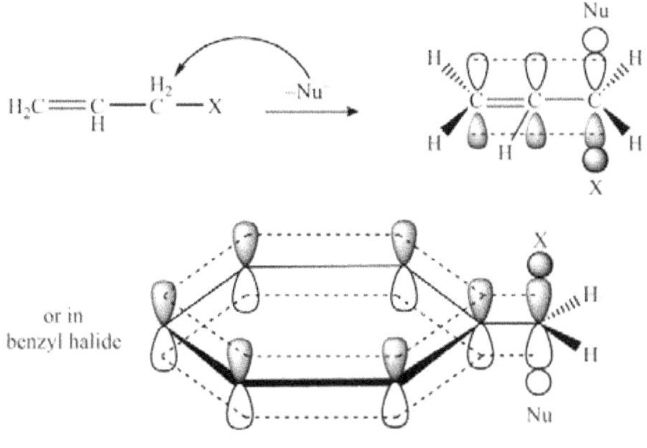

Therefore, we can conclude that allyl and benzyl halides are more reactive than alkyl halides in SN_1 as well in SN_2 pathways.

3. Steric effects: Since the alkyl carbocations are primarily obtained from alkyl halides with tetrahedral geometry, a link between the steric relief and carbocation formed can be established. During the formation of carbocations in such cases, the carbon-carbon bond angles change from 109°28' to 120°. Therefore, the carbon with bulky groups around is expected to get more relief from this carbocationic conversion. The stability order of alky carbocations on the basis of steric effect is given below.

Furthermore, the steric bulk around or near the central carbon also affects the rate of bimolecular nucleophilic substitution (SN_2) but in a negative way. This is because the bulky groups disfavor the formation of transition state resulting in a slower reaction rate.

4. Presence of heteroatom at α-carbon: The rate of SN_1 reactions is greatly enhanced by the presence of heteroatom at α-carbon because it stabilizes the carbocation formed during the first step.

5. Substitution at bridgehead carbon: The nucleophilic substitution at bridgehead carbons is extremely disfavoured both in SN$_1$ as well as in SN$_2$ pathways. The reason for low reactivity at bridgehead carbon in the SN$_1$ pathway is the 'non-planar' structure of carbocation involved.

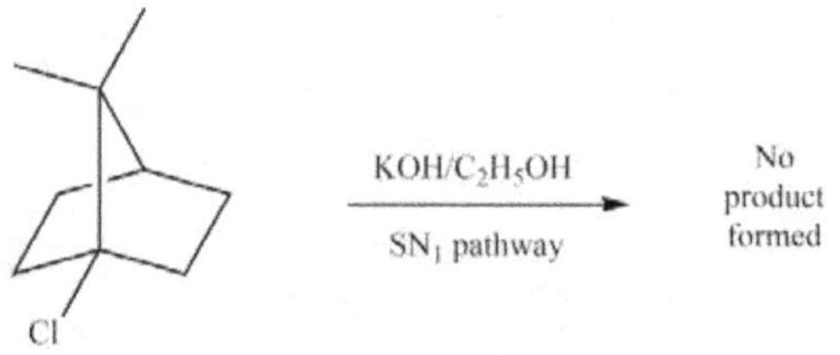

$$\xrightarrow[\text{SN}_1 \text{ pathway}]{\text{KOH/C}_2\text{H}_5\text{OH}} \quad \text{No product formed}$$

However, the low reactivity at bridgehead carbon in the SN$_2$ pathway is due to different reasons i.e. disfavorability of backside attack of nucleophile which an essential request of the same.

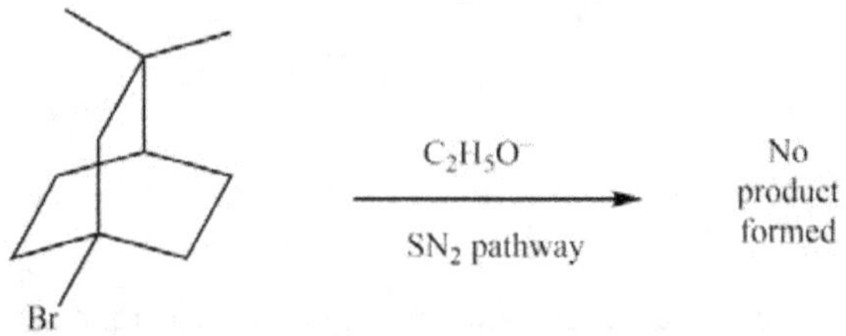

$$\xrightarrow[\text{SN}_2 \text{ pathway}]{\text{C}_2\text{H}_5\text{O}^-} \quad \text{No product formed}$$

6. Carbonyl groups at α-carbon in alkyl halides: The rate of SN$_2$ reactions is greatly enhanced if some carbonyl group is situated at the α-carbon in alkyl halides. This is because the electron-withdrawing effect raises the stability of the transition state.

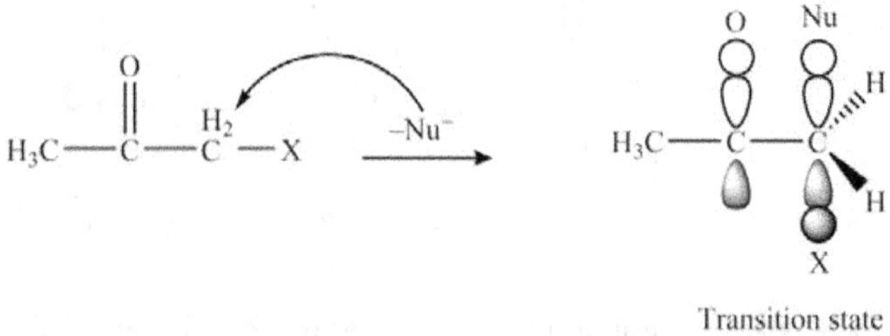

Transition state

However, a reverse effect will be observed for SN$_1$ pathway which is obviously due to the destabilization of carbocation formed.

$$\text{H}_3\text{C} - \text{X} \qquad\qquad \text{H}_3\text{C} - \text{O} - \text{CH}_2\text{X} \qquad\qquad \text{C}_6\text{H}_5 - \overset{\overset{\displaystyle O}{\|}}{\text{C}} - \overset{\text{H}_2}{\text{C}} - \text{X}$$

| Reactivity of substrate in SN$_2$ pathway | 202 | 920 | 100000 |

> *Effect of Nucleophile on the Reactivity of Aliphatic Nucleophilic Substitution*

Since the rate in SN_1 reactions is determined by the 1st step (formation of carbocation), and the nucleophile attacks in the second step; the rate of SN_1 reactions is pretty much independent of nucleophilic strength. However, the rate of SN_2 and SN_2' depends upon the strength of the attacking nucleophile due to second-order kinetics. The effect of the nucleophile on the reactivity of aliphatic nucleophilic substitution reactions can be divided into the following categories.

1. Formal charge: The nucleophiles with higher negative charge have more strength than corresponding conjugate acids. In other words, if the conjugate acid of a nucleophile is also a nucleophile, it will be of less strength always. For instance, the amide is a stronger nucleophile than ammonia; and OH^- is stronger than water.

2. Basic character: The strength of different nucleophiles follows the same order as basicity if the attacking atom comes from the same period. For instance, consider the following order.

$$NH_2^- > RO^- > OH^- > R_2NH > ArO^- > NH_3 > \text{pyridine} > F^- > H_2O > ClO_4^-$$

Furthermore, it is also worthy to recall that nucleophilicity is kinetically controlled but basicity is thermodynamically controlled.

3. Solvation level: The strength of different nucleophiles increases as the attacking comes from more down the group. Now, although the basic character of such nucleophiles decreases down the group, the increasing nucleophilicity can be rationalized in terms of higher solvation of smaller anions making them less suitable for attack. For instance, consider the following order.

$$I^- > Br^- > Cl^- > F^-$$

The effect is more prominent in polar protic solvents where smaller nucleophiles are more solvated than usual. Furthermore, it is also worthy to note that this effect becomes inefficient for neutral nucleophile because they don't get much solvated. However, they also show an increase in nucleophilic strength down the group which can be explained in terms of the hard-soft acid-base principle. In other words, the reason for the increasing strength of neutral nucleophiles is that a softer nucleophile prefers an alkyl carbocation (soft acid) over the proton (soft acid).

4. Attacking liberty: As the nucleophile becomes more free to attack, the rate of reaction increases. For instance, consider the case of $(EtOOC)_2CBu^-Na^+$ in C_6H_6, where the addition of Na^+-solvating compound makes the $(EtOOC)_2CBu^-$ ion free to attack, and increasing rate consequently.

Nevertheless, it must be kept in mind that these rules do not hold at all conditions always because there are many factors like steric influences. For instance, the tert-butoxide ion (Me_3CO^-) much weaker nucleophile though it is an even stronger base than OH^- or OEt^-; which can be attributed to the fact that its huge bulk hinders it from approaching a substrate more closely. Finally, we may conclude that only rough estimates can be made about the order of nucleophilic strengths and there is no absolute order of either leaving group and nucleophilicity.

> ➤ *Effect of Leaving Group on the Reactivity of Aliphatic Nucleophilic Substitution*

A good leaving group is that becomes stable after the detachment from substrate. The effect of leaving group on the reactivity of aliphatic nucleophilic substitution reactions can be divided into the following categories.

1. Basic character: The betterment of the leaving-group is inversely proportional to the strength of its basic character. In other words, weak bases are good leaving and vice-versa. For instance, consider the order of goodness for halides as leaving group.

$$I^- > Br^- > Cl^- > F^-$$

This is just the opposite trend of acidic character order of corresponding their hydro acids i.e., $HI > HBr > HCl > HF$.

2. Ester conversion: The power of a leaving group can be improved by converting it into its ester, which is why brosylates or tosylates (sulphonic esters) are more popular as leaving group than halides.

3. Ring strain: The leaving-group power can also be increased via ring strain. For instance, simple ethers do not undergo breaking while protonated ethers show cleavage conditions are strenuous; however, epoxides undergo bond cleavage effortlessly whereas protonated epoxides show even more tendency.

Tosylates

Brosylates

Furthermore, it is also worthy to note that the nature of leaving the group not only affects the reaction speed but can also change the mechanism of the same from associative to dissociative and vice-versa.

> ***Effect of Reaction Medium on the Reactivity of Aliphatic Nucleophilic Substitution***

The effect of the reaction medium on the reactivity of aliphatic nucleophilic substitution reactions can be divided into the following categories.

1. Solvent polarity: The rate of reaction in SN_1 transformations increases with the increase in solvent polarity and vice-versa. This is because the polar solvent stabilizes carbocation intermediate greatly, which in turn decreases the activation energy required for the change.

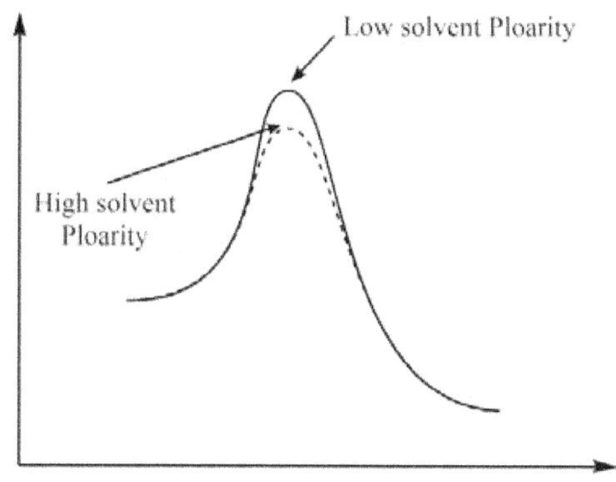

Reaction coordinate diagram

Furthermore, the increasing solvent polarity also raises the rate of the associative pathway (SN_2) which obviously because of the better dissolution of nucleophile involved.

2. Protic-aprotic type: It has been observed that polar protic solvents decrease the nucleophilic strength and enhances the stability of the anionic leaving group; consequently, supporting the dissociative pathway (SN_1).

On the other hand, protic solvents slow down the rate of SN_2 reaction by solvating the nucleophile involved; whereas, aprotic solvents enhance the rate by freeing the nucleophile via the binding cation mechanism.

❖ Ambident Nucleophiles and Regioselectivity

Ambidentate nucleophiles may simply be defined as the nucleophilic species that have more than one atom for the attack to result in different structural isomers; nevertheless, if one isomer is preferred over the other, the phenomenon is called as regioselectivity.

In this section, we will discuss the different types of ambident nucleophiles; and then will try to explain the factors governing the regioselectivity arising from these nucleophiles.

➢ *Different Types of Ambident Nucleophiles*

The different types of ambident nucleophiles which are very common in synthetic organic chemistry are discussed below.

1. [−CO−CR−CO−]⁻-type ions: The ions of this type are obtained by the elimination of a proton from β-keto esters, malonic esters, or b-diketones, and resonance hybrid structures are needed to depict them.

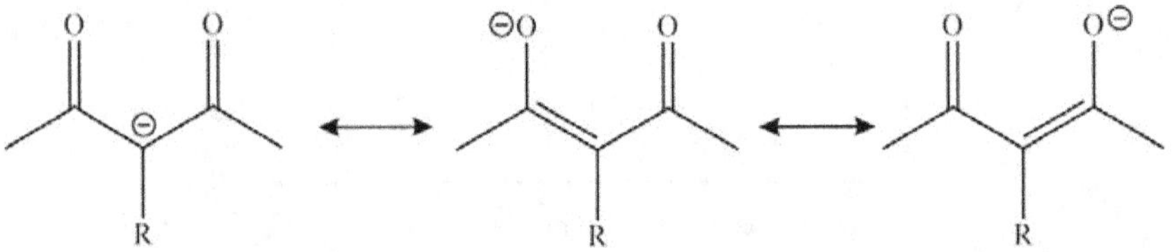

The possibility of nucleophilic attack arises at the saturated carbon through the oxygen or carbon atoms resulting in O-alkylation or C-alkylation, respectively. However, if the ions are unsymmetrical in nature, three different products can be obtained because either oxygen can show an attacking tendency. Furthermore, the O- or C-acylation can be achieved if the substrate has a carbonyl group.

2. [CH₃CH−CH₂−CO−]-type molecules: Upon treating with 2 molar equivalents of a strong a base, these types of molecules can release two protons forming a dicarbanion.

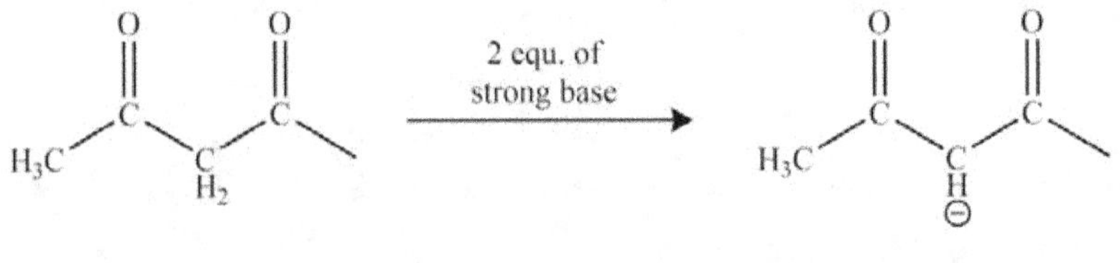

Now although the two atoms bearing negative charge are carbon, these nucleophiles are still called as ambident because they are non-equivalent in nature.; and the possibility of attack by oxygen is also there. Now because the hydrogen at carbon attached to one carbonyl group is less acidic than that of a carbon bonded to two, the CH₂ group should be more basic than the CH group, and therefore, more suitable for the attacks.

All this results in a useful generalization that if there is a stronger acidic group but we need to eliminate a proton from some other carbon to create a nucleophilic site, the removal of both is possible; however, the attack will happen at the weakly acidic position because this conjugate base of the more weak acid. Alternatively, one more acidic proton is needed to be removed if we want to attack a more acidic position.

3. CN⁻ Ion: This nucleophilic ion can give rise to isocyanides or nitriles depending upon the mode of the chemical bond between carbon and the attacking group.

4. NO₂⁻ Ion: This nucleophilic ion can give rise to nitro compounds or nitrites depending upon the mode of the chemical bond between the carbon and attacking group.

5. Phenoxide Ions: Ions of this type are analogous to enolate anions and can undergo a chemical change to give rise to O-alkylation or C-alkylation.

6. Aliphatic nitro compound's derivatives: In these types of reactions, a carbanion of kind $O_2N{-}R2C^-$ is formed which can be alkylated at carbon or oxygen atom. Furthermore, it is also worthy to remember that the alkylation at oxygen results in the nitronic esters, but these esters would break down to give an aldehyde or ketone or oximes upon simple heat treatment.

> ### *Factors Dictating the Nature of Regioselectivity*

Following the definition of regioselectivity (one product is favored over the other), various factors dictating this preferential attack are discussed below.

1. Mechanism of the reaction: According to HSAB (hard-soft–acid-base) principle, the hard acids prefer to bind with hard bases whereas soft acids prefer to bind with soft bases. Therefore, we can say that ambident nucleophile uses it more electronegative atom to participate in SN_1 mechanism because the carbocation formed is a hard base; whereas it will use it less electronegative atom to participate in SN_2 mechanism because the carbon center of a molecule is soft acid. In other words, we can say that supporting the SN_2 character of a reaction motivates the nucleophile to attack via a less electronegative group and the vice-versa is also true.

2. Nature of the cation: Since a positive counterion exists for each negatively charged nucleophile, the SN_1 mechanism will be supported if this ion helps the elimination of leaving group (such as Ag^+). On the other hand, if this is a more common type like K^+ or Na^+, then SN_2 pathway will be supported.

3. Nature of the solvent: The nature of the solvent also affects the site of the attack by making nucleophile more or less free. A freer nucleophile prefers to attack via its more electronegative atom whereas a more engaged nucleophile prefers to attack via the less electronegative atom. The atom with a higher electronegativity atom is better solvated in protic solvents by the formation of hydrogen bonds. However, neither atom of the nucleophile is significantly solvated in polar aprotic solvents and these solvents are very efficient in solvating different cations making more electronegative atom suitable for attack. Therefore, the higher electronegative end of the nucleophile shows low engagement form both cation as well as from solvent in polar aprotic solvents.

5. Steric effects: It is quite a well-known fact that sterics play a very important role in deciding the final product of a chemical reaction across the domain. If one site surrounded by more bulky groups, is less likely to be attacked in an associative pathway than a less crowded site, and vice-versa is also true. Furthermore, the product with more steric hindrance is less likely to be formed.

4. Electron withdrawing effect: The attachment of more electron-withdrawing groups make the carbon more electron-deficient and suitable for the attack via more electronegative site, whereas the attachment of more electron-donating groups makes the carbon less electron-deficient and suitable for the attack via a less electronegative site.

For instance, the shift for the attack from carbon to oxygen takes place for the alkylation of acetylacetone's sodium salts.

❖ Phase Transfer Catalysis

The phase-transfer catalysis may simply be defined as a special form of heterogeneous catalysis that supports the movement of a reactant from one phase into another phase where the reaction occurs.

Ionic reactants are often soluble in an aqueous phase but insoluble in an organic phase in the absence of the phase-transfer catalyst (PTC). The function of a phase transfer catalyst resembles detergent to solubilize the salts into the organic-phase. In other words, the phase-transfer catalysis accelerates the reaction rate. By using this type of catalysis, we can get faster rates with better yields, fewer byproducts, eliminate the need for dangerous or expensive solvents that will dissolve all the reactants in one phase, eliminate the need for expensive raw materials and minimize waste-related complications. In this section, we will discuss the prominent mechanisms and applications of phase transfer catalysis.

➢ *Mechanisms of Phase Transfer Catalysis*

The mechanism of phase-transfer catalysis can primarily be divided into two categories where the acting routes are different but the final effects are similar. In other words, the anion is made to enter into the organic phase and permitted to be fairly free to react with the substrate in both types.

1. Quaternary phosphonium or ammonium salts: This mechanism can be understood by taking the example of the reaction between sodium cyanide and 1-chlorooctane where no substitution product (1-chlorooctane) is formed even after heating (with constant stirring) the mixture for weeks. However, even we add a very small amount of a suitable "quaternary ammonium salt", sufficient amount of 1-chlorooctane is obtained in less than two hours.

The reason for no reaction when "quaternary ammonium salt" is absent is that CN^- ions cannot pass from aqueous phase to organic phase in sufficient concentration all alone, leavening behind Na^+ ions because this would disturb the electrical neutrality of both phases; and Na^+ ions have no motivation go into organic phase since highly hydrated in the aqueous medium. However, "quaternary ammonium or phosphonium salts" are added, the quaternary ammonium (R_4N^+) and quaternary phosphonium ions (R_4P^+) are produced which do have the ability to pass through the interface between two phases. This is because 'R' groups in R_4N^+ and R_4P^+ are quite bulky causes poor solvation in the aqueous phase. Therefore, when they cross the phases' interface, they also carry CN^- ions with them to keep the electrical neutrality of both phases. Once the CN^- ions reach sufficient concentration, they start reacting with RCl to yield RCN and halide anions (Cl^-). The whole process can be depicted as given below.

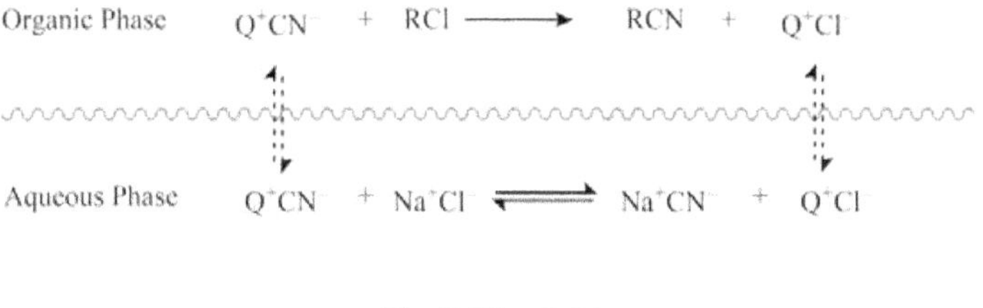

$$Q^+ = R_4N^+ \text{ or } R_4P^+$$

2. Crown Ethers and Other Types of Cryptands: It is quite a well-known fact that some cryptands have the ability to surround certain cations. For instance, when KCN salt is mixed with dicyclohexano-18-crown-6, a new salt is obtained which has the same anion much larger cation as shown below.

dicyclohexano-18-crown-6

This new cation has a much smaller positive charge density than K^+; and therefore, it gets very poorly hydrated. Furthermore, the new cryptate salt is quite soluble in many solvents including organic-types; therefore, we can add this new salt directly to the organic phase without the need for an aqueous phase. Many cryptands have been employed to make salts with OAc^-, F^-, I^-, Br^-, and CN^- to enhance the rates of nucleophilic substitution reactions.

Finally, we can conclude that "quaternary phosphonium or ammonium salts" as well as "cryptands" are quite capable of moving the anion from aqueous to organic phase; however, it has been found that even if sodium and potassium salts were soluble in the organic phase, they don't very fast nucleophilic substitution because they exist as ion pairs with corresponding anions, and are not free to attack the substrate efficiently. On the other hand, R_4N^+ or R_4P^+ and cryptate cations ions don't pair with anions very effectively, letting them, free to attack the substrate.

> ➤ *Applications of Phase Transfer Catalysis*

Although the application domain is quite large, some of the main applications of phase transfer catalysis in synthetic organic chemistry are given below.

i) Phase transfer catalysis is widely used in industries for commercial production.

ii) Phase transfer catalysis is used in the alkylation of phosphothioates to produce pesticides.

iii) Chiral quaternary ammonium salts are used as phase transfer catalysts for the asymmetric alkylations.

iv) Phase transfer catalysis is not bound to systems with hydrophobic and hydrophilic reactants and is employed in liquid-solid and liquid-gas reactions sometimes.

v) Phase-transfer catalysts are especially valuable in green chemistry by allowing the use of water, and therefore, the requirement for organic solvents is lowered.

❖ Problems

Q 1. What are aliphatic nucleophilic substitution reactions? Name various types.

Q 2. What are the major differences between SN_1 and SN_2 reactions?

Q 3. Describe neighboring group participation? How does σ-bond can also place the same role?

Q 4. Define anchimeric assistance. Discuss with suitable examples.

Q 5. Discuss the stability of classical and non-classical carbocations.

Q 6. What are meso-phenonium ions? How are they different from C_2-symmetric phenonium ions?

Q 7. Give major rearrangement reactions of classical carbocations.

Q 8. Write a short note on the $SN_i{'}$ mechanism with special emphasis on its relative study with normal SN_i.

Q 9. Discuss the effect of substrate structure on the reactivity of nucleophilic substitution reactions.

Q 10. What are nucleophiles? How does their strength affect the rate of nucleophilic substitution reactions?

Q 11. Discuss ambident nucleophiles?

Q 12. State and explain regioselectivity.

Q 13. What is phase transfer catalysis?

Q 14. Discuss the applications of NMR spectroscopy in the structure determination of carbocations.

Q 15. What is the 'single electron transfer (SET)' mechanism? Explain with suitable example.

❖ Bibliography

1. M.S. Singh, *Reactive Intermediates in Organic Chemistry*, John Wiley & Sons, Inc., New Jersey, USA, 2014.

2. H. Zimmerman, *Quantum Mechanics for Organic Chemists*, Academic Press, New York, USA, 1975.

3. J. Clayden, N. Greeves, S. Warren, *Organic Chemistry*, Oxford University Press, Oxford, UK, 2012.

4. R. L. Madan, O*rganic Chemistry*, Tata McGraw Hill, New Delhi India, 2013.

5. M. B. Smith, *March's Advanced Organic Chemistry: Reactions, Mechanisms, and Structure*, John Wiley & Sons, Inc., New Jersey, USA, 2013.

6. D. Klein, *Organic Chemistry*, John Wiley & Sons, Inc., New Jersey, USA, 2015.

7. C. A. Coulson, B. O'Leary, R. B. Mallion, *Hückel Theory for Organic Chemists*, Academic Press, Massachusetts, USA, 1978.

CHAPTER 7

Aliphatic Electrophilic Substitution

❖ Bimolecular Mechanisms – SE$_2$ and SE$_i$

The electrophilic substitution in the aliphatic compounds is just similar to the aliphatic nucleophilic substitution, except for the fact that here an electrophile displaces a functional group rather than an electrophile. In this section, we will discuss the two major types of electrophilic substitutions; SE$_2$ (substitution electrophilic bimolecular) and SE$_i$ (substitution electrophilic internal) mechanisms.

➢ SE$_2$ (Substitution Electrophilic Bimolecular) Mechanism

The bimolecular electrophilic substitution (SE$_2$) reactions may simply be defined as the chemical changes where a stronger electrophile displaces a weaker one in an aliphatic substrate.

This mechanism is quite analogous with the SN$_2$ route excepting the mode of attack. In the SN$_2$ mechanism, a stronger nucleophile replaces a weaker one via the backside attack due to its inter-cloud repulsion with the leaving group; however, in the SE$_2$ route, the attacking electrophile may come from the front, as well as from the backside because it is only using is vacant orbital towards substates causing little to no repulsion. So, the SE$_2$ mechanism can be divided into SE$_2$-front and SE$_2$-back based upon the front and back attacks, respectively.

Illustrative reaction: The general reaction showing both types of electrophilic attacks (with their corresponding products), is shown below.

Mechanism involved: The proposed mechanism for the reaction given above says that the two electrons of the carbon-electrophile bond reside in the central orbital. Ingold proposed that the electronic distribution responsible for different stereochemistry of products is a function of bond-extension's magnitude and the extent of bond ionicity of the transition state.

It is obvious that the transition state needs to have high ionicity and good bond extension potential for SE_2-back reactions so that the carbon's orbital is sufficiently spread on both ends, resulting in the inverted configuration in the case of a chiral substrate. On the other hand, if there is a very little bond extension and low ionicity in the transition state, the electron-pair of the original bond pretty much retains its position and gives rise to the retention of the configuration, and we get SE_2-front case.

Where E^+ is the attacking electrophile whereas L+ is the leaving group. Furthermore, it is also worthy to note that organometallic compounds have exceptional susceptibility towards electrophilic substitution.

Salient Features: The main features of the mechanism involved in electrophilic substitution bimolecular or SE_2 type reactions are given below.

i) SE_2 reactions follow second-order kinetics with the rate law

$$Rate = k[RX][E]$$

Where k is the rate constant. The symbol $[RX]$ and $[E]$ represent the molar concentration of the substrate and attacking electrophile, respectively.

ii) If the substrate is chiral, then the SE_2 mechanism can lead to the inversion, as well as, retention of the configuration; depending upon the mode of attack (front or back).

iii) The rate of the substitution becomes independent of the concentration of the attacking electrophile if its concentration is extremely high in comparison to the substrate.

iv) Stereochemical studies can be employed to differentiate between SE_2-front and SE_2-back.

v) The SE_2 reactions are favored by the more polar C−L bond.

➤ ***SE_i (Substitution Electrophilic Internal) Mechanism***

The internal electrophilic substitution (SE_i) reactions may simply be defined as the chemical changes where a stronger electrophile displaces a weaker one in an aliphatic substrate by assisting its departure.

This mechanism is also very analogous with the SN_2 route excepting the mode of attack. In the SN_2 mechanism, a stronger nucleophile replaces a weaker one via the backside attack due to its inter-cloud repulsion with the leaving group; however, in the SE_i route, the attacking electrophile comes from the front and assists the departure of leaving group by forming a bond with it.

Illustrative reaction: The general reaction showing this type of electrophilic attack (with the corresponding product) is shown below.

Retention
of configuration

Mechanism involved: The proposed mechanism for the reaction given above says that the two electrons of the carbon-electrophile bond reside in the central orbital. It is observed that if there is a very little bond extension and low ionicity in the transition state, the electron-pair of the original bond pretty much retains its position and gives rise to the retention of the configuration, and we get SE_i case (like SE_2-front).

Retention
of configuration

Salient Features: The main features of the mechanism involved in electrophilic substitution internal or SE_i type reactions are given below.

i) SE_i reactions follow second-order kinetics with the rate law

$$Rate = k[RX][EZ]$$

Where k is the rate constant. The symbol $[RX]$ and $[EZ]$ represent the molar concentration of the substrate and attacking electrophile, respectively.

ii) Like SE_2-front, SE_i reactions result in the retention of the configuration.

iii) The SE_2 reactions are favored by the more polar C−L bond.

❖ The SE_1 Mechanism

The unimolecular electrophilic substitution (SE_1) reactions may simply be defined as the chemical change in which a stronger electrophile displaces a weaker one in an aliphatic substrate via the formation of a carbanion.

This mechanism is quite analogous with the SN_1 route excepting the nature of intermediate. In the SN_1 mechanism, a stronger nucleophile replaces a weaker one via the formation of a carbocation intermediate; however, in the SE_1 route, before the attacking electrophile attack, the intermediate formed after the dissociation of electrofuge is a carbanion.

Illustrative reaction: The general reaction showing this type of electrophilic attack (with its corresponding product) is shown below.

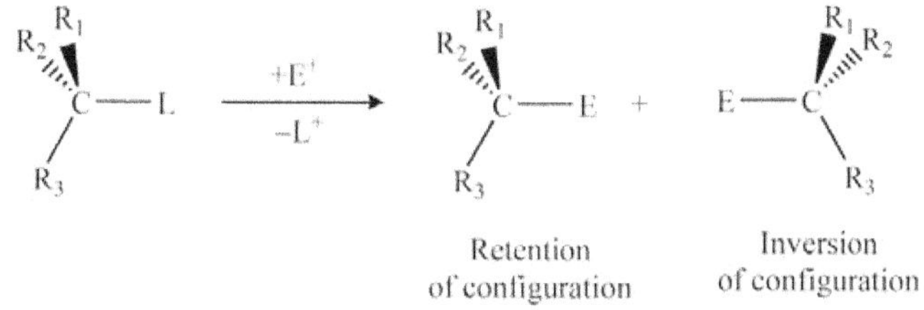

Retention
of configuration   Inversion
of configuration

Mechanism involved: The proposed mechanism for the reaction given above involves a three-step route which must be discussed before we give the salient features of the same.

i) Heterolysis in substrate: This is the slowest, and therefore, is the rate-determining step that gives rise to a carbanion.

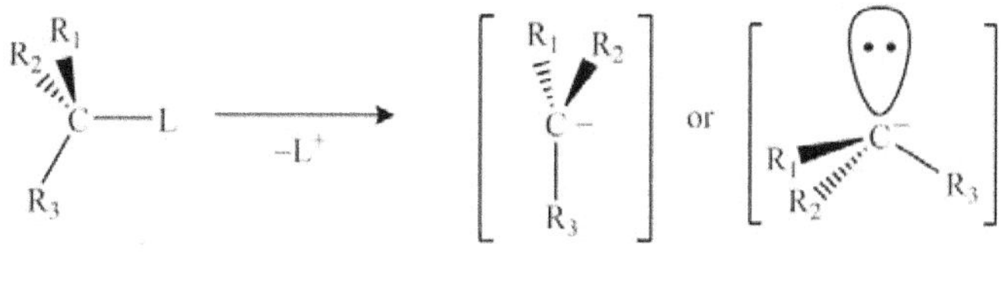

ii) Electrophilic attack: This is a very fast step and involves the combination of attacking electrophile with the carbanion produced in the previous step.

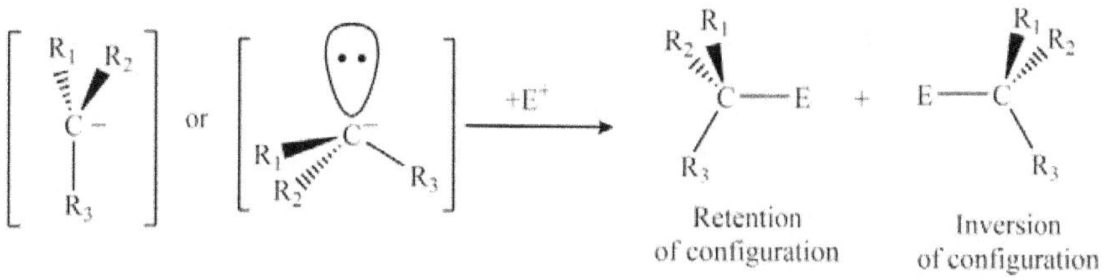

Retention
of configuration   Inversion
of configuration

The stereochemistry of SE_1 reactions is quite complicated to rationalize because of the configuration of intermediatory carbanions obtained via the first step of heterolysis. Generally, we consider carbanions planar (sp^2 hybridization) or pyramidal (sp^3 hybridization), or an in-between configuration. As far as the energy is concerned, pyramidal geometry is more advantageous because lone pair will stay in sp^3 hybridized orbital. Furthermore, a pyramidal carbanion can retain its structure in the course of substitution to result in the retention of the final configuration. However, it does not always go this way because a pyramidal carbanion has been shown to results in racemization due to 'pyramidal inversion'; amines and R_3C^- carbanions are typical examples.

Figure 1. The pyramidal inversion of carbanion.

On the other hand, if the carbanion is of trigonal planar geometry, the electrophile can attack from both sides to give rise to racemized yield. So, we can conclude that racemization is the characteristic feature of the SE_1 route. However, it is quite tedious to determine how the racemization actually occurred; via pyramidal inversion or planar carbanions.

Salient Features: The main features of the mechanism involved in electrophilic substitution unimolecular or SE_1 type reactions are given below.

i) SE_2 reactions follow first-order kinetics with the rate law

$$Rate = k[RX]$$

Where k is the rate constant. The symbol $[RX]$ represents the molar concentration of the substrate.

ii) If the substrate is chiral, then this always leads to the racemization of the final product.

iii) Unlike SN_1-type, SE_1 reaction can also occur at bridgehead carbon because the intermediate (carbanion in this case) need not to be planar.

iv) The rate of the substitution increases as the steric bulk around the carbon center decreases.

v) The SE_2 reactions are favored by the more polar C–L bond.

❖ Electrophilic Substitution Accompanied by Double Bond Shifts

If the substrate in electrophilic substitution is allylic in nature, the final product may undergo rearrangement, which is quite similar to the allylic rearrangements in nucleophilic substitutions. There are two main routes for this behavior to occur; one is analogous to the SE_1 pathway (leaving group is detached first) giving a resonance-stabilized allylic carbanion which attacks the electrophile E, the second one involves the initial attack on E by the π-bond to yield a carbocation which then which loses X forming new alkene unit.

➤ *Base-catalyzed Double Bond Migration*

The first pathway is the base-catalyzed double bond migration, where an equilibrium mixture of isomers is obtained with stable configuration as the major product. The reaction occurs in two steps, in which the step is the abstraction of a proton by the base to yield a resonance stabilized carbanion, which in turn, is attacked by electrophile (proton in this case) to give rise to a more stable species. The typical reaction portraying mechanism is given below.

It is also worthy to note that terminal and non-conjugated alkenes can easily be converted into internal and conjugated olefins using this route, proving its significance in synthetic organic chemistry.

➤ *Acid-catalyzed Double Bond Migration*

The second pathway is the acid-catalyzed double bond migration, where an equilibrium mixture of isomers is obtained with a stable configuration as the major product. The reaction initiates with the attack of E on the π-bond to yield a carbocation which then loses L forming a new alkene unit. The typical reaction portraying mechanism is given below.

Just like base-catalyzed double bond migration, this route can also be used to convert terminal and non-conjugated alkenes into internal and conjugated olefins.

❖ Effect of Substrates, Leaving Group and the Solvent Polarity on the Reactivity

The reactivity of aliphatic electrophilic substitution reactions is affected by many factors that can be better understood via experimental data and theoretical treatment combined. In this section, we will discuss some major factors that greatly influence the electrophilic substitution's rate in aliphatic compounds like substrate structure, leaving group and reaction medium.

➢ *Effect of Substrate Structure*

The electron-donating groups of the substrate decrease the rate of SE_1 reactions whereas electron-withdrawing groups show an opposite trend. This declining rate in the SE_1 pathway with electron-donating groups is quite normal for a reaction-type where the proton's dissociation is the rate-determining step (like in the case of acidic character). Jensen and Davis proved that the reactivity of alkyl groups is similar in SE_2-back as that for the SN_2 pathway, which can be attributed to the backside attack and steric hindrance, simultaneously.

Reactivity of alkyl groups for SE_2-back pathway

Furthermore, it has also been observed that the rate of front-mode electrophilic substitution in aliphatic compounds increases as the branching in the substrate increases increased, which can be attributed to the electron-releasing effect of the alkyl groups that makes the electron-deficient transition state more stable. However, it is also worthy to note that β-branching will reduce the substitution rate in SE_2-front because of the steric hindrance.

➢ *Effect of Leaving Group*

The ease of electrofuge's detachment in both types (SE_1 and SE_2) increases with the increasing polarity of the C−X bond. Nevertheless, if the leaving group is metallic in nature and metal has a valence greater than one, then any group attached to the metal center will affect its electrofugal ability. For instance, consider the case of $Me_3C-Hg-W$ (organomercurials) where the rate of reaction has decreased. The reason lies in the fact that although Hg and W have less electronegativity than carbon (which is why the C−Hg bond is polar), the C−Hg bond becomes less polar due to the higher electronegativity W than Hg. In other words, carbon will have a lower negative charge in the C−Hg bond when W is attached to Hg because tungsten will support Hg to hold the shared pair more firmly. Therefore, −HgMe will be a better leaving group than −HgCl.

$$Hg-t\text{-}Bu > Hg-i\text{Pr} > HgEt > HgMe$$

Leaving-group order

Furthermore, it is also worthy to note that carbon-based leaving groups support the SE_1 mechanism, whereas SE_2 or SE_i mechanisms are favored by metal-based leaving groups.

D DALAL
INSTITUTE

> ### *Effect of Solvent Polarity*

Just like the case of aliphatic nucleophilic substitution reactions, the raise in solvent polarity boosts the chances of the SE_1 pathway by supporting the ionization because of the better solvation of carbanions. However, if SE_2 and SE_1 reactions are competing with each other in parallel propagation, then less polar solvents favor the SE_2 pathway and polar solvents favor the SE_1 mechanism. Finally, If the nucleophilic character of the solvent is very small, the electrophile with properly placed assisting functionality might support the reaction; and therefore, motivating the reaction towards the SE_i pathway; otherwise, solvent polarity has little to no effect upon SN_i reactions.

❖ Problems

Q 1. Define electrophilic substitution reactions.

Q 2. What is the fundamental difference between SE_2 and SE_i mechanisms?

Q 3. Discuss the step-to-step mechanism of SE_1 reactions.

Q 4. How can electrophilic substitution occur via double bond shift?

Q 5. Discuss the effect of substrate structure and leaving group on the reactivity of electrophilic substitution in aliphatic compounds.

Q 6. How does the nature of nucleophiles affect the rate of aliphatic electrophilic substitution?

Q 7. Write down a short note on the solvent polarity's effect on electrophilic substitution reactions in aliphatic systems.

❖ Bibliography

1. M. B. Smith, *March's Advanced Organic Chemistry: Reactions, Mechanisms, and Structure*, John Wiley & Sons, Inc., New Jersey, USA, 2013.

2. D. Klein, *Organic Chemistry*, John Wiley & Sons, Inc., New Jersey, USA, 2015.

3. C. A. Coulson, B. O'Leary, R. B. Mallion, *Hückel Theory for Organic Chemists*, Academic Press, Massachusetts, USA, 1978.

4. M.S. Singh, *Reactive Intermediates in Organic Chemistry*, John Wiley & Sons, Inc., New Jersey, USA, 2014.

5. H. Zimmerman, *Quantum Mechanics for Organic Chemists*, Academic Press, New York, USA, 1975.

6. J. Clayden, N. Greeves, S. Warren, *Organic Chemistry*, Oxford University Press, Oxford, UK, 2012.

7. R. L. Madan, *Organic Chemistry*, Tata McGraw Hill, New Delhi India, 2013.

CHAPTER 8

Aromatic Electrophilic Substitution

❖ The Arenium Ion Mechanism

Electrophilic aromatic substitution (EAS) is the organic reaction in which an atom that is attached to an aromatic system (typically hydrogen) is replaced by an electrophile. This is quite possible in aromatic systems because there is π-electron density above and below the plane which is easily available for attacking electrophile; and nucleophilic attack is opposed because of π-cloud shields the carbon from such invasions.

Illustrative reaction: The general reaction showing the electrophilic substitution in aromatic compounds is shown below (E is electrophile).

Mechanism involved: The proposed mechanism for the reaction given above involves three steps which must be discussed before we give salient features of the same.

i) Attack of the electrophile on aromatic ring forming carbocation intermediate: In this step, the electrophile attacks the aromatic ring to form a resonance stabilized carbocations.

iii) Departure of the leaving group: In this step, the leaving group detaches itself from the aromatic ring to give rise to the final product.

Salient Features: The main features of the mechanism involved in electrophilic aromatic substitution type reactions are given below.

i) The aromatic electrophilic substitution (EAS) reactions follow second-order kinetics with the rate law as give below.

$$Rate = k[RX][E]$$

Where k is the rate constant. The symbol $[RX]$ and $[E]$ represent the molar concentration of the substrate and attacking electrophiles, respectively.

ii) The substituents like $-XY$, where X has loan pairs of electrons and no conjugated double bond in extended part (Y), increase the electron density at the ring; and therefore, are strongly activating. Furthermore, these types of groups donate those loan pair of electrons to the π-system, creating a negative charge density at para and ortho sites; and therefore, become ortho/para-directing via resonance. In other words, these groups make these positions more susceptible to an electron-deficient electrophile.

iii) The substituents like $-X=Y$, where three is conjugated double bond (w.r.t ring), decrease the electron density at the ring; and therefore, are strongly deactivating. Furthermore, these types of groups accept those electrons from the π-system, creating a positive charge density at para and ortho sites; and therefore, become meta-directing via resonance. In other words, these groups make m-positions more susceptible towards an electron-deficient electrophile.

iv) Alkyl groups as substituents increase the electron density at the ring via hyperconjugation; and therefore, are strongly activating. In other words, these groups make o- and p-positions more susceptible towards an electron-deficient electrophile.

iv) Although the halogens as substituents also have loan pairs of electrons, they decrease the electron density at the ring; and therefore, are strongly deactivating. This is because these types of groups donate those loan pair of electrons to the π-system via resonance (creating a negative charge density at para and ortho sites), but also withdraw electron density via inductive effect which is more dominant. In other words, these groups make these positions (o- and p- to the attached halogen) more susceptible towards the electron-deficient electrophile.

v) The addition of an entering group to a site in an aromatic compound that already has a substituent group (other than H), the attacking group may replace that substituent group but may also itself get expelled or moved to another position in a subsequent step, called '*ipso*-substitution'.

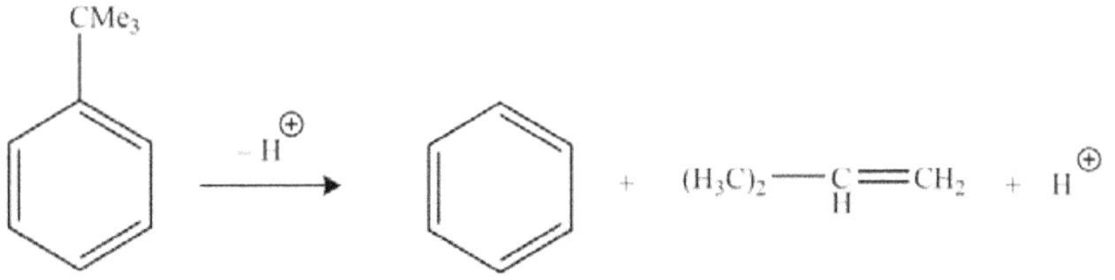

❖ Orientation and Reactivity

The orientation and reactivity in monosubstituted benzene for electrophilic substation reaction can be divided into four major types as discussed below.

➢ *Ring Activator with o-, p-Directing Influence*

The substituents like −XY, where X has loan pairs of electrons and no conjugated double bond in extended part (Y), increase the electron density at the ring; and therefore, are strongly activating. Furthermore, these types of groups donate those loan pair of electrons to the π-system, creating a negative charge density at para and ortho sites; and therefore, become ortho/para-directing via resonance. Nevertheless, one might ask that what if these groups are electron-withdrawing via inductive effect. The answer would be the same (i.e., ring activated) because the electron donation via mesmeric effect is much dominant than the inductive effect in these types of groups. Some typical examples of this type of group are given below with NH_2 as reference.

$$-NH_2, -NR_2, -NHR, -OH, -OR, -OCOR, -SR, -NHCOR$$

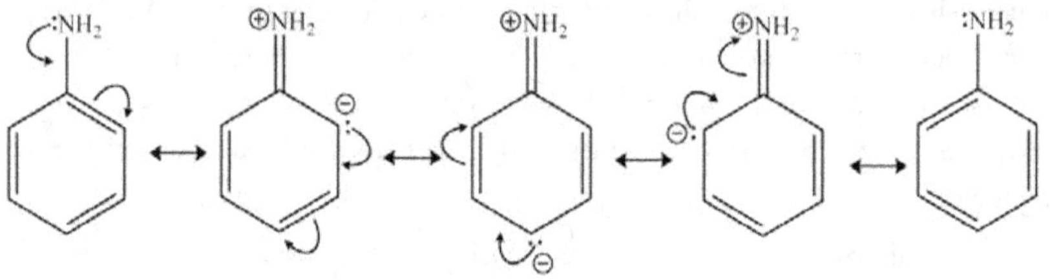

In other words, we can conclude that these groups make these positions more susceptible to an electron-deficient electrophile.

➢ *Ring Deactivator with m-Directing Influence*

The substituents like −X=Y, where three is conjugated double bond (w.r.t ring), decrease the electron density at the ring; and therefore, are strongly deactivating. Furthermore, these types of groups accept those electrons from the π-system, creating a positive charge density at para and ortho sites; and therefore, become meta-directing via resonance. Some typical examples of this type of group are given below.

$$-NO_2, -CN, -CHO, -COOH, -COOR, -CONH_2, -SO_3H$$

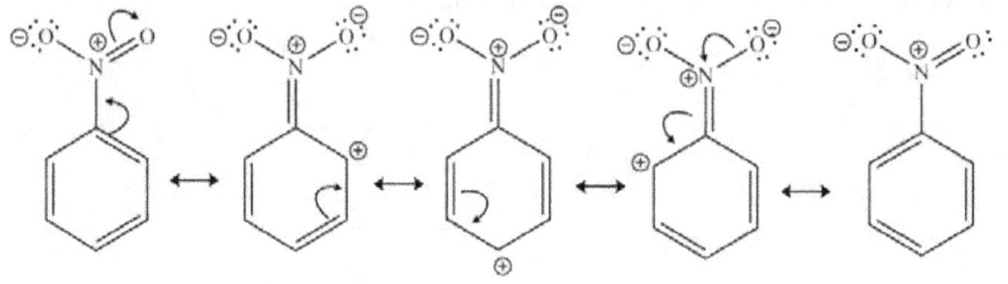

In other words, we can conclude that these groups make *m*-positions more susceptible towards an electron-deficient electrophile.

➢ *Ring Activator with o-, p-Directing Influence*

Alkyl groups as substituents increase the electron density at the ring via hyperconjugation; and therefore, are strongly activating. In other words, these groups make o- and p-positions more susceptible towards an electron-deficient electrophile. Furthermore, these groups are also electron releasing via inductive effect making them highly ring-activating species. Some typical examples of this type of substituents are given below with CH_3 as reference.

$$-NH_2, -NR_2, -NHR, -OH, -OR, -OCOR, -SR, -NHCOR$$

In other words, these groups make o- and p-positions more susceptible towards an electron-deficient electrophile.

➢ *Ring Deactivator with o-, p-Directing Influence*

Although the halogens as substituents also have loan pairs of electrons, they decrease the electron density at the ring; and therefore, are strongly deactivating. This is because these types of groups donate those loan pair of electrons to the π-system via resonance (creating a negative charge density at para and ortho sites), but also withdraw electron density via inductive effect which is more dominant. Some typical examples of these types of groups are given below with CH_3 as the reference.

$$-Cl^-, -Br^-, -I^-, -F^-$$

In other words, these groups make these positions (*o-* and *p-* to the attached halogen) more susceptible towards the electron-deficient electrophile.

❖ Energy Profile Diagrams

The general mechanism for the aromatic electrophilic substitution involves two main steps which must be discussed before we give an energy profile diagram of the same. The electrophile attacks the aromatic ring to form resonance stabilized carbocations, followed by the detachment of leaving group.

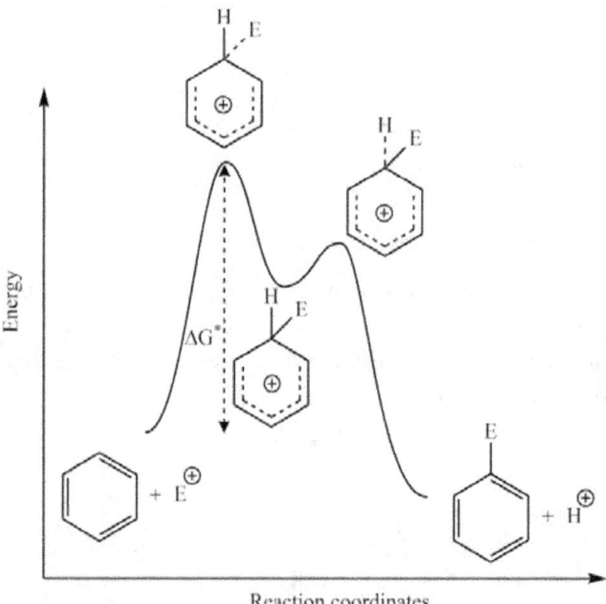

A typical energy level profile of various intermediates and transition states in the course of electrophilic substitution reactions in aromatic compounds is shown below.

Figure 1. Energy profile diagram for a typical aromatic electrophilic substitution.

Finally, it is also worthy to note that the rate of electrophilic substitution in already substituted aromatic compounds depends upon the height of the potential barrier which will be different for different types of attack i.e., *o-*, *m-* or *p*-attacks.

❖ The Ortho/Para Ratio

If there is an *o*-, the *p*-directing substituent on the benzene ring, the prediction of the relative amount of both products is quite difficult. This is because the attack at these sites is dictated by many factors like mathematical probability, charge density, steric effects. However, certain conclusions can still be made by considering only one factor at a time keeping others constant. In this section, we will discuss the factors affecting the ratio of o- and p-products one by one.

➤ *Statistical Factor*

One might say the relative ratio of o- and p-products should be 2:1 because there are two ortho sites but only one *p*- site. However, it is rarely obtained because the actual situation isn't that simple. For instance, consider the chlorination of methylbenzene where statistical treatment will suggest 2:1 o-/p- ratio but the actual ratio varies from 62:38 to 34:66.

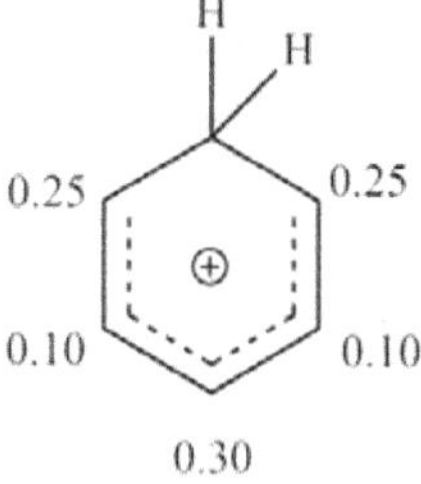

➤ *Charge Density*

Besides statistical factors, the distribution of charge density at different carbons is of great importance in deciding the attack, which sometimes can completely outrank the former factor. For instance, the charge densities in the phenonium ion are shown below.

It is obvious from the above diagram that a para substituent would show a more significant stabilizing effect on the neighboring carbons than an ortho group. This implies that the *p*-substituted product should be more than 1/3, and the *o*-substituted product should be less than 2/3, provided that other factors like steric hindrance same in both attacks. In hydrogen exchange reactions, the ratio of partial rate factor for o-position to the partial rate factor for o-position was found to be 0.865 which quite comparable to the ratio of charge densities (i.e., 0.25/0.30 = 0.833). Similar results were obtained (o-/p- = 67:33, though the effect was small) if the benzene had m-directing groups.

➤ *Steric Factors*

The attack at *p*-position is also supported by the factor steric hindrance which says that bulky will face more crowding at o- sites; and therefore, will try to avoid that approach. For instance, the nitration of toluene and tert-butylbenzene results in 58% o-product and 16% o-product, respectively.

73%   16%

➤ *Loan Pair on o-, p-Directing Group*

If there is a loan pair on the o-, p-directing group, the formation of *p*- product is favored at the cost of the ortho. This can be understood by considering the following example.

It is obvious that the A intermediate has an o-quinoid structure, whereas B intermediate has a p-quinoid structure; since o-Quinones are less stable than the p-Quinones, B should also be more stable than A, and hence, is expected to contribute more towards the resonance hybrid.

➤ *Substrate's Enclosing*

Although it seems impossible but exclusively p-products can be obtained enclosing the substrate molecules in a cavity where only the p- site is exposed to the electrophile. For instance, the chlorination of Anisole in solutions containing a cyclodextrin p-/o- ratio of 21.656; however, only 1.48 p-/o- ratio was obtained when cyclodextrin was absent.

Anisole

❖ *ipso*-Attack

Besides the attack at *o*-, *m*-, or *p*-positions, the incoming electrophile can also attack the substituted carbon itself called *ipso*-attack. For instance, consider the case when the NO_2^+ group attacks at substituted benzene as given below.

It is obvious that the carbonium ion formed after the attack can result in five different products via five different pathways.

Path I: The NO_2^+ can be lost from arenium resulting in the initial compounds. Since there is no net chemical change, this pathway is quite difficult to detect.

Path II: The NO_2^+ electrophile can show the 1, 2-rearrangement, and losing a proton afterward; and therefore, would result in a product which is equivalent to the product given by o-attack. Nevertheless, the amount of o-product cannot be predicted exactly though it is quite reasonable. Therefore, an ambiguity arises about the *o*-, *m*-, or *p*-positions because some of the 'calculated' o-activity may actually be the result of the ipso-attack followed by this pathway.

Path III: In this case, Z^+ is lost (instead of NO_2^+) from the arenium ion, resulting in a product that is equivalent to a simple electrophilic substitution where leaving group is not proton.

Path IV: A nucleophilic attack can happen at the carbonium ion, which may result in cyclohexadiene which is equivalent to 1,4-addition to the aromatic ring.

Path V: The Z^+ electrophile can show the 1, 2-rearrangement, and losing a proton afterward; and therefore, would result in a product that is equivalent to the product given by o-attack or Path II.

❖ Orientation in Other Ring Systems

The unsubstituted fused ring systems don't have all the sites equivalent like they were in benzene. Therefore, the electrophile will prefer some sites over the others even if single carbon isn't substituted. For instance, the 1st position in naphthalene is more reactive towards electrophilic substitution than the 2nd site. This is because the carbonium ion formed after the attack at 1st carbon has more resonating structures than we would have obtained for the carbonium ion formed by the attack at 2nd carbon.

Nevertheless, the β-position may also become of the highest priority under special experimental conditions. For instance, the sulfonation of naphthalene at 160° gives rise to a β-substituted product; whereas if the same reaction is carried out at 80°C, we get an β-substituted product.

less stable thermodynamically, more resonance stblized

more stable thermodynamically, less resonance stblized

Furthermore, it is also worthy to note that excessive resonance stabilization of arenium ions makes polycyclic conjugated ring systems more reactive toward electrophilic substitution than benzene; and the orientation can be rationalized on the basis of stability of same arenium ions. The same statement is true for heterocyclic ring systems.

❖ Quantitative Treatment of Reactivity in Substrates and Electrophiles

The quantitative treatment of reaction rate in case of electrophilic substitution with reference to substrate structure and attacking electrophile is given below.

➢ *Quantitative Treatments of Reactivity in the Substrate*

Unlike nucleophilic substitution where there is only one leaving group, there are many hydrogens that can leave in electrophilic substitution reactions; and therefore, the quantification of rate-ratios isn't that simple in the later category. For instance, we know the ratio of the total rate of acetylation of toluene to the total rate of acetylation of benzene, but have little to no idea about the rate ratios at individual positions. To do so, we need to carefully analyze the proportion of isomers obtained (for kinetically controlled reactions).

The partial rate factor (for a particular group and a certain reaction) may simply be defined as the rate of substitution at a single site relative to a single site in benzene.

To understand the definition more clearly, consider the case mentioned earlier i.e., acetylation of toluene. The magnitudes of partial rate factors for o-, m- and p-sites are 4.5, 4.8, and 749, respectively; which implies that the overall rate of acetylation of toluene at p-site is 749 times faster than what it is at a single site in benzene molecule. However, since there are six positions available in benzene, it will only be 125 (749/6) times faster than the total rate of benzene's acetylation. Furthermore, it is also very important to note that the group is considered as a position-activator if the partial rate factor comes out to be greater than unity, and vice-versa is also true. Hence, we can conclude that the methyl group is definitely a p-position activator since the partial rate factor for the same is 749 >1. Also, the magnitudes of partial rate factors may vary from reaction to reaction, or in the same reaction at different experimental conditions.

Now, after obtaining the partial rate factors, the ratio of possible isomers (when two or more substituents are attached on the cycle which have presumably independent effects) can be forecasted. For instance, if we consider the case of m-xylene, the theoretical values of partial rate factors at each site can be obtained by multiplying those from methylbenzene as shown below. Now we can easily calculate the ratio of different isomeric products arising from the acetylation of m-Xylene using the following expression.

$$x_i = f_i \Big/ \sum_{i=1}^{i=n} f_i$$

Where x_i and f_i are the are mole fraction of isomer arising from the ith type position and 'partial rate factor' for the ith site, respectively.

For instance, the mole fraction of 1-(2,6-dimethylphenyl) ethan-1-one (i.e., 2-isomer) will be obtained by dividing the partial rate factor of the same by the sum of partial rate factors of all positions i.e.

$$x_2 = \frac{20}{20 + 3375 + 23 + 3375} = 0.002944$$

To get the isomeric proportion on the percentage scale, it will be 100 i.e., 0.002944×100 =0.2944. Similarly, we can also obtain the percentage of mole fraction for other isomers.

Table 1. Experimental and calculated isomeric mole fractions (%) in the Acetylation of m-Xylene.

Site	Experimental	Theoretical
2nd	0	0.29
4th and 6th (same)	97.5	99.36
5th	2.5	0.34

Finally, the total rate-ratio for the acetylation of *m*-xylene to benzene can also be obtained theoretically sum of all 'partial rate factors' by six i.e.,

$$Rate\ ratio = \frac{20 + 3375 + 23 + 3375}{6} = 1132$$

It is pretty much obvious that that though the isomeric mole fractions are quite reasonable, the theoretical ratio (1132) is quite different from the actual one 347. This is because the real situation isn't that simple and many other factors like steric hindrance also play an important role.

> ➢ *Quantitative Treatment of Reactivity of the Electrophile*

Different electrophiles have different magnitudes of reactivity; for instance, the NO_2^+ ion can react with benzene and also with aromatic rings having deactivating substituents, but the diazonium ions react only with aromatic rings having strongly activating substituents. The Hammett equation can be used to analyze this preferred behavior of attacking electrophiles. To do so, recall the general form of Hammett equation for the cases of *m*- and *p*-XC_6H_4Y (X a variable substituent, and Y is the 'reaction spot' not group) i.e.

$$\log \frac{k}{k_0} = \sigma\rho = \log \frac{K}{K_0} \tag{1}$$

where k and k_0 are the rate constants for the substituent X and X = H; ρ and σ are the constants for reaction conditions and substituent X, respectively. The symbol K and K_0 are the equilibrium constant for the group X and X = H. Furthermore, it should also be noted that since the Hammett equation is for monosubstituted benzene rings, the substrate becomes a simple benzene ring if X = H because 'Y' the reaction spot and not any ring-attached group.

However, the equation given above is for overall rate constants (I mean, for all the available sites); and therefore, if we want to use this equation for single-site comparison, we will need to convert overall rates into partial rate factors first. This can be done by dividing k_0 by 6 (benzene has six equivalents 'reaction spots' i. e. Y), and by dividing k by 2 and 1 for m- and p- electrophile-attack, respectively (C_6H_5X have two meta and one para site). Now one might ask why we would need to convert overall rate constants into 'partial rate factors, then the answer is the same as we have discussed in the previous section; to find isomeric proportions. Now the isomeric proportions proposed by this equation were quite accurate if X is an electron withdrawing in nature. Nevertheless, large deviations were observed X is electron-donating in nature. In other words, partial rate factors derived from k values given by equation (3) weren't very reliable if group X are electron-donating in nature; and therefore, a modification was needed.

$$f_m = k_m/2$$

$$f_0 = k_0/6$$
All sites are equally reactive

$$f_p = k_p/1$$
One p- site

An American chemist, H.C. Brown, solved the problem by introducing modified σ-values labeled as $σ^+$, where the plus symbol represents a positive charge appearing in the transition state. Substituents with negative and positive $σ^+$-values position activating and deactivating, respectively. In other words, the modified equation can be used to rationalize the aromatic substitution at rings with electron-donating, as well as, electron-withdrawing groups.

Now the question arises, what does the parameter ρ correlates because σ (or $σ^+$) values are almost exclusively correlated with the reactivity of substrate structure in electrophilic substitution. The answer is, it is connected with the reactivity of attacking electrophile and how it affects the overall reaction-stability. Nevertheless, besides the electrophile, ρ-values may also change with the reaction's experimental conditions. A smaller negative ρ-value implies a less reactive electrophile and vice-versa. Furthermore, it should also be kept in mind that this is true only for meta- and para-sites because the Hammett equation isn't applicable to ortho sites.

Brown developed his idea on the basis of the fact that the reactivity of a particular species shows an inverse variation with selectivity. He observed that all the electrophiles can be classified on the basis of their choice of the attack on benzene vs toluene and ortho- vs para positions (in toluene).

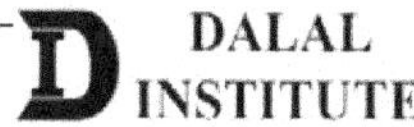

Table 2. Comparative reaction rates and products' ratio in some typical electrophilic substitutions on benzene and toluene.

Reaction-type	m-isomer	p-isomer	$k_{toluene}/k_{benzene}$
Bromination	0.3	66.8	605
Chlorination	0.5	39.7	350
Benzoylation	1.5	89.3	110
Nitration	2.8	33.9	23
Mercuration	9.5	69.5	7.9
Isopropylation	25.9	46.2	1.8

It is obvious from the above data that if an electrophile prefers to attack toluene over benzene, it will also prefer to attack p-site to give major product, and vice-versa. Brown formulated this observation into the following formula.

$$S_f = \log \frac{f_p}{f_m} \tag{2}$$

Where S_f is the selectivity and $f_{p,m}$ are the partial rate factors. The conclusion from equation (2) is that more reactive species tend to attack meta-position rather than para, and vice-versa is also true. Furthermore, it also possible to modify the Hammett-brown equation using the above result to prove the following.

$$\log f_p = \left(\frac{\sigma_p^+}{\sigma_p^+ - \sigma_m^+} \right) S_f \tag{3}$$

and

$$\log f_m = \left(\frac{\sigma_m^+}{\sigma_p^+ - \sigma_m^+} \right) S_f \tag{4}$$

and

$$S_f = \left(\sigma_p^+ - \sigma_m^+ \right) \rho \tag{5}$$

Partial rate factors and ρ-values obtained by employing equation (3–4) were quite comparable with experimental data.

❖ Diazonium Coupling

The diazonium coupling may simply be defined as the reaction between a diazonium salt and an aromatic compound to give rise to an azo compound.

In this type of aromatic electrophilic substitution, the aryldiazonium cation acts as the electrophile whereas the activated arene behaves as a nucleophile; and the resulting diazonium compound is also aromatic in most cases.

Illustrative Reaction: The general chemical reaction showing the diazonium coupling in aromatic compounds is shown below.

Mechanism Involved: The mechanism involved in the diazonium coupling reaction that gives rise to Orange-I dye is shown below.

More resonating structure

Orange I

Some of the most important applications of diazonium coupling in industries include the synthesis of dyes, pigments, and lakes.

DALAL INSTITUTE

❖ Vilsmeier Reaction

The Vilsmeier (or Vilsmeier-Haack) reaction may simply be defined as the chemical transformation of a substituted amide with phosphorus oxychloride and an electron-rich arene to result in an aryl ketone or aldehyde.

The reaction was invented by Anton Vilsmeier and Albrecht Haack, and therefore, is also named after them. The common examples of the Vilsmeier-Haack reaction include the reaction of benzanilide and dimethylaniline with phosphorus oxychloride to give an unsymmetrical diaryl ketone.

Illustrative Reaction: One of the most popular reactions of this type is the reaction of anthracene with N-methylformanilide, also using phosphorus oxychloride, gives 9-anthracenecarboxaldehyde.

Mechanism Involved: The reaction of a substituted amide with phosphorus oxychloride gives a substituted chloroiminium ion, also called the Vilsmeier reagent.

The initial product is an iminium ion, which is hydrolyzed to the corresponding ketone or aldehyde during workup.

❖ Gattermann-Koch Reaction

The Gattermann (or Gattermann-Koch) reaction may simply be defined as a chemical transformation in which aromatic compounds are formylated by a mixture of carbon monoxide (CO) and hydrogen chloride (HCl) in the presence of a Lewis acid catalyst such as $AlCl_3$.

The reaction was invented by Ludwig Gattermann and Julius Arnold Koch, and therefore, is also named after them. Furthermore, it is also important to note that this reaction is a variant of the Gattermann reaction where HCN is replaced by CO.

Illustrative Reaction: The general chemical reaction showing the Gattermann-Koch reaction in aromatic compounds is shown below.

Mechanism Involved: The mechanism involved in the Gattermann-Koch reaction in aromatic compounds is shown below.

It should also be noted that this reaction cannot be applied to phenol and phenol ether substrates which were common in simple Gattermann reaction. Initially was thought that formyl cation first forms formyl chloride as an intermediate but nowadays it is assumed that the formyl cation directly.

❖ Problems

Q 1. Define arenium ion. Discuss the organic reaction mechanism involving them.

Q 2. What are ring activators and deactivators? Explain with suitable examples.

Q 3. Draw and discuss the typical energy level diagram for the electrophilic substation.

Q 4. Define ortho/para ratio. What factors affect this ratio?

Q 5. State and explain *ipso*-attack.

Q 6. Discuss the directive influence of substituents in ring systems other than benzene.

Q 7. Explain the quantitative treatment of reactivity in substrates and electrophiles.

Q 8. What is diazonium coupling?

Q 9. Discuss the mechanism of the Vilsmeir reaction.

Q 10. Write a short note on the Gattermann-Koch reaction.

❖ Bibliography

1. M. B. Smith, *March's Advanced Organic Chemistry: Reactions, Mechanisms, and Structure*, John Wiley & Sons, Inc., New Jersey, USA, 2013.

2. H. Zimmerman, *Quantum Mechanics for Organic Chemists*, Academic Press, New York, USA, 1975.

3. J. Clayden, N. Greeves, S. Warren, *Organic Chemistry*, Oxford University Press, Oxford, UK, 2012.

4. R. L. Madan, *Organic Chemistry*, Tata McGraw Hill, New Delhi India, 2013.

5. D. Klein, *Organic Chemistry*, John Wiley & Sons, Inc., New Jersey, USA, 2015.

6. C. A. Coulson, B. O'Leary, R. B. Mallion, *Hückel Theory for Organic Chemists*, Academic Press, Massachusetts, USA, 1978.

7. M.S. Singh, *Reactive Intermediates in Organic Chemistry*, John Wiley & Sons, Inc., New Jersey, USA, 2014.

CHAPTER 9

Aromatic Nucleophilic Substitution

❖ The ArSN₁, ArSN₂, Benzyne and S$_R$N₁ Mechanisms

An aromatic nucleophilic substitution in organic chemistry may simply be defined as a chemical reaction where the nucleophile displaces a good leaving group, such as a halide, on an aromatic ring. The aromatic nucleophilic substitution can primarily occur via three different routes as given below.

> #### ➤ *ArSN₁ or Aryl Cation Mechanism*

The unimolecular nucleophilic substitution on aromatic rings is mainly given by aromatic diazonium salts. The typical reaction of such type is given below.

Illustrative reaction: The typical reaction involving nucleophilic substitution in aromatic compounds is shown below.

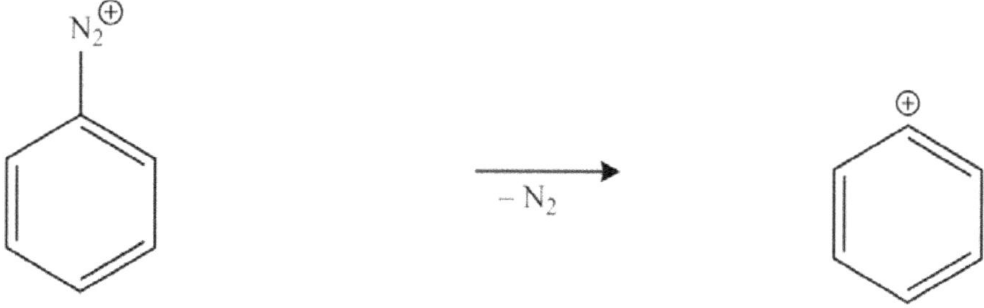

Mechanism involved: The proposed mechanism for the reaction given above involves two steps which must be discussed before we give salient features of the same.

i) Formation of aryl cation: Now although the aryl carbocation is highly unstable, its formation is still favored due to the high stability of dinitrogen (i.e., good leaving group).

Now although the aryl carbocation is highly unstable, its formation is still favored due to the high stability of dinitrogen (i.e., good leaving group).

ii) Attack by the Nucleophile:

Now since the faces of the carbocations formed are homotopic, the Nu⁻ can attack from either side to give the same product.

Salient Features: The main features of the mechanism involved in aromatic nucleophilic substitution unimolecular or $ArSN_1$ type reactions are given below.

i) $ArSN_1$ reactions follow first-order kinetics with the rate law

$$Rate = k[RX]$$

Where k is the rate constant and $[RX]$ represents the molar concentration of the substrate.

ii) The presence of $+R$ groups at *ortho* and *para* positions raises the reactivity of the substrate and vice-versa.

> ➤ ***ArSN₂ or Addition-Elimination Mechanism***

The bimolecular nucleophilic substitution on aromatic rings is most common among the class. The typical reaction of such type is given below.

Illustrative reaction: The typical reaction involving this type of mechanism is given below.

Mechanism involved: The proposed mechanism for the reaction given above involves two steps which must be discussed before we give salient features of the same.

i) ipso-addition of the nucleophile:

Now although an ion is no longer an aromatic species; however, it is relatively stable due to the delocalization of the negative charge over 3 carbon atoms by the pi system.

ii) Elimination of the leaving group:

Salient Features: The main features of the mechanism involved in aromatic nucleophilic substitution bimolecular or $ArSN_2$ type reactions are given below.

i) $ArSN_2$ reactions follow second-order kinetics with the rate law

$$Rate = k[RX][Nu]$$

Where k is the rate constant. The symbol $[RX]$ and $[Nu]$ represent the molar concentration of the substrate and attacking nucleophiles, respectively.

ii) The reactivity increases as the leaving group gets better.

iii) The rate of the substitution increases as the $-I$ or $-R$ effect of the groups attached *o-* and *p*-positions increases.

iv) The reactivity is also proportional to the electronegativity of the heteroatom (if any) in the ring.

v) The $ArSN_2$ reactions are favored in polar aprotic solvents.

> ➢ **Aryne (Benzyne) or Elimination-Addition Mechanism**

The elimination-addition mechanism involves a highly unstable intermediate called benzyne (dehydrobenzene). A typical reaction of such type is given below.

Illustrative reaction:

Steps involved: The proposed mechanism for the reaction given above involves two steps which must be discussed before we give salient features of the same.

i) First step is the elimination of proton ortho to the substituent present and formation of benzyne:

ii) Attack of amide ion on the benzyne intermediate:

iii) Abstraction of the proton from ammonia:

Salient Features: The main features of the mechanism involved in aromatic nucleophilic substitution via benzyne are given below.

i) At least one hydrogen must be present at ortho position in the inactivated aryl halide.

ii) The incoming group may or may not occupy the position vacated by the leaving group i.e. cine substitution.

> **Substitution Radical Nucleophilic Unimolecular ($S_{RN}1$)**

Radical-nucleophilic aromatic substitution or $S_{RN}1$ in organic chemistry is a type of substitution reaction in which a certain substituent on an aromatic compound is replaced by a nucleophile through an intermediary free radical species.

Illustrative reaction:

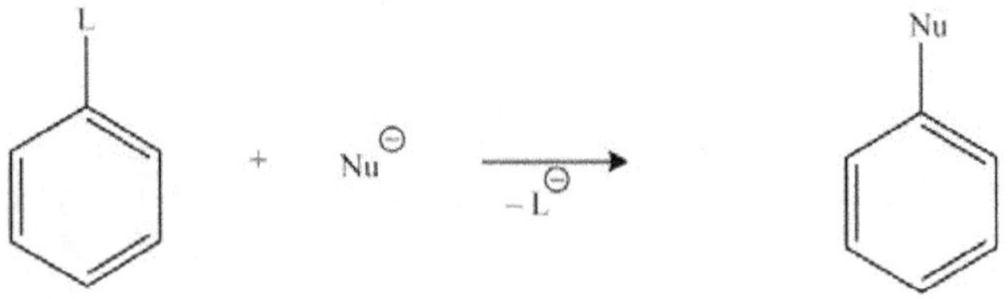

Mechanism involved: The proposed mechanism for the reaction given above involves two steps which must be discussed before we give salient features of the same.

i) Formation of radical anion: The aryl halide accepts an electron from a radical initiator to form a radical anion.

ii) Transformation of radical anion into aryl radical:

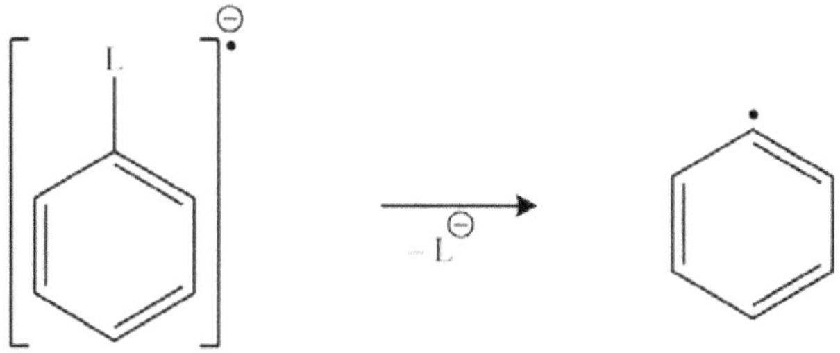

iii) Attack of the nucleophile on the aryl radical:

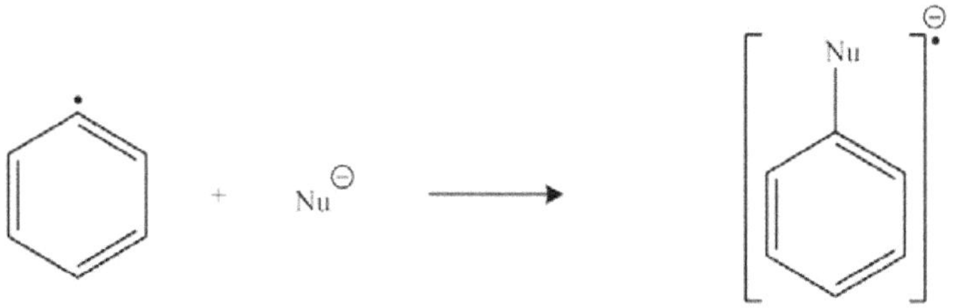

iv) Transfer of electron to new aryl halide:

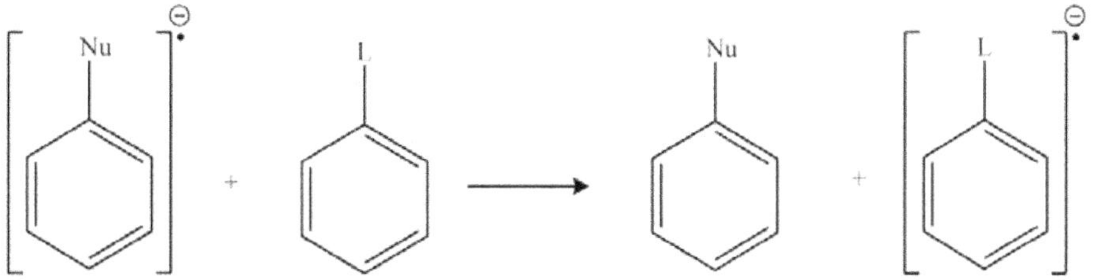

Salient Features: The main features of the mechanism involved in $S_{RN}1$ (substitution radical nucleophilic unimolecular) type reactions are given below.

i) $S_{RN}1$ reactions follow first-order kinetics with the rate law

$$Rate = k[RX]$$

Where k is the rate constant and $[RX]$ represents the molar concentration of the substrate.

ii) The phenyl radical can also abstract any loose proton to form arene in a chain termination reaction to yield the final product.

❖ Reactivity – Effect of Substrate Structure, Leaving Group and Attacking Nucleophile

In this section, we will discuss the effect of substrate structure, leaving the group and attacking nucleophiles on the reactivity of nucleophilic substitution in aromatic compounds.

➤ *Effect of Substrate Structure on the Reactivity of Aromatic Electrophilic Substitution*

Just like the case of aromatic electrophilic substitution, substrate structure also affects the reactivity of aromatic nucleophilic substitution w.r.t orientation as well as ring-activation. However, orientation-effect was of more importance in the electrophilic case four or five hydrogens were able that act as leaving groups in comparison to the nucleophilic-case where the typical number of leaving group one. Consequently, substrate reactivity is primarily studied w.r.t other molecules rather than the same species.

Aromatic nucleophilic substitution reactions are typically opposed by electron attracting groups but enhanced by electron-withdrawing groups mainly at ortho and para positions to the leaving group, which is just the reverse order for electrophilic substitutions. Therefore, all the groups can be arranged in ascending or descending order of their activating (or deactivating) abilities.

Groups with nitrogen atoms activating in nature w.r.t to o- and p-position with N_2^+ as the strongest activator. Also, $-NO_2$ is the most common activating group; whereas 2,4-dinitrophenyl halides and 2,4,6-trinitrophenyl halides are considered as the most common substrates. Furthermore, substrates without activating groups are largely useless to serve in $ArSN_{1,2}$ pathways, which can be attributed to the presence of 2 antibonding electrons in the ring. If attached groups are electron-withdrawing, they can activate the reaction by withdrawing electron density, and therefore, will stabilize the transition states (or intermediate). Aromatic electrophilic substitutions of type $ArSN_{1,2}$ are also supported if a transition metal is connected to the aromatic ring. Finally, the Hammett equation can also be modified for aromatic electrophilic substitution with the difference of σ^- instead of σ^+.

As far as the benzyne pathway is concerned, the reactivity w.r.t to the substrate can be factored in categories; one is the direction the aryne forms in, and the second is the presence of groups at ortho or para positions to the leaving group.

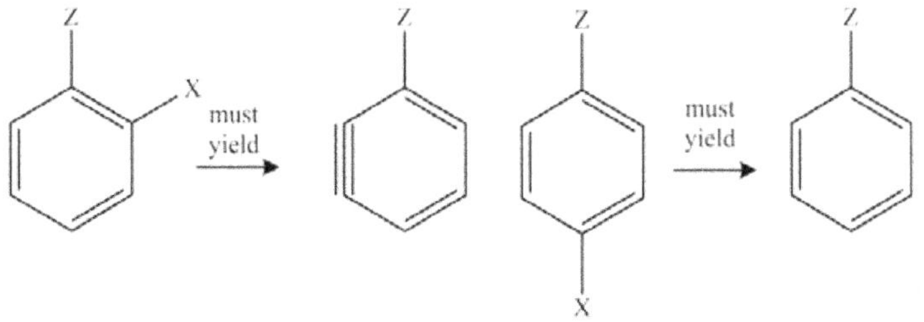

However, if the aryne substrate is *m*-substituted, the nucleophilic substitution can occur via two different routes as shown below.

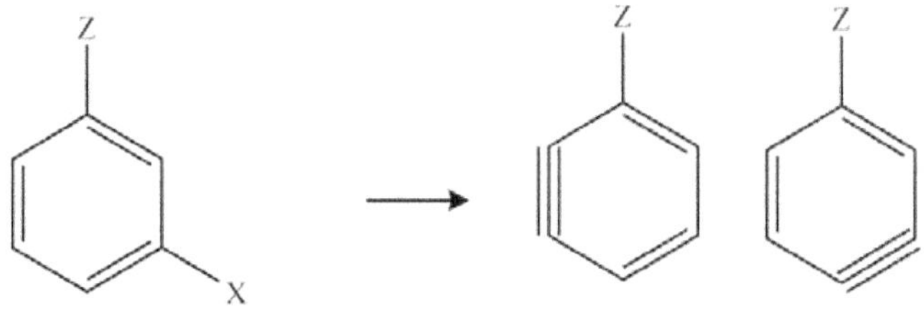

These types of attacks occur via the removal of more acidic hydrogen; and because the acidity is correlated to the magnitude of field effect of the $-Z$ group, we can conclude that Z with more electron-attracting character will support the removal of the o-hydrogen whereas the Z with electron-donating character will elimination of the *p*-hydrogen atom. On the other hand, the other route says that though the aryne attacked at two sites, the favored site for the nucleophile to attack will be the one that gives rise to a stabler carbanion (which is also a function of Z's field effect).

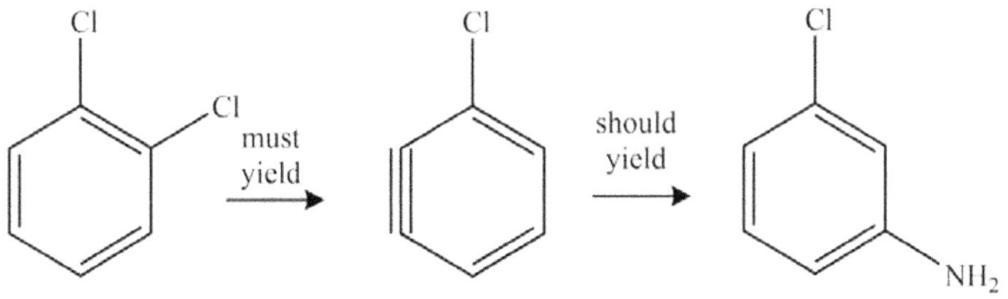

In other words, we may conclude that the carbanion with the negative charge closer to Z will be more stable than others.

> ### *Effect of Leaving Group on the Reactivity of Aromatic Electrophilic Substitution*

Typical leaving groups for nucleophilic substitutions in aromatic compounds are X^- (halides), sulfonate, sulfate, NR_3^+, etc., which also act as a leaving-group for nucleophilic substitution in aliphatic compounds. Nevertheless, some groups are common; for instance, OAr, SO_2R, NO_2, OR, and SR, which do not act as leaving-group in aliphatic compounds but exclusively in aromatic systems. The typical order of leaving group power in aromatic nucleophilic substitution is given below.

$$F > NO_2 > OTs > SOPh > Cl > Br > I > N_3 > NR_3^+ > OAr, OR, SR, NH_2$$

It is obvious from the order given above that F and NO_2 are exceptionally good leaving groups in the aromatic nucleophilic substitution. Nonetheless, it should also be kept in mind that a better leaving group doesn't always lead to the preferred product because the nature of attacking nucleophile also decide the final departed result. For example, Cl is better a better leaving group than OR but the attack of NH_2^- on $C_6Cl_5OCH_3$ always results in $C_6Cl_5NH_2$ which is contrary to expectations.

Routinely, the formation of an inorganic ester can also make OH act like a leaving group. It can also be seen that leaving group order give here different than aliphatic nucleophilic substitution because the first step (rate-determining) in the present case is assisted by prevalent $-I$ effects.

> ### *Effect of Attacking Nucleophile on the Reactivity of Aromatic Electrophilic Substitution*

Just like the order of leaving groups, the manifestation of a universal nucleophilicity order is very hard; though an approximation can still be made.

$$NH_2 > Ph_3C > PhNH^- > ArS > RO > R_2NH > ArO > OH > ArNH_2 > NH_3 > I > Br > Cl > H_2O > ROH.$$

The nucleophilic strength also depends upon the base strength and shows an increase we select the attacking atom more down the group. Nevertheless, like other physical concepts, some exceptions are always present like a stronger basic character of OH than ArO but weaker nucleophilicity.

It is also worthy to remember that even though the nucleophilicity order is not invariant, it still finds application in a wide range of synthetic and practical applications including the assignment of electrophilicity parameters in the case of electron-short heteroarenes.

❖ The von Richter, Sommelet-Hauser, and Smiles Rearrangements

In this section, we will discuss some common types of rearrangement reactions that involve aromatic electrophilic substitution.

> ### *Von Richter Reaction*

The von Richter reaction may simply be defined as the chemical transformation where aromatic nitro compounds react with KCN in aqueous ethanol to yield cine substitution product by a carboxyl group.

This reaction was invented by a German chemist Victor von Richter in 1871; and therefore, it is also named after him; and It is practically unimportant because of low yield and by-products formation.

Illustrative reaction: Common example is the conversion of *p*-bromonitrobenzene into m-bromobenzoic acid.

Mechanism involved: The most widely accepted mechanism for the von Richter reaction was given by Rosenblum in 1960 when he employed ^{15}N labeling experiments.

In the first step, the carbon ortho to the nitro group is attacked by cyanide; which is followed by ring-closing through nucleophilic invasion at the cyano group; finally resulting in the rearomatization of the imidate intermediate. The opening of the cycle via nitrogen-oxygen bond-breaking gives rise to an ortho-nitroso benzamide that recyclizes to yield a compound with a nitrogen-nitrogen bond. The elimination of H_2O results in a cyclic azoketone, which undergoes nucleophilic invasion by OH^- to result in a tetrahedral intermediate. This intermediate breakdowns with the removal of the azo group to give an aryldiazene with an o-carboxylate group, which squeezes out dinitrogen gas to be able to have the anionic form of the benzoic acid.

> ### *Sommelet-Hauser Rearrangement*

The Sommelet-Hauser rearrangement may simply be defined as the rearrangement reaction of certain benzyl quaternary ammonium salts where the reagent used is sodium amide (or alkali amide) and the reaction results in the N,N-dialkylbenzylamine with a new alkyl substituent in the aromatic o-position.

Now because the final product is a benzylic tertiary amine, it can further undergo alkylation followed by reoccurring rearrangement, and then repeating the process until the blockage of o-site.

Illustrative reaction: The common type of this type of rearrangement is benzyltrimethylammonium iodide that rearranges in the presence of $NaNH_2$ to give the o-methyl derivative of N,N-dimethylbenzylamine.

Mechanism involved: The widely accepted mechanism for the Sommelet-Hauser mechanism is depicted below clear the propagation picture.

The benzylic methylene hydrogen is acidic and deprotonation occurs to give the benzylic ylide, which is in equilibrium with another ylide formed via deprotonation of one ammonium methyl substituent. Nevertheless, the second ylide is available in much minor quantity, it shows a 2,3-sigmatropic rearrangement as it has a more reactive character than the initial one, and will show subsequent aromatization to give rise to the end product.

> ### *Smiles Rearrangements*

The Smiles rearrangement may simply be defined as an intramolecular aromatic nucleophilic substitution (ArSN) reaction, where the breaking of a C–X single bond and creation of a new C–C or C–X bond take place via ipso substitution.

This reaction was invented by a British chemist Samuel Smiles; and therefore, it is also named after him. Its dependence upon leaving group, nucleophile, and the cycle-size of transition state makes it suitable for arene functionalization.

Illustrative reaction: The general reaction showing this type of transformation (Smiles rearrangement) is shown below.

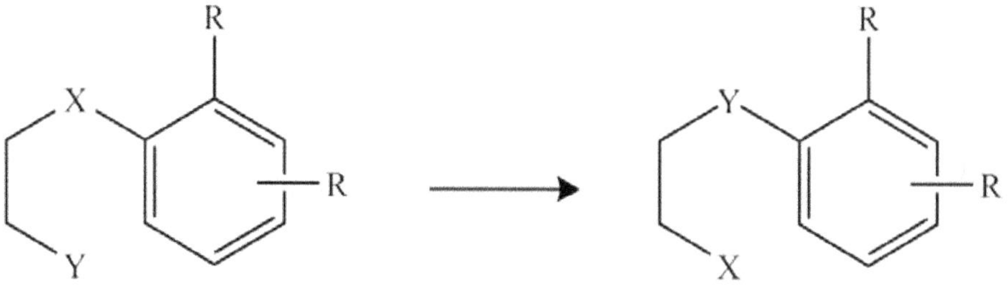

where X represents a sulfide, a sulfone, an ether, or any group capable of displacing from a negatively charged arene. On the other hand, Y represents a group that is capable to act as a strong nucleophile (like alcohol, thiol, or amine).

Mechanism involved: The widely accepted mechanism for the Smiles mechanism is depicted below clear the propagation picture.

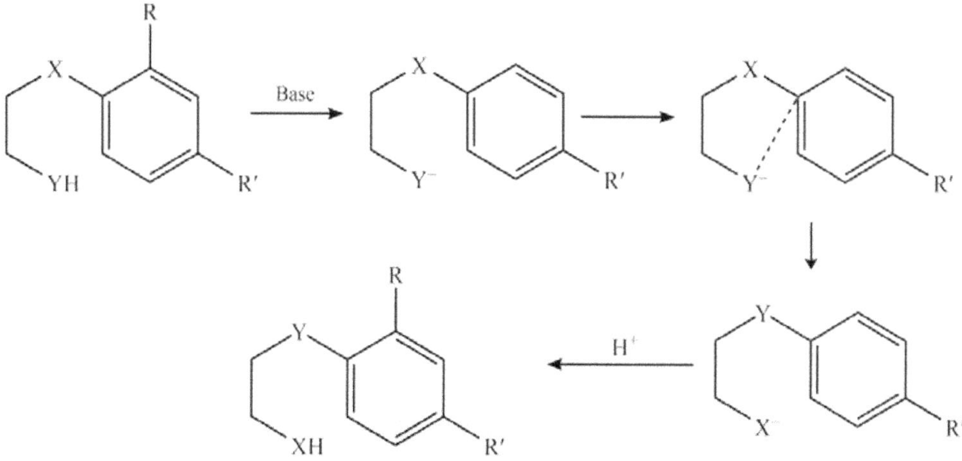

Just like other aromatic nucleophilic substitutions, an electron-withdrawing group at ortho position is also required for activation. However, in Truce-Smiles rearrangement, no additional activation is required because the income g nucleophile is enough strong (i.e., organolithium).

❖ Problems

Q 1. What is aromatic nucleophilic substitution? Explain with special reference to the benzyne mechanism.

Q 2. Give five points of differences between $ArSN_1$ and $ArSN_2$ mechanism.

Q 3. Discuss the effect of Substrate Structure, Leaving Group, and Attacking Nucleophile on the overall Reactivity of aromatic nucleophilic substitution reactions.

Q 4. Define von Richter reaction. Also, explain its mechanism.

Q 5. Discuss Smiles rearrangement in detail.

Q 6. What is the Sommelet-Hauser reaction?

Q 7. Give SRN_1 mechanism of nucleophilic substitution reactions.

❖ **Bibliography**

1. M. B. Smith, *March's Advanced Organic Chemistry: Reactions, Mechanisms, and Structure*, John Wiley & Sons, Inc., New Jersey, USA, 2013.

2. H. Zimmerman, *Quantum Mechanics for Organic Chemists*, Academic Press, New York, USA, 1975.

3. J. Clayden, N. Greeves, S. Warren, *Organic Chemistry*, Oxford University Press, Oxford, UK, 2012.

4. R. L. Madan, *Organic Chemistry*, Tata McGraw Hill, New Delhi India, 2013.

5. C. A. Coulson, B. O'Leary, R. B. Mallion, *Hückel Theory for Organic Chemists*, Academic Press, Massachusetts, USA, 1978.

6. M.S. Singh, *Reactive Intermediates in Organic Chemistry*, John Wiley & Sons, Inc., New Jersey, USA, 2014.

7. D. Klein, *Organic Chemistry*, John Wiley & Sons, Inc., New Jersey, USA, 2015.

CHAPTER 10

Elimination Reactions

❖ The E_2, E_1 and E_1CB Mechanisms

No organic reaction is capable of giving 100% yield of a single product only, including nucleophilic substitutions. The reason for this type of behavior is the fact that nucleophilic substitution reactions are in direct competition with the elimination reactions; and if the reagent and substrate carefully with a fine-tuning of experimental conditions, elimination yield can even surpass the nucleophilic substitution product. In this section, we will discuss some important elimination mechanisms like E_2, E_1, and E_1CB-type.

➤ E_2 (Bimolecular Elimination) Mechanism

An E_2 elimination reaction is a type of organic reaction in which two substituents are removed from a molecule in a one-step concerted mechanism. The numbers refer not to the number of steps in the mechanism, but rather to the kinetics of the reaction which means that E_2 is a bimolecular (second-order) reaction.

Illustrative reaction: The reaction of ethoxide ion with ethyl bromide falls into this category because the rate of reaction depends only upon the concentration of the substrate as well as of reagent.

$$H_3C-CH_2-Br \;+\; C_2H_5O^{\ominus} \longrightarrow H_2C{=}CH_2 \;+\; Br^{\ominus} \;+\; C_2H_5OH$$

Bromoethane   Ethene   Ethanol

Mechanism involved: The proposed mechanism for the reaction given above involves one step which must be discussed before we give salient features of the same.

$$Br-CH_2-CH_2-H + C_2H_5O^{\ominus} \xrightarrow{\text{slow}} Br{\sim}C{\sim}C{\sim}H{\sim}OC_2H_5$$

Transition State

$$H_2C{=}CH_2 \;+\; Br^{\ominus} \;+\; C_2H_5OH$$

It is worthy to recall that even though the reaction is shown as a two-step process, it is actually concerted in nature where the bond breaking and bond-making occur simultaneously.

Salient Features: The main features of the mechanism involved in bimolecular elimination reactions are given below.

i) E_2 reactions follow second-order kinetics with the rate law

$$Rate = k[RX][B]$$

Where k is the rate constant. The symbol $[RX]$ and $[B]$ represent the molar concentration of the alkyl halide and base, respectively.

ii) E_2 typically takes place with primary alkyl halides but is possible with some secondary alkyl halides.

iii) Because of the formation of a π-bond in the course of the E_2 mechanism, the two leaving groups (often a halogen and hydrogen) must be antiperiplanar. The reason for this is the fact that the antiperiplanar transition state will have lower energy (staggered-conformation) than a synperiplanar transition state (eclipsed conformation). In other words, the E_2 reactions are favored by staggered conformation whereas the E_1 by eclipsed one.

iv) E_2 typically uses a strong base. It must be strong enough to remove the weakly acidic hydrogen.

v) E_2 competes with the SN_2 reaction mechanism if the base can also act as a nucleophile (true for many common bases).

> ### E_1 *(Unimolecular Elimination) Mechanism*

An E_1 elimination reaction is a type of organic reaction in which two substituents are removed from a molecule in a two-step mechanism. The numbers refer not to the number of steps in the mechanism, but rather to the kinetics of the reaction which means that E_1 is a unimolecular (first-order) reaction.

Illustrative reaction: The reaction of NaOH with tert-butyl bromide falls into this category because the rate of reaction depends only upon the concentration of the substrate.

$$H_3C \!-\! \underset{\underset{\displaystyle CH_3}{|}}{\overset{\overset{\displaystyle CH_3}{|}}{C}} \!-\! Br \;+\; \overset{\ominus}{OH} \;\longrightarrow\; H_3C \!-\! \underset{\underset{\displaystyle CH_3}{|}}{C} \!=\! CH_2 \;+\; \overset{\ominus}{Br} \;+\; H_2O$$

Mechanism involved: The proposed mechanism for the reaction given above involves two steps which must be discussed before we give salient features of the same.

i) *Generation of carbocation:* The aryl halide accepts an electron from a radical initiator to form radical anion.

$$H_3C \!-\! \underset{\underset{\displaystyle CH_3}{|}}{\overset{\overset{\displaystyle CH_3}{|}}{C}} \!-\! Br \;\rightleftharpoons\; H_3C \!-\! \underset{\underset{\displaystyle CH_3}{|}}{\overset{\overset{\displaystyle CH_3}{|}}{C}} \!\sim\! Br \;\overset{Slow}{\rightleftharpoons}\; H_3C \!-\! \underset{\underset{\displaystyle CH_3}{|}}{\overset{\overset{\displaystyle \oplus}{C}}{}} \!-\! CH_3 \;+\; \overset{\ominus}{Br}$$

Transition State I

ii) Removal of proton:

Transition State II

Salient Features: The main features of the mechanism involved in unimolecular elimination reactions are:

i) E1 reactions follow first-order kinetics with the rate law

$$Rate = k[RX]$$

Where k is the rate constant. The symbol $[RX]$ represents the molar concentration of the substrate.

ii) The E_1 reactions generally occur with tertiary alkyl halides but are possible with some 2° alkyl halides.

iii) The E_1 reaction typically takes place in the presence of a weak base or the base can be absent at all.

iv) E_1 reactions compete with SN1 reactions because they share the same intermediate.

> ➢ **E_1CB (Conjugate Base Elimination) Mechanism**

The E_1CB elimination reaction is a type of chemical transformation where the elimination occurs in the presence of a strong base, and the hydrogen to be removed is comparatively acidic, while the leaving group (such as -OH or -OR) is a relatively poor one.

Illustrative reaction: An example of the E_1CB reaction mechanism in the degradation of a hemiacetal under basic conditions.

Mechanism involved: The proposed mechanism for the reaction given above involves two steps which must be discussed before we give salient features of the same.

i) Proton abstraction: The aryl halide accepts an electron from a radical initiator to form a radical anion.

ii) Detachment of leaving group:

Salient Features: The main features of the mechanism involved in unimolecular elimination conjugate-base reactions are:

i) E_1CB reactions follow first-order kinetics with the rate law

$$Rate = k[RX]$$

Where k is the rate constant. The symbol $[RX]$ represents the molar concentration of the substrate.

ii) The substrate must have acidic hydrogen on its β-carbon and a relatively poor leaving group on the α-carbon.

❖ Orientation of the Double Bond

In this section, we will discuss the effect of the orientation of the double bond (regio- and stereochemistry) on the reactivity of the elimination reaction.

➤ *Regiochemistry of the Double Bond*

The possibility of regioselectivity in elimination reactions arises if more than one carbon have β-hydrogens. For instance, the sec-butyl halide (two types of β-hydrogen) can result in either 1- or 2-butene whereas $PhCH_2CH_2Br$ is not able to do so (only $PhCH=CH2$).

Therefore, the rules that dictate the major-minor products must be discussed for any stereochemical rationalization first.

Rule 1: The double bond will not shift to a bridgehead carbon irrespective of the mechanism-type until the ring size becomes quite big. This rule is also known as Bredt's rule because it was given Julius Bredt in 1902, and codification was completed in 1924.

Will not form

Will be formed

Rule 2: If there is a possibility of a newly formed double bond to get into conjugation with some previously present multiple bond or aromatic ring, it will always do so even if it leads to unfavorable stereochemistry.

Rule 3: The directional shift of the double bond in E_1 reactions is always decided by the relative stabilities of resulting products; which is because the initial departure of the leaving group leaves both possibilities open. In other words, the double bond moves primarily toward the more substituted carbon (Zaitsev's rule); which can be explained on the basis of the heat of hydration or hyper-conjugative structure. For instance, 2,3-dimethyl-2-pentene is formed from 3-bromo-2,3-dimethylpentane rather than either 3,4-dimethyl-2-pentene or 2-ethyl-3-methyl-1-butene.

2,3-dimethylpent-2-ene   3-bromo-2,3-dimethylpentane   3,4-dimethylpent-2-ene

Since the departure of the leaving group occurs first, the Zaitsev's rule dictates the orientation of double bond in E_1 reaction irrespective of the leaving group's nature (neutral or positive); however, in case it cannot be said for E_2 reactions, where the orientation of the double bond and the departure of the leaving group takes place simultaneously. Nevertheless, the non-Zaitsev product can also be the major product even in the case of E_1 eliminations because of reduced steric hindrance, or the formation of ion-pair.

Rule 4: It is quite a well-known fact that a trans β proton is required for the anti E_2 mechanism to be active; which is accessible only in one direction creating only one double-bond-shifting possibility. However, it is limited to the cyclic systems because the molecule may free rotation about carbon-carbon single bond (if the steric hindrance isn't very high). On the other hand, if two or more carbons have trans-β hydrogens, two types of products can be obtained; sometimes Zaitsev's product (double bond shifting toward the more substituted carbon), sometimes Hofmann's product (double bond shifting toward the least substituted carbon).

Zaitsev
product

Hofmann
product

It has also been observed that Zaitsev's rule is followed in all substrates if the compound has uncharged nucleofuges (leaving as negative ions like Cl$^-$); whereas Hofmann's rule is followed if the compound has charged nucleofuges (leaving as neutral ions like NR_3^+) provided that the substrate is acyclic, otherwise Zaitsev's product (i.e., if the leaving group is connected to a benzene ring). Now because Zaitsev's rule gives rise to the thermodynamically stable product, its outranking by Hofmann's rule in some compounds should also be explained. This change of double bond orientation in acyclic systems can be rationalized in the terms of two different factors.

Less acidic supports Zaitsev
product

Bulky group
supports Hofmann
product

The first one is that the β-hydrogen becomes less acidic due to the presence of the alkyl group, which in turn favors Hoffman's rule. The second one is the fact that positively charged groups are generally larger in size than the neutral group, and therefore a CH_3 group (less substituted) is more prone to attack than a primary or secondary carbon; in other words, steric effects dictate the final product.

Rule 5: It has been observed that the Hofmann orientation is significantly favored over the Zaitsev product in syn-E$_2$ eliminations.

Rule 6: The regioselectivity is of less importance as far as the E$_1$CB-type reactions are concerned because this pathway is typically found in the systems with an electron-withdrawing group at β-site, attracting double bond movement towards itself.

Rule 7: It is a well-known fact that E$_2$C reactions are susceptible to Zaitsev orientation, and this preference is of great importance as far as commercial production is concerned. However, the non-Zaitsev product is also obtained in some cases where conjugation with the aromatic ring is obtainable.

> *Stereochemistry of the Double Bond*

If CHAB–CGGX or CH3–CABX type compounds undergo elimination, the resulting alkene cannot show cis-, trans-isomers. However, the CH₂E–CABX CHEG–CABX type compounds do have the ability to give rise to cis- and trans- isomers after undergoing elimination. For instance, consider the following transformation.

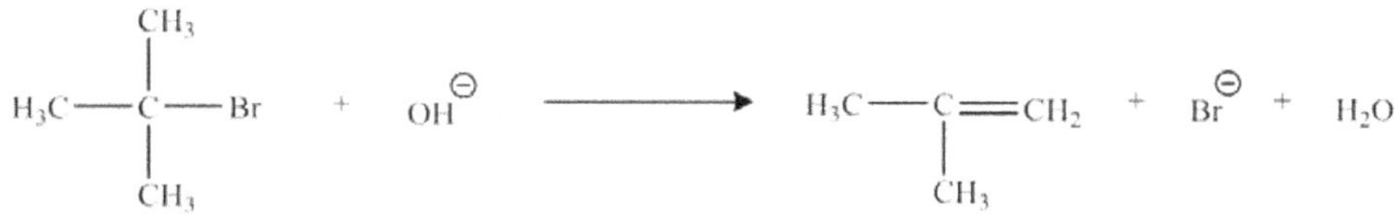

$$H_3C-\underset{\underset{CH_3}{|}}{\overset{\overset{CH_3}{|}}{C}}-Br \;+\; OH^{\ominus} \longrightarrow H_3C-\underset{\underset{CH_3}{|}}{C}=CH_2 \;+\; Br^{\ominus} \;+\; H_2O$$

It has been observed that the threo- and erythro-compound gives rise to trans- and cis- alkene; respectively. Furthermore, two conformations can be obtained for the transition state in the case of compound II, where two isomers can be synthesized. Nevertheless, the eclipsing effects will dictate the major-minor one; like the Zaitsev elimination of 2-bromopentane leads to the trans-isomer as major. This because confirmation A (Et is in between H and Br) is more stable than conformation B (Et is in between Me and Br), and the effect becomes more dominating as the groups get bigger.

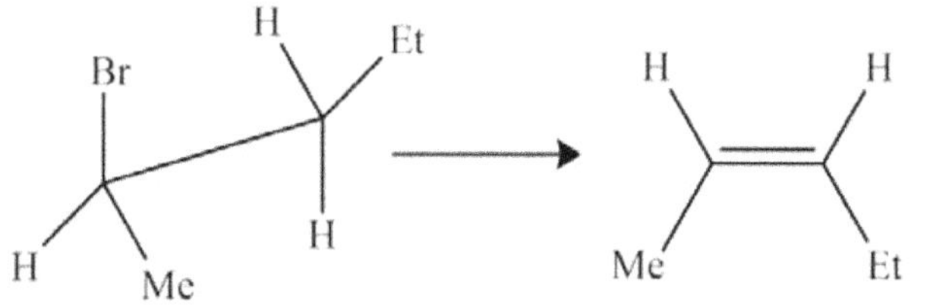

Minor product

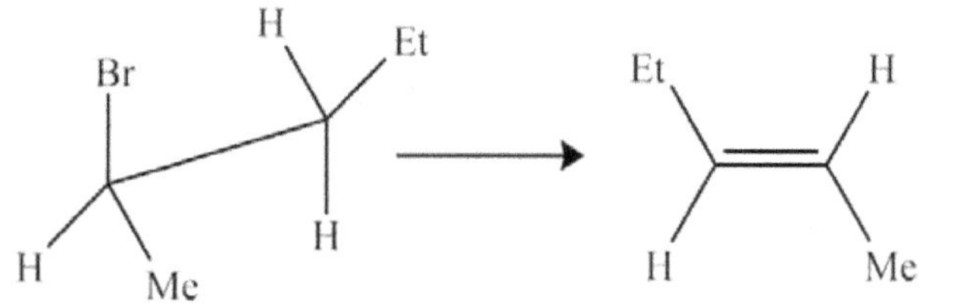

Major product

Moreover, the ratio of cis/trans isomers in the anti-E_2 reaction is also dictated by the solvent, nature of the leaving group, the substrate, and the attacking base. The complete picture of these effects is still not clear as far as the stereochemistry of the double bond is concerned. For instance, E_1-system with a bigger D-E pair opposite to the smaller AB pair is more stable if the carbocation free to rotate; and therefore, we should get the corresponding alkene. On the other hand, E_2-like products will be formed if the carbocation formed is not totally free; and the same is true for E_1CB reactions.

❖ Reactivity – Effects of Substrate Structures, Attacking Base, the Leaving Group and The Medium

In this section, we will discuss the effects of substrate structures, attacking base, the leaving group, and the nature of the medium on the reactivity of elimination reactions.

> ### *Effect of Substrate Structure on the Reactivity of Elimination Reactions*

The effects of substrate structures on the reactivity of elimination reactions can be divided into the following categories.

1. Effect on the reaction Rate: Different Groups bonded to the α-carbon (C with nucleofuge i.e., C–X) or β-carbon (C that loses proton i.e., C–H) can primarily exercise four types of influences as discussed below.

i) The emerging double bond can be stabilized or destabilized by these groups.

ii) The groups attached at β-carbon can affect the acidity of β-proton by stabilizing or destabilizing the emerging negative charge.

iii) The groups attached at α-carbon can affect the stability of the emerging positive charge.

iv) The groups attached at α- and β-carbon can exert eclipsing effects (steric effects).

The first and fourth types of effects can affect all three kinds of elimination mechanisms, with steric effects as most dominant in E_2 reactions. Also, the second and third kinds of effects cannot be applied to E_1 and E_1CB, respectively. If the C=C bond formation is the rate-determining, the presence of C=C or aromatic ring enhances the reaction rate in all mechanisms. Finally, the presence of electron-withdrawing groups at β-position raises the acidity of leaving hydrogen but has little to no effect at α-sites provided that no multipole bond conjugation available; making CN, Br, Cl, NO_2, Ts, SR, and CN suitable E_2-kind reactions.

2. Effect on E_1 vs E_2 vs E_1CB: Since the presence of aryl or alkyl group at α-carbon can stabilize the carbocation via resonance or inductive effect, A shift towards the E_1 pathway should be observed in the same. Also, the same shift can also be carried out by Alkyl groups at β-position by decreasing the acidity of the hydrogen atom. Nevertheless, the presence of aryl groups at β-carbon will push the towards E_1CB pathway by carbanion's stabilization. Conclusively, it has been observed that the presence of any electron-withdrawing group at β-site always pushes the reaction E_1CB pathway. Finally, it should also be remembered that E_2C reactions are also favored by the presence of alkyl groups at α-sites.

$H_3C\!-\!\underset{\underset{\displaystyle CH_3}{|}}{\overset{\overset{\displaystyle CH_3}{|}}{C^+}}$ $H_3C\!-\!CH^+$

Stablized via hyperconjugation   Stablized via Resonance
or inductive effect   (support E_1)
(support E_1)

3. Effect on elimination vs substitution: In bimolecular elimination reaction, the rate of reaction increases as the branching increases. This behavior can easily be rationalized in terms of statistical and steric factors. In other words, the increased α-branching leads to more base-attackable hydrogens, and increased steric hindrance opposes the attack at the carbon simultaneously. Moreover, the increased α-branching also supports unimolecular elimination over unimolecular nucleophilic substitution. The E_2 pathway is also favored over SN_2 when branching at the β-carbon is raised because of the suppression of the latter. Similarly, The E_1 pathway is also favored over SN_1 when the branching at the β-carbon is raised because of the steric factors. If the leaving group has a charge on it, the branching at the β-carbon will slow down the rate of E_2 reactions (Hofmann's rule). Also, the presence of electron-withdrawing groups at the β-site supports E_2-pathway with the simultaneous shift towards the E_1CB route but (upsurging the elimination/substitution ratio).

> ➢ *Effect of Attacking Base on the Reactivity of Elimination Reactions*

The effects of attacking base on the reactivity of elimination reactions can be divided into the following two categories.

1. Effect on E_1 vs E_2 vs E_1CB: The outside base isn't required in E_1 reaction under typical conditions because itself can act as the base. Therefore, the reaction pathway shifts from E_1 to E_2 when the outside bases is mixed. Furthermore, adding more outside stronger base will shift the pathway even towards E_1CB. Nevertheless, weak bases are also capable of yielding elimination reactions with some particular substrate-types. Bases (besides organic) yielding normal E2 reactions are given below.

$$NH_2^-, OR^-, OAr^-, CO_3^{2-}, LiAlH_4, CN^-, H_2O, NR_3, OH^-, OAc^-, I^-$$

It should also be noted that not all the bases are useful as far the practical synthetic route is concerned; for instance, NH_2^-, OR^- and OH^- are valuable for normal E_2 reaction; whereas the bases like OAc^-, Cl^- and RS^- are useful in preparing quaternary salts.

2. Effect on elimination vs substitution: Besides supporting E_2 over E_1, strong bases also assist elimination over substitution reactions. The concentrated solution of strong bases in a nonionizing solvent not only favors E_2 but also helps them to outrank the SN_2 pathway. On the other hand, dilute basic solutions in ionizing solvents not only favor E_1 but also help them to outrank the SN_1 pathway. It was also deduced from the nucleophilic substitution studies that stronger bases aren't necessarily stronger nucleophile; and therefore, a weaker nucleophile but stronger base will prefer elimination over substation. Nevertheless, weak bases can also lead to elimination if the solvent used is polar and aprotic.

> ➢ *Effect of Leaving Group on the Reactivity of Elimination Reactions*

The effects of leaving-group on the reactivity of elimination reactions can be divided into the following categories.

1. Effect on the general reactivity: Despite the different nature of the pathways, the leaving groups in both cases behave pretty much similar. Leaving groups in E_2 are NO_2, F, Cl, Br, I, NR_3^+, OHR^+, SO_2R, PR_3^+, SR_2^+, OSO_2R, OOR, OCOR, OOH, and CN; leaving groups in E_1 are NR_3^+, OSO_2R, SR_2^+, OH_2^+, OHR^+, Br, I, OCOR, Cl, and N_2^+.

2. Effect on E_1 vs E_2 vs E_1CB: Since the better leaving groups make the ionization easier, they will move the pathway towards E_1 reactions, which can be confirmed via ρ-values. Furthermore, positively charged or poor leaving groups will shift the pathway towards the E_1CB reactions, which can be attributed to the increased acidity β-protons that arises from the strong field effects of electron-withdrawing nature. Finally, it has also been observed that good leaving groups support E_2C reaction.

3. Effect on elimination vs substitution: Since the reaction pathway (elimination or substitution) is decided only after the departure of the leaving group in the reactions following first-order kinetics, the leaving group will not be able to prefer elimination over substitution, and vice-versa. Nevertheless, if ion-pair formation had taken place, the leaving group will affect the final product. Therefore, the elimination to substitution ratio (e/s) is largely independent of a halide as leaving group with only a minor raise in elimination to $Cl < Br < I$. On the other hand, the substitution pathway is strongly favored if the leaving group is like OTs. For instance, n-$C_{18}H_{37}Br$ treated with t-BuOK results in 85% elimination, whereas n-$C_{18}H_{37}OTs$ gives rise to 99% substitution under similar experimental conditions. Conversely, leaving groups with a positive charge will increase the elimination yield.

> ➢ *Effect of Medium on the Reactivity of Elimination Reactions*

The effects of the medium on the reactivity of elimination reactions can be divided into the following categories.

1. Effect on E_1 vs E_2 vs E_1CB: It is quite a well-known fact that reaction-rate increases with solvents polarity increases if the intermediates involved are ionic in nature. Furthermore, it has also been observed that the rate of E_1 and E_1CB pathways are also supported by increasing solvent's polarity and ionic strength, even if the leaving group is neutral in nature. Lastly, aprotic polar solvents encourage E_2C reactions with some particular substrates.

2. Effect on elimination vs substitution: The SN_2 pathway is favored at the cost of E_2 one if the solvent polarity is increased. For instance, KOH results in more elimination in alcohol but favors substitution in water as a solvent; which can partially be explained via the charge-dispersal phenomenon. On the other hand, the SN_1 pathway is encouraged over E_1 in most of the solvent mediums. Nevertheless, in the polar solvents with low nucleophilic character, the E_1 pathway become more prominent than the usual (like dipolar aprotic mediums). Finally, it has also been observed in the gas-phase studies (i.e., no medium) that when MeO^- reacts with1-bromopropane exclusively via the elimination-route even if the substrate used in the process are only primarily substituted.

3. Effect of temperature: It has been proven again and again that raising the reaction temperature almost always supports the elimination over substitution despite the order of the reaction (first or second-order). This behavior can simply be attributed to the higher activation energies of eliminations than those of substitutions, arising from larger bonding-alteration.

❖ Mechanism and Orientation in Pyrolytic Elimination

The pyrolytic elimination or E_i (elimination internal/intramolecular) mechanism is a special kind of elimination reaction where two vicinal groups on an alkane framework leave simultaneously through a cyclic transition state to form an alkene with a syn-elimination, and that is why they also called as pericyclic syn-or thermal syn elimination.

➤ *Mechanism in Pyrolytic Elimination*

The pyrolytic elimination is a unique type of elimination because it is activated thermally and does not need additional reagents unlike regular eliminations where an acid, a base, or charged intermediates is needed; and as the name suggests, it is often found in pyrolysis.

Illustrative reaction: one of the most common examples of pyrolytic elimination reaction is shown below for more clear understanding.

Mechanism involved: The proposed mechanism for the reaction given above involves one step which must be discussed before we give salient features of the same.

The elimination must be syn and the atoms coplanar for five and four-membered transition states, but coplanarity is not needed in the case where six-membered transition states are involved.

Salient Features: The main features of the mechanism involved in elimination internal (or intramolecular elimination) reactions are given below.

i) E_i reactions follow first-order kinetics with the rate law

$$Rate = k[RX]$$

Where k is the rate constant. The symbol $[RX]$ represents the molar concentration of the substrate.

ii) Elimination internal is thermally activated and does not need additional reagents unlike typical eliminations pathways where an acid or base is required, or sometimes involve charged intermediates.

iii) The elimination mode must be syn and the atoms must be coplanar for five and four-membered transition states; nevertheless, the coplanarity is not necessary for six-membered transition states.

iv) The rate of the reactions is not affected by the use of free-radical inhibitors.

> ### Orientation in Pyrolytic Elimination

Just like in normal elimination reactions, Bredt's rule is also applicable in the case of pyrolytic elimination. Nevertheless, conjugated systems are preferred non-conjugated systems (if allowed sterically) if a double bond is available. Furthermore, some more conclusive remarks regarding orientation in pyrolytic elimination are also of great importance.

1. The pyrolytic elimination requires a β-hydrogen in cis position; and therefore, the double bond will have only one direction to move in cyclic systems with only cis-hydrogen. Nevertheless, the condition of the leaving groups to be cis isn't required in six-membered transition states (due to non-coplanarity). Consequently, the hydrogen must be at the equatorial site if the leaving group is present at the axial position because the transition state cannot be comprehended with both at the axial positions. On the other hand, the leaving group will become able to create a transition state with β-hydrogen if it occupies an equatorial site. Conclusively, we can say if the leaving group is at the axial site, the double bond formation will not take place in the carbethoxyl group's direction due to the lack of equatorial hydrogen. Therefore, compound A will result in 100% C; whereas 50% of each type of alkene will be obtained if an equatorial leaving group is present.

2. It has been observed that the more stable alkene product dominates (Zaitsev's rule) in many cases, particularly with cyclic reactants. For instance, more of Hofmann product was expected menthyl acetate due to the presence of cis-β hydrogen on both sides, but the experimental yield is opposite i.e., 65% Zaitsev and 35% Hofmann product.

3. In many cases, it has also been observed that steric effects also dictate the elimination's direction since minimum steric interactions are favorable in both transition state and ground state of the substrate.

4. If all the three effects mentioned above are absent, the orientation dictation will be statistical in nature, and therefore, will be controlled by the number of β-hydrogens (Hofmann's rule). For instance, 60% 1-butene and 40% 2-butene were obtained from sec-butyl acetate where the hydrogens present are also in the 3:2 ratio.

❖ Problems

Q 1. Define elimination reactions.

Q 2. Discuss E1 and E2 mechanisms. How they are different from SN_1 and SN_2 pathways?

Q 3. State and explain E_1CB pathway of elimination reactions.

Q 4. What is pyrolytic elimination? How would you explain the orientation effect in the same?

Q 5. Discuss the reactivity of elimination reactions with special reference to substrate structure and attacking base.

Q 6. How does a leaving group affect the rate of elimination reactions?

Q 7. Write a short note on the reactivity of elimination reactions with special reference to reaction medium.

❖ Bibliography

1. M. B. Smith, *March's Advanced Organic Chemistry: Reactions, Mechanisms, and Structure*, John Wiley & Sons, Inc., New Jersey, USA, 2013.

2. H. Zimmerman, *Quantum Mechanics for Organic Chemists*, Academic Press, New York, USA, 1975.

3. M.S. Singh, *Reactive Intermediates in Organic Chemistry*, John Wiley & Sons, Inc., New Jersey, USA, 2014.

4. D. Klein, *Organic Chemistry*, John Wiley & Sons, Inc., New Jersey, USA, 2015.

5. J. Clayden, N. Greeves, S. Warren, *Organic Chemistry*, Oxford University Press, Oxford, UK, 2012.

6. R. L. Madan, *Organic Chemistry*, Tata McGraw Hill, New Delhi India, 2013.

7. C. A. Coulson, B. O'Leary, R. B. Mallion, *Hückel Theory for Organic Chemists*, Academic Press, Massachusetts, USA, 1978.

CHAPTER 11

Addition to Carbon-Carbon Multiple Bonds

❖ Mechanistic and Stereochemical Aspects of Addition Reactions Involving Electrophiles, Nucleophiles and Free Radicals

We know that addition reactions in organic chemistry are the chemical transformations where two or more molecules combine to yield a usually single but bigger molecule called an adduct. Since these addition reactions are restricted to chemical compounds with multiple bonds, molecules with carbon-carbon multiple bonds (alkenes, alkynes, or many cyclic species like benzene derivatives or cyclo-alkene/alkynes), or with carbon-heteroatom multiple bonds (like carbonyl C=O or imine C=N derivatives) are suitable candidates. Furthermore, these addition reactions can be classified into polar addition (electrophilic and nucleophilic) and non-polar addition (free radical and cycloaddition) reactions. Nevertheless, in this section, we will only discuss the mechanistic and stereochemical aspects of electrophilic, nucleophilic, and free radical addition to the carbon-carbon multiple bonds.

> #### Electrophilic Addition to Carbon-Carbon Multiple Bond

We know from the wave-mechanical treatment that space below and above the chemical bond is quite rich in electron density due to π-overlap; which makes the carbon-carbon multiple bonds very susceptible to electrophilic attacks. The general reaction showing the electrophilic attack on carbon-carbon multiple bonds is shown below.

Now first we will discuss the mechanism responsible for this transformation and then we will study the stereochemical aspects of the same.

Mechanism: Since the reaction between the reagent and substrate requires them to get close to each other first, some attractive force is needed to do so. This can be achieved by considering the attacking reagent as a species that can be fragmented into electrophile (E^+) and nucleophile (Nu^-).

Now because the double bond is a Lewis base (and nucleophile), it will attract the electrophilic part of the attacking reagent towards itself, forming π-complex. One might ask the since we have a nucleophilic part too in the attacking reagent then why we don't call the nucleophilic addition; the answer would be that the electrophilic part attacks first, and therefore, dictates almost everything. Also, we can not assign the electrophile to any specific carbon because the empty orbital of the attacking electrophile is overlapping with π-bond and not with any particular atomic orbital. However, this π-complex so formed will get convert into carbocation with real sigma bonds as shown below.

π-Complex   Carbocation

Furthermore, if the attacking electrophile is having a lone pair of electrons, which can be donated to neighboring carbon, a three-membered cyclic cation will be obtained which can be represented via three resonating structures as shown below.

I   II   III

Now depending upon the relative stability of three resonating structures, the intermediatory carbocation becomes "more cyclic" or even acyclic at the extreme. In other words, the intermediate carbocation will be cyclic if structure II is more stable (and hence more contributing) and will be acyclic if the structure I and III are more stable (and hence more contributing). This cyclic (or acyclic) cation is then attacked by the nucleophilic part of the attacking reagent to give rise to the final product.

Attack on classical carbocation   Attack on cyclic cation

The whole process of the electrophilic attack on the carbon-carbon multiple bonds can be fragmented into two steps as shown below.

1st Step:

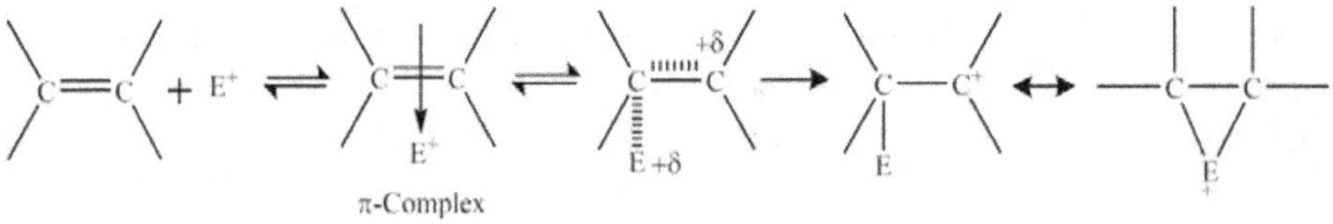

π-Complex

2nd Step:

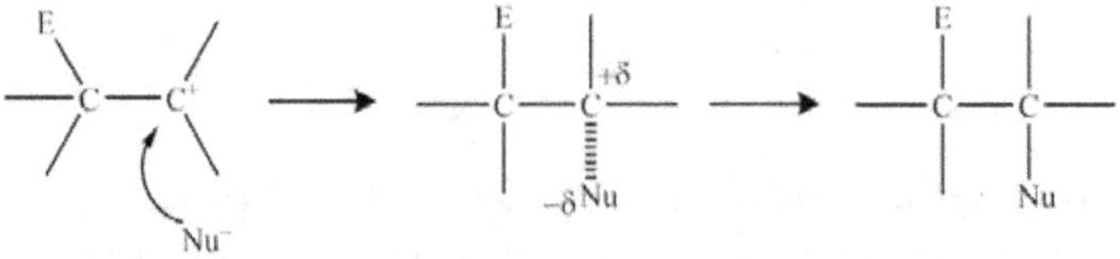

Stereochemistry: The stereochemistry of electrophilic addition to carbon-carbon multiple bonds is affected by two primary factors as discussed below.

i) The electrophile can attach itself to the double bond on the same or different side of the nucleophile i.e., syn- or anti-additions, respectively.

ii) In addition to the geometrical profile of addendum E^+ and Nu^-, the stereochemistry of the final product is also decided by the configuration of the addition product i.e., the orientation w.r.t rest of the molecule.

In other words, the electrophilic addition at the carbon-carbon multiple bonds can be cis- (syn) or trans- (anti), and may or may not be stereospecific. The only way for the nucleophile to attack is from backward if the intermediate is a cyclic cation; resulting in a syn addition product. Furthermore, if the reagent forms a 4-membered ring intermediate (instead of three), the addition will still be 'syn'.

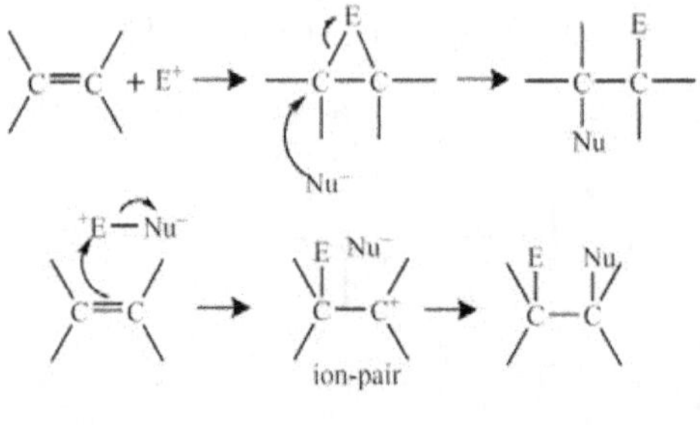

ion-pair

Conversely, if the classical carbocations dominate as intermediate and are having a sufficiently longer lifespan, they can show rotation about carbon-carbon single to yield a non-stereospecific product. However, if the classical carbocationic intermediate is short-lived, the nucleophile coming after the electrophilic attack may generate an ion-pair leading to a syn-addition product.

To find wheater the addition is syn or anti for a certain reagent (E−Nu), we need to use a substrate of form $abC=Cab$ where $a \neq b$ but E may or may not be the same as Nu. At this point, two scenarios can be realized, one when E ≠ Nu, and the other is when E = Nu; and we will discuss them one by one.

Case-I (E ≠ Nu):

If the addition is syn but on cis-compound, we will get a (*dl*)-erythro form via such type of transformation as shown below.

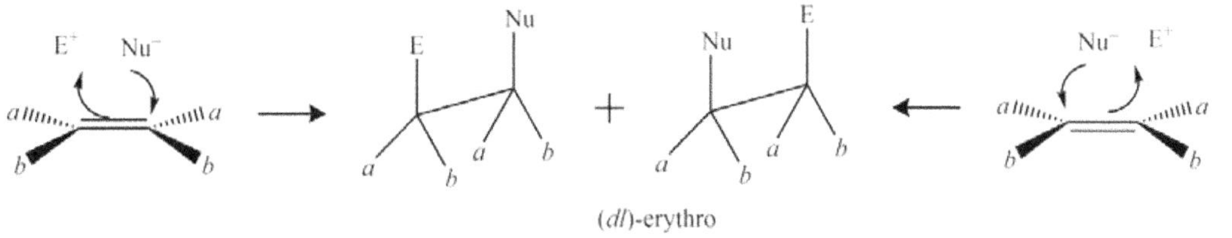

(*dl*)-erythro

On the other hand, if the addition is syn but on trans-compound, we will get a (dl)-threo form via such type of transformation as shown below.

(*dl*)-threo

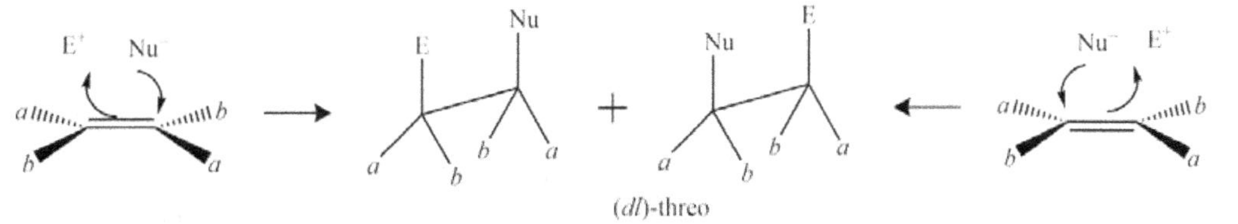

Similarly, If the addition is anti but on cis-compound, we will get a (*dl*)-threo form via such type of transformation as shown below.

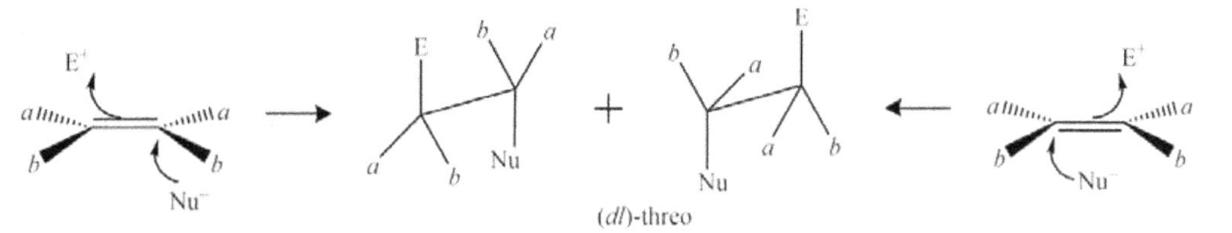

(*dl*)-threo

On the other hand, if the addition is anti but on trans-compound, we will get a (dl)-erythro form via such type of transformation as shown below.

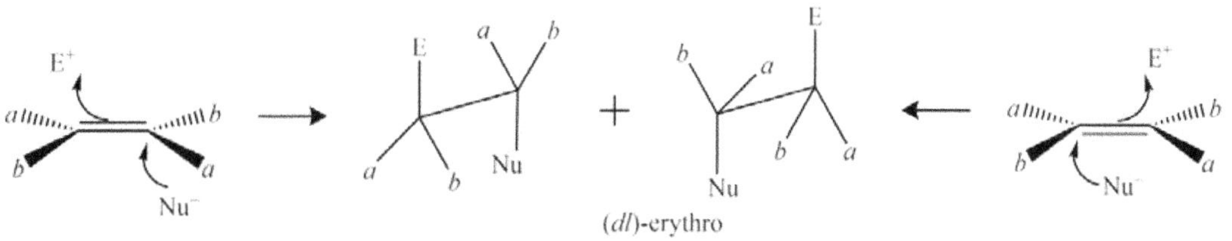

(*dl*)-erythro

Case-II (E = Nu):

In these types of cases, the (*dl*)-erythro forms will become meso-products; whereas the threo-forms will remain the same as shown below.

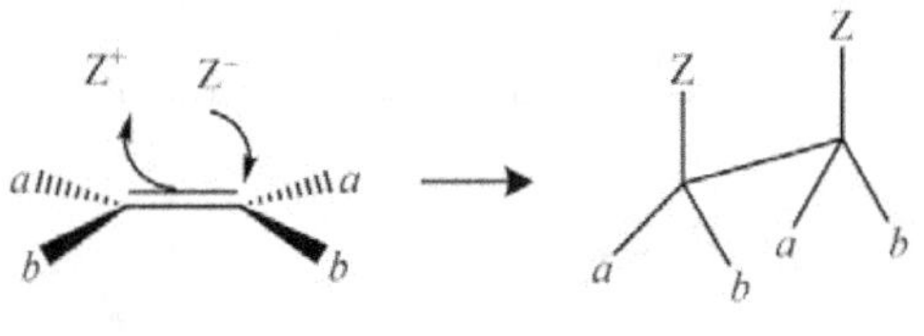

meso form

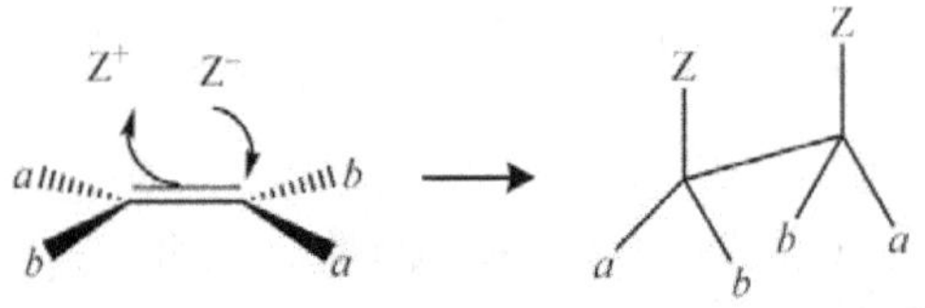

(*dl*)-threo form

Therefore, we may conclude that if the configuration of both the product and substrate are identified, the reaction pathway along with the addition mode can simply be predicted.

> ### Nucleophilic Addition to Carbon-Carbon Multiple Bond

The nucleophilic addition in organic chemistry is an addition reaction where an organic compound with an electrophilic multiple bond reacts with an attacking nucleophile in such a way that the multiple bond is broken. It is different from the electrophilic additions because it involves the group, to which atoms are being attached, accepts electron pairs; whereas in electrophilic addition, the group, to which atoms are being attached, donates electron pairs. The reaction of nucleophilic attack on C-C multiple bonds is shown below.

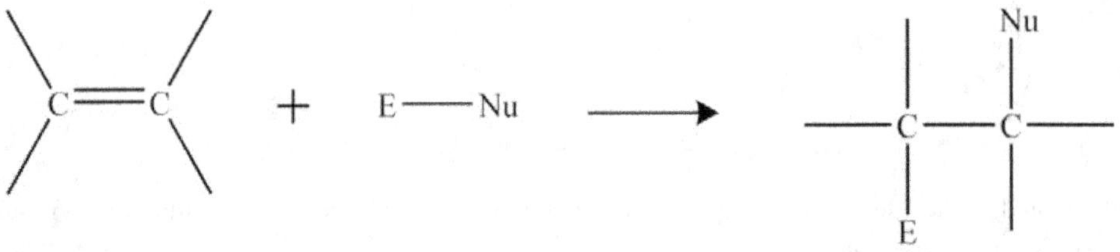

Substrate   Reagent   Product

Now first we will discuss the mechanism responsible for this transformation and then we will study the stereochemical aspects of the same.

Mechanism: The mechanism of nucleophilic addition to the carbon-carbon multiple bonds follows a two-step pathway as shown below.

Step I: The driving force for the addition to the alkenes is the generation of a nucleophile X^- that creates a covalent bond with an electron-deficient unsaturated system $-C=C-$ (first step); and the negative charge on nucleophile is shifted to the C–C bond.

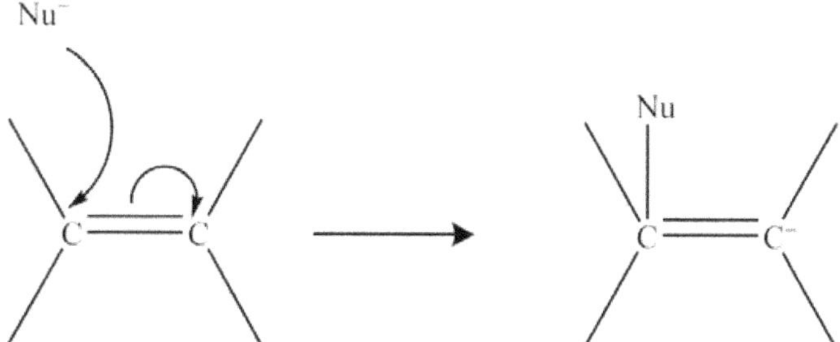

Step II: In the second step, the carbanion binds with the electrophilic part of the reagent (E^+) that is electron-deficient to form another covalent bond. Simple alkenes are not vulnerable to a nucleophilic attack due to the non-polar nature of the bond.

Stereochemistry: The stereochemistry of nucleophilic addition to carbon-carbon multiple bonds is affected by two primary factors as discussed below.

i) The nucleophile can attach itself to the double bond on the same or different side of the electrophile i.e., syn- or anti-additions, respectively.

ii) In addition to the geometrical profile of addendum Nu^- and E^+, the stereochemistry of the final product is also decided by the configuration of the addition product i.e., the orientation w.r.t rest of the molecule.

In other words, the nucleophilic addition at the carbon-carbon multiple bonds can be cis- (syn) or trans- (anti), and may or may not be stereospecific. The carbanion intermediate can be attacked in syn or anti mode to yield different products. To find wheater the addition is syn or anti for a certain reagent (E–Nu), we need to use a substrate of form $abC=Cab$ where $a \neq b$ but E may or may not be the same as Nu. At this point, two scenarios can be realized, one when $E \neq Nu$, and the other is when $E = Nu$; and we will discuss them one by one.

Case-I (E ≠ Nu):

If the addition is syn but on cis-compound, we will get a (*dl*)-erythro form via such type of transformation as shown below.

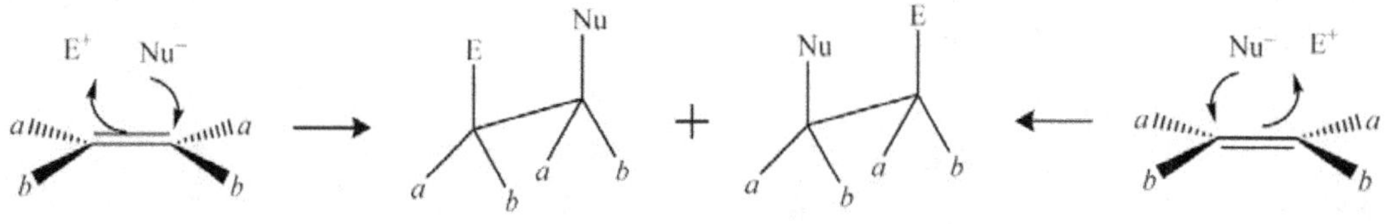

(dl)-erythro

On the other hand, if the addition is syn but on trans-compound, we will get a (dl)-threo form via such type of transformation as shown below.

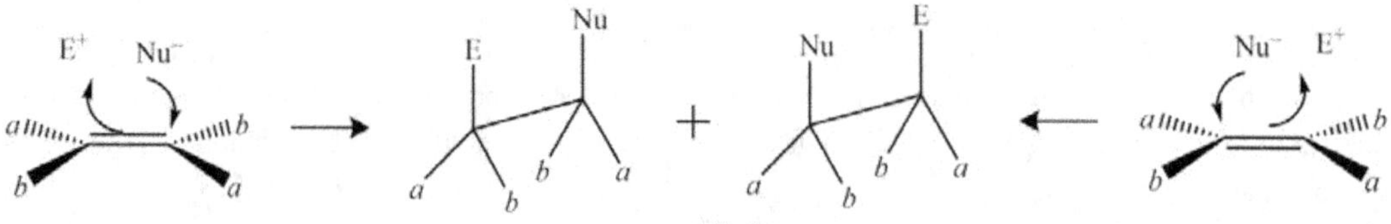

(dl)-threo

Similarly, If the addition is anti but on cis-compound, we will get a (*dl*)-threo form via such type of transformation as shown below.

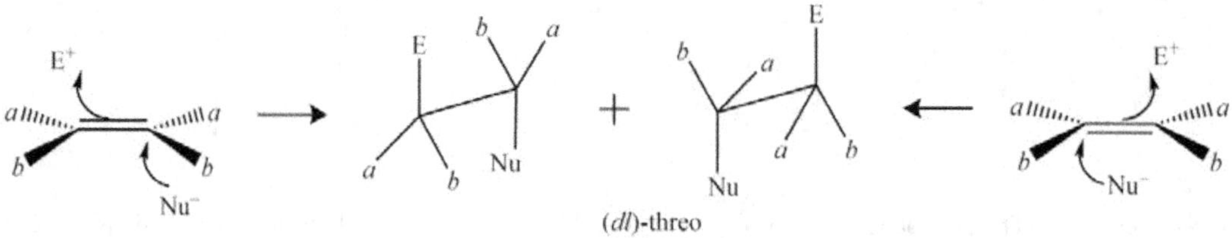

(dl)-threo

On the other hand, if the addition is anti but on trans-compound, we will get a (dl)-erythro form via such type of transformation as shown below.

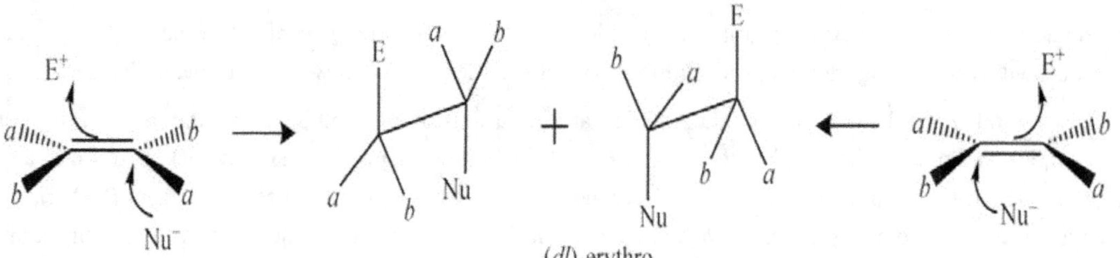

(dl)-erythro

Case-II (E = Nu):

In these types of cases, the (*dl*)-erythro forms will become meso-products; whereas the threo-forms will remain the same as shown below.

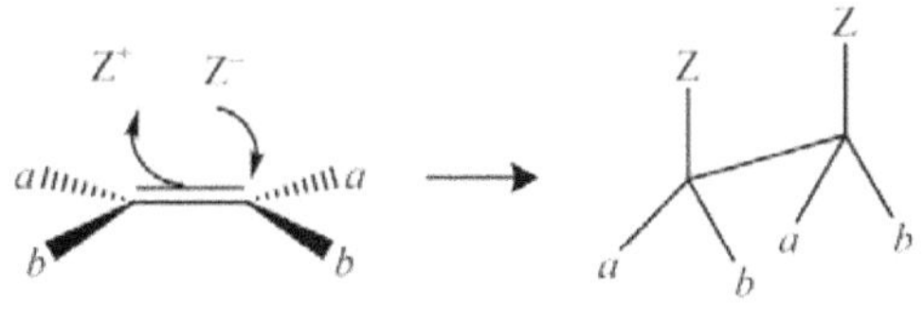

meso form

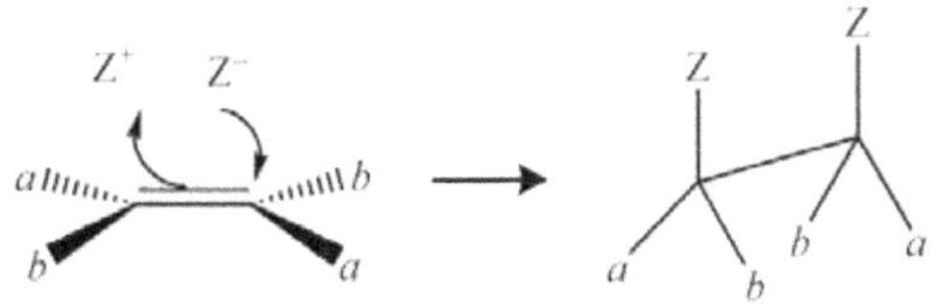

(dl)-threo form

Therefore, we may conclude that if the configuration of both the product and substrate are identified, the reaction pathway along with the addition mode can simply be predicted.

> ### Free Radical Addition to Carbon-Carbon Multiple Bond

Besides electrophiles and nucleophiles, the reactive species that can initiate addition reactions to carbon-carbon multiple bonds are free radicals. All this started with the regioselectivity HBr additions where the product from Markovnikov Rule wasn't the 'major' product suggesting some other route than the normal electrophilic addition. Further research in this field showed that the reason for the anti-Markovnikov product is the contamination of the reactants by peroxide; which in turn initiated an entirely different pathway called the free-radical mechanism. Nevertheless, if extremely pure HBr is added to pure 1-butene, 1-bromobutane (Markovnikov product) was the main yield. The reaction of nucleophilic attack on carbon-carbon multiple bonds is shown below.

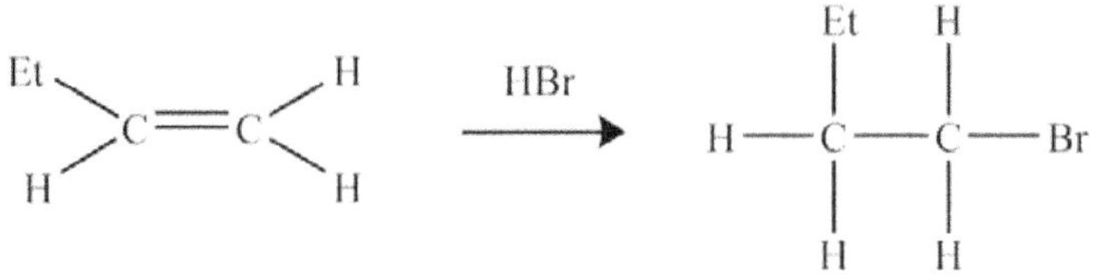

Free-radical reactions depend on a reagent having a (relatively) weak bond, allowing it to homolysis to form radicals (often with heat or light). Reagents without such a weak bond would likely proceed via a different mechanism.

Mechanism: The mechanism of radical addition to the carbon-carbon multiple bonds follows a three-step pathway as shown below.

Initiation: In this step, a catalytic amount of organic peroxide is needed to abstract the acidic proton from HBr and generate the bromine radical.

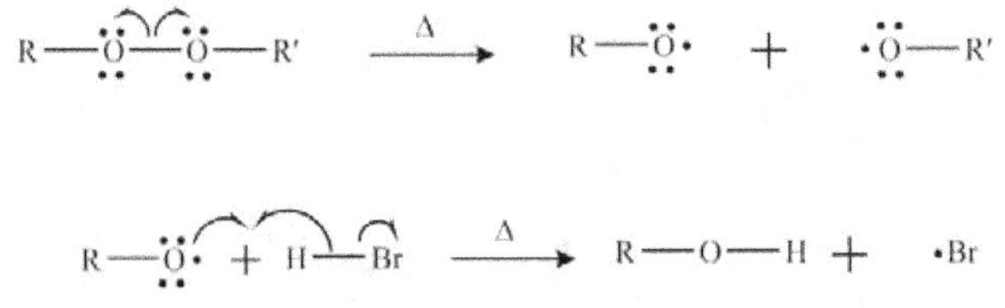

Propagation: In this step, the radical initiated addition to carbon-carbon multiple bonds propagates via the attack of free radicals on the substrate.

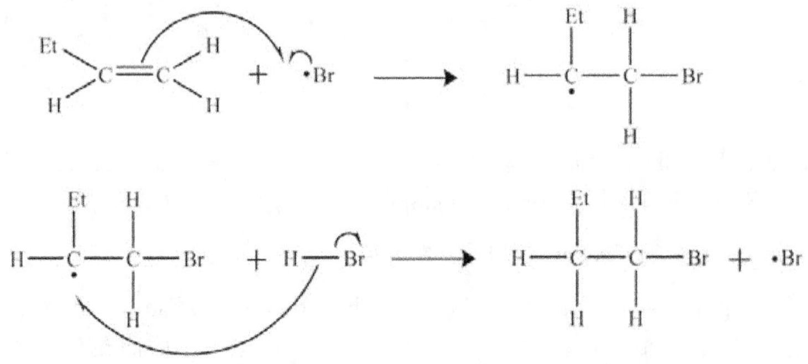

Termination: In this step, the radical initiated addition to carbon-carbon multiple bonds terminates via the attack of free radicals on the substrate.

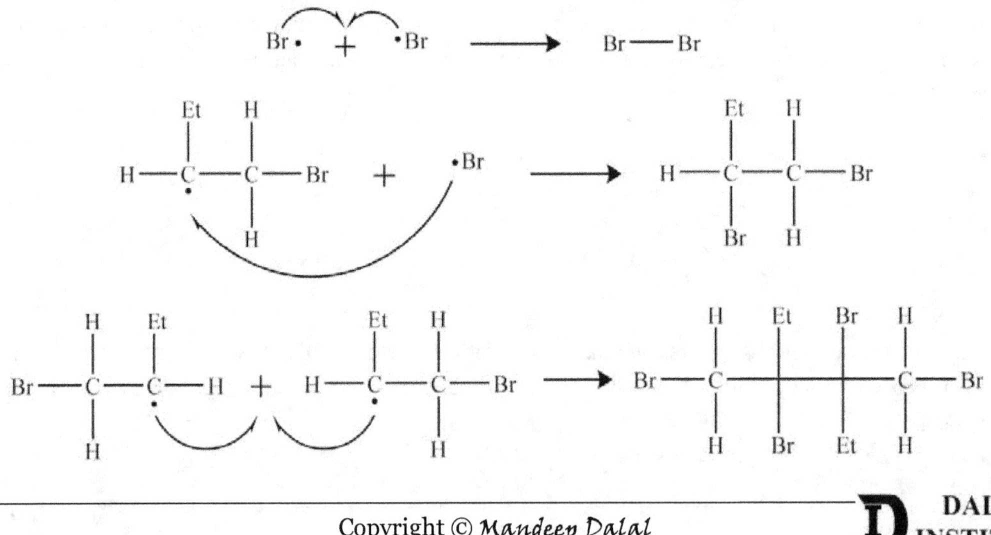

Stereochemistry: The addition of HBr to acyclic cis- or trans-olefins at room temperature results in 20% cis and 80% trans product, indicating that the radical addition of HBr to acyclic olefins is stereoselective but not stereospecific.

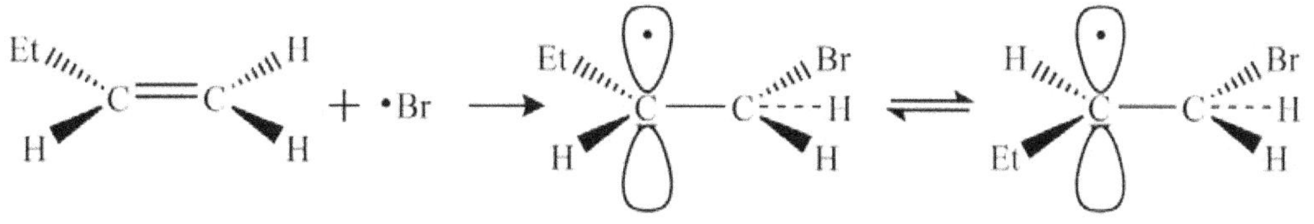

This can be rationalized in terms of rotation about C–C single bond in bromo-alkyl radical that can give rise to many conformations.

Conversely, if the reaction is carried out at −80°C, the trans addition resulted in 90% meso product whereas the cis addition resulted in 100% of (*dl*)-pair. All this can be rationalized in terms of a bridged molecular geometry. For cis-addition, we have

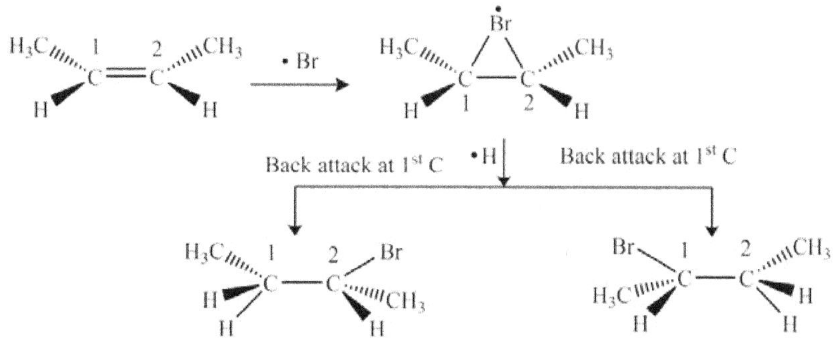

Back attack at 1st C   •H   Back attack at 1st C

Similarly, for trans addition, we have

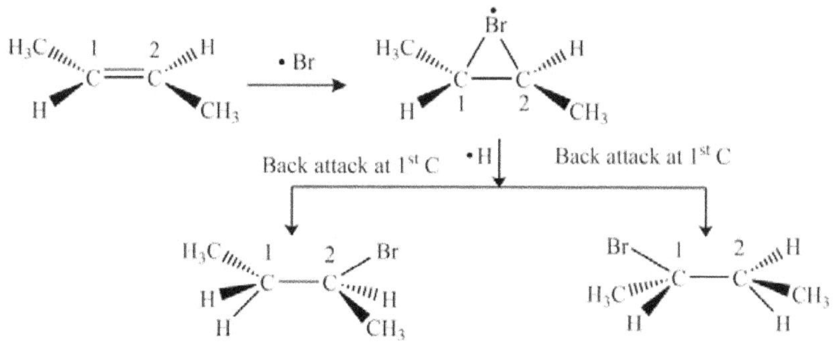

Back attack at 1st C   •H   Back attack at 1st C

It is also worthy to note that the free-radical addition does not occur with the molecules HCl or HI because both reactions are extremely endothermic and are not chemically favored.

❖ Regio- and Chemoselectivity: Orientation and Reactivity

In this section, we will study the orientation (or regio-selectivity) and reactivity (chemo-selectivity) of addition to carbon-carbon multiple bonds.

➤ *Orientation of Addition to Carbon-Carbon Multiple Bonds*

The structural orientation will not affect the final product if either the regent or the alkene is symmetrical in nature. On the other hand, if the alkene and attacking reagent both are unsymmetrical, two different products can be obtained as shown below.

$$H_3C - \overset{H}{\underset{H}{C}} = \overset{}{\underset{H}{C}} - CH_3 \; + \; HBr \longrightarrow H_3C - \overset{H}{\underset{\underset{Br}{|}}{C}} - \overset{H_2}{C} - CH_3$$

Symmetrical   Unsymmetrical   Single Product Possible

$$H_3C - \overset{}{\underset{H}{C}} = CH_2 \; + \; Br_2 \longrightarrow H_3C - \overset{H}{\underset{\underset{Br}{|}}{C}} - \overset{H_2}{C} - Br$$

Unsymmetrical   Symmetrical   Single Product Possible

In such a case, one of the products will be major and the other one will be minor depending upon their relative yield. In other words, the structural orientation is nothing but the preference that the double gives during its shift to decide which carbon to bind electrophile and which one to the nucleophile. The regioselectivity of electrophilic addition can be rationalized via two different rules as given below.

1. Markovnikov's Rule: The problem of structural orientation or regioselectivity in electrophilic addition was solved by a Russian chemist, Vladimir Markovnikov, in 1870 by giving an empirical rule called Markovnikov's rule. This rule states that when a polar reagent (like protic acid HX) is added to unsymmetrical alkenes, the electronegative part (i.e., halide) binds to the carbon with more alkyl groups; whereas the electropositive part (i.e., hydrogen) binds to the carbon with more hydrogens.

$$H_3C - \overset{}{\underset{H}{C}} = CH_2 \; + \; HBr \longrightarrow H_3C - \overset{H}{\underset{\underset{Br}{|}}{C}} - CH_3$$

Unsymmetrical   Unsymmetrical   Markovnikov Product

The theoretical basis for Markovnikov's Rule is the creation of a stable carbocation in the course of addition. The addition of the H⁺ to one of the carbons in alkene gives rise to a positive charge on another carbon, yielding an intermediate carbocation.

If the carbocation is highly substituted, its stability will increase due to hyperconjugation and induction. The addition reaction's major product will be the one that is formed from the intermediate with the highest stability. So, the major product of the HX-addition (X is atom greater electronegativity than H) to an alkene has the H atom in the less substituted site and X in the more substituted site. Nonetheless, the other substituted product from less stable carbocation will still be yielded as minor having the opposite conjugate attachment of X.

2. Anti-Markovnikov's Rule: If the addition pathway carbon-carbon double bond doesn't involve an intermediatory carbocation, the regioselectivity will not be dictated by Markovnikov's rule; for instance, the free radical addition. Such additions are labeled as anti-Markovnikov additions because the halogen binds to the less substituted carbon (i.e., the reverse of Markovnikov addition).

$$(H_3C)_3\overset{\oplus}{N}-\underset{H}{C}=CH_2 \;+\; HBr \;\longrightarrow\; (H_3C)_3\overset{\oplus}{N}-\overset{H_2}{C}-\underset{|}{CH_2}$$
$$Br$$

$$F_3C-\underset{H}{C}=CH_2 \;+\; HBr \;\longrightarrow\; F_3C-\overset{H_2}{C}-\underset{|}{CH_2}$$
$$Br$$

The anti-Markovnikov rule can be demonstrated via the addition of HBr to isobutylene in the presence of hydrogen peroxide or benzoyl peroxide. The addition of hydrogen bromide to substituted alkenes was archetypal in the study of free-radical addition. Chemists in the early era found that the reason for the inconsistency in the ratio of Markovnikov to anti-Markovnikov products was because of the presence of peroxides (which are free radical ionizing substances). They thought that the O$-$O bond in the peroxide is comparatively weak; and therefore, light or heat, or sometimes even acting on its own, this bond gets broken to give two radicals species. These radicals can then interact with hydrogen bromide to give a Br radical, which in turn, reacts with the C$-$C double bond.

Now because the Br atom is comparatively bigger, more likely it will encounter and react with the least substituted carbon which can be attributed to less static interactions between the two. Additionally, just like a positively charged species, the radical will be more stable if the unpaired electron is in the more substituted site. The intermediary radical is then stabilized by the hyperconjugation effect. In the more substituted sites, a greater number of carbon-hydrogen bonds are aligned with the electron-deficient molecular orbital of the radical. All this implies that there are superior hyperconjugation effects, so that site will be more favorable. In such a case, the terminal carbon of the substrate will give rise to the primary addition product rather than the secondary one.

> ➤ *Reactivity of Addition to Carbon-Carbon Multiple Bonds*

The double bond's reactivity towards the electrophilic addition increases with electron-donating groups and decreases with electron-withdrawing groups. For instance, consider the following order of reactivity.

$$CH_3CH=CH_2 > ClCH_2CH=CH_2 > Cl_2CHCH=CH_2 > CCl_3CH=CH_2$$

The above-given order gets revered if addition type becomes nucleophilic it will be supported by electron-withdrawing groups rather than electron-donating i.e.

$$CH_3CH=CH_2 < ClCH_2CH=CH_2 < Cl_2CHCH=CH_2 < CCl_3CH=CH_2$$

Also, the favourability of nucleophilic addition becomes significant only if three or four strongly electron-withdrawing groups like F_2C-CF_2 and $(NC)_2C-C(CN)_2$ are present. The substituents' effect is so dominant that polyhalo (or polycyano) alkenes almost always react via the nucleophilic pathway, whereas simple alkenes prefer to go via the electrophilic-addition route. Many reagents like ammonia which can only act as nucleophiles attack only react substrates vulnerable for such attacks. Conversely, the 'electrophilic only' type reagents simply refuse to add to the substrate like $F_2C=CF_2$. Furthermore, there are some reagents that add via electrophilic pathway if the substrate is simple alkene, but changes their mode of addition to nucleophilic if the substrate a polyhalo-alkene. For instance, HF and Cl_2 normally add via electrophilic attack, but it has been shown that Cl_2 and HF add to $(N\equiv C)_2C=CHC\equiv N$ and $F_2C=CClF$ via Cl^- and F^-, respectively.

Very fast   Slow

Reactivity Toward Bromine in Acetic Acid

Substrates like the form $-C=C-Z-$ (Z = CHO, COR, etc.), where the substituent is conjugated with the double bond, almost always react via nucleophilic addition. The order of activation by the substituent Z is given below.

$$NO_2 > COAr > CHO > COR > SO_2Ar > CN > CO_2R > SOAr > CONH_2 > CONHR$$

The nucleophilic attack on substituted alkenes can be attributed to the reduced electron density which attracts the electron-rich species. Nevertheless, some alkenes do react via an electrophilic pathway even after substituted with electron-withdrawing groups.

The rationale given above isn't totally suitable when comparing addition to carbon-carbon double vs triple bonds. For instance, even though the triple bond is richer in electron density than the double bond, it is less susceptible to electrophilic attack, and usually prefers to react via nucleophilic addition. It has been observed that reagents that form bridged intermediate prefer to add to double bond the triple one. Also, the rate ratio of alkenes to alkyne is reduced when electron-withdrawing groups are attached.

The higher susceptibility of triple bonds to nucleophilic attack can be attributed to the firm attachment of electrons in the triple bond due to smaller carbon-carbon bond length, which in turn make the electron density less available for any such attack. Alternatively, the lower susceptibility of triple bonds to electrophilic attack can be explained in terms of the accessibility of the empty orbital in the alkyne. In other words, it has been shown theoretically that bent alkynes have a π^* orbital of lower energy than the π^* orbital of simple alkenes; and therefore, linear alkynes can get a bent during transition states (electrophile addition) but alkene cannot. Also, bridged-ion intermediates arising from electrophilic addition to triple bonds will be more strained than their double bond counterparts; and therefore, slowing the rate of electrophilic addition. Nevertheless, triple bonds conjugated to the Z group favor the nucleophilic addition more aggressively.

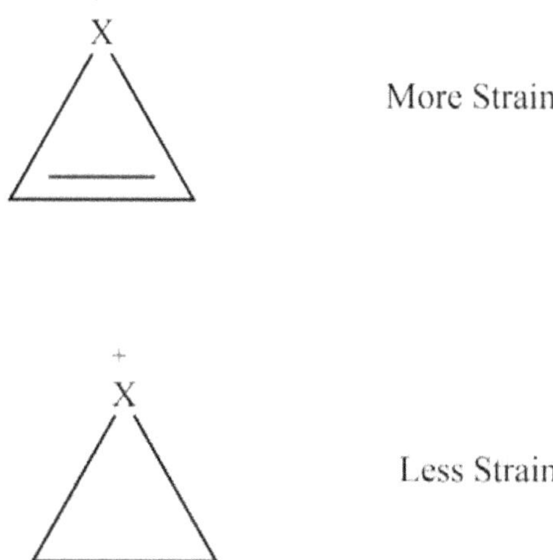

As expected, the attachment of alkyl groups typically increases the electrophilic addition's rates because of increased electron density; thought the order might change depending upon whether the intermediate formed is an open carbocation or a cyclic cation. If the first step is slowest (rate-determining) in electrophilic additions, like in the case of brominations, the rates for different substituted alkenes are dictated by the corresponding ionization potentials only and steric effects play little to no role.

No special types of substrates are required for free radical additions and the presence of a reactive free radical species predominantly dictates the overall rate. In the absence of initiator, reagents HBr or RSH) prefer to attack via ionic pathway; however, the mechanism changes to free radical addition as the in the radical initiator is mixed. Nucleophilic and electrophilic radicals behave more or less like nucleophiles and electrophiles, respectively; and the rate is affected accordingly. Nevertheless, it isn't expected but The rate of reaction of nucleophilic radical attack is faster with alkenes than with alkynes. Finally, it is also worthy to note that the steric effect might get an important role in some particular cases like catalytic hydrogenation where substitution decreases the reaction rate due to adsorption on the catalyst surface.

❖ Addition to Cyclopropane Ring

It is quite a well-known fact that cyclopropane rings behave in a similar manner as double bonds as far the reactivity is concerned. Therefore, like carbon-carbon multiple bonds, the cyclopropane ring can also undergo addition giving rise to open chain products as shown below.

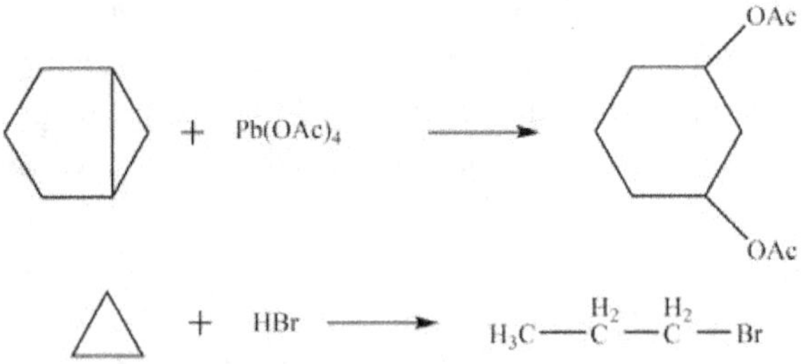

Now although the attack at cyclopropanes can occur via polar as well as non-polar additions, the electrophile type is the most important to discuss. The final product of electrophilic addition to substituted cyclopropanes is primarily dictated by Markovnikov's rule with rare exceptions.

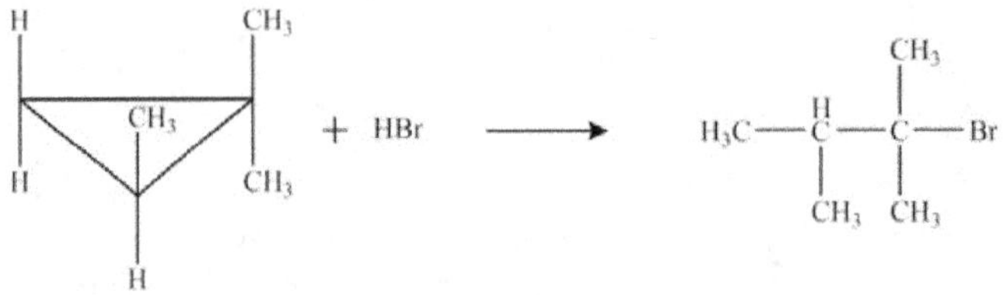

The stereochemical configuration of the final product can be analyzed in terms of the electrophilic as well as nucleophilic part of the attacking reagent. The electrophilic position can give rise to retention, inversion, or a mixture of two; whereas the nucleophilic position almost always gives rise to inversion. Three primary mechanisms that the electrophilic addition can adapt are given below.

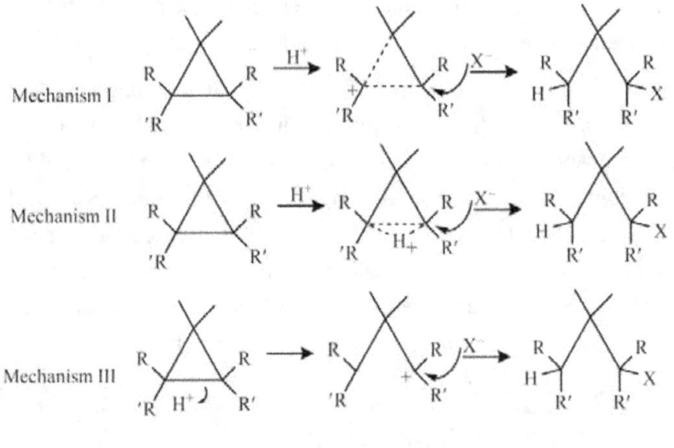

It is obvious that a cyclopropane ring system with one cornered carbon protonated is involved in the first mechanism; for instance, 7-norbornenyl and 2-norbornyl cations. On the other hand, the cyclopropane ring system with one protonate edge is involved in the second mechanism. The third mechanism involves an SE_2 type attack of H^+ to result in a classical cation, which subsequently reacts with the nucleophilic part of the attacking regent. It is also important to recall the fact that despite the depiction of configuration retention at carbon in all three cases, the 1st and 3rd routes are capable of giving inversion also. Since the cherry-picking of the mechanisms is not always possible, all or some of the cases take place at the same time. It has been found that Br^+ and Cl^+ react primarily via the second pathway; whereas D+ and Hg^{2+} react via the first pathway. Furthermore, density functional analysis has shown that edge-protonated is less stable than the corner-protonated; and the third pathway is usually opposed or less favorable. The reagents like Br_2 and Cl_2 can add to cyclopropanes via the free radical pathway (according to Markovnikov's rule) when the sample is irradiated with ultra-violet light. These free radical additions are stereospecific w.r.t only one carbon as shown below.

The addition to the cyclopropane ring may also occur in conjugated mode if the cyclopropyl ring is conjugated with a multiple bond.

❖ Hydrogenation of Double and Triple Bonds

The double and triple bonds in organic compounds can easily be reduced by heterogeneous catalysis. One of the most common examples of such type of addition to multiple bonds is the process of hydrogenation of alkene or alkyne. These reactions involve the attachment of two hydrogen atoms across the double or triple bond. Since a σ-bond is stronger than the π one, the hydrogenation of a multiple bond is thermodynamically favored (exothermic reaction). The molecule's stability can also be quantified in terms of its heat released during its hydrogenation.

Illustrative Reaction:

cyclohexene → cyclohexane

ethyne → ethene → ethane

Mechanism Involved: It is also worth noting that the hydrogenation of multiple bonds does not proceed without the addition of a catalyst because the final product is thermodynamically favorable but the reactants are kinetically stable. To illustrate this, the reaction coordinate diagram for the hydrogenation of alkenes and alkynes is given below.

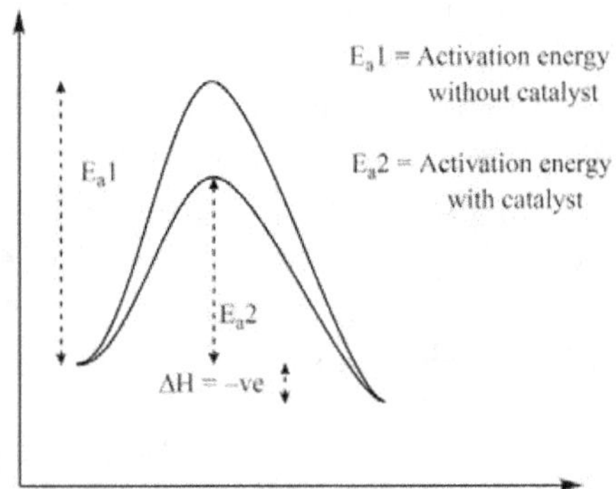

Figure 1: The reaction coordinate diagram for the catalytic hydrogenation of alkenes or alkynes.

The H–H bond in dihydrogen breaks in the presence of a metal catalyst and each hydrogen atom gets attached to the surface heterogeneous metal catalyst via a metal-hydrogen bond. The participating alkene also gets absorbed on the catalyst's surface. At this stage, an H atom is moved to the participating alkene, via a new C–H bond, followed by the movement of the second hydrogen atom via another C–H bond. Furthermore, since the hydrogens and alkene are on a flat surface of the metal catalyst, the two hydrogens being attached must do so via syn addition (i.e., at the same face of the double bond).

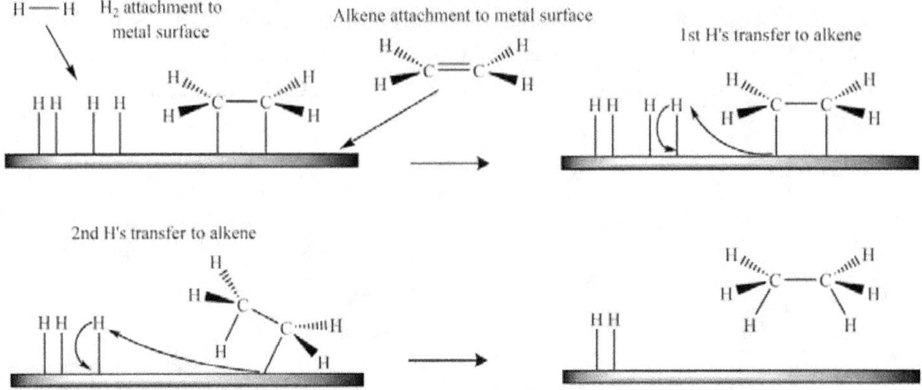

The most commonly used catalysts for alkene- or alkyne hydrogenation are platinum in the form of PtO_2, insoluble metals like palladium in the form of Pd-C, and Ni in the form of Ra-Ni.

❖ Hydrogenation of Aromatic Rings

It is not an easy task to hydrogenate aromatic rings even If we use precious metal catalysts, and require higher pressures and temperatures will still be needed. Nonetheless, after initiating the hydrogenation of the benzene ring, it won't stop at partial hydrogenation and will give rise to cyclohexane. This can be attributed to the fact that once it is converted to cyclohexadiene, (endothermic step), the aromatic will be lost, and therefore, the subsequent hydrogenation-steps will become exothermic and will take place at a much faster transformation rate.

At normal temperatures, commonly used catalysts are Pt and Rh, while Ru catalysts or Raney-Nickel need much higher pressures and temperatures. Furthermore, the Raney-Nickel catalyst is used for mass-level hydrogenations including temperature and pressures of $150^{\circ}C$ and 100-200 atm, respectively. Another important catalyst is 'Rh over Alumina' which is needs very mild experimental conditions. It is also worthy to note that this catalyst does not make $C-O$ bonds undergo hydrogenolysis, which is another useful aspect as far as practicality is concerned.

Polycyclic aromatic rings, like phenanthrenes and naphthalenes, can also undergo hydrogenation under suitable experimental conditions, and can fully- or partially be hydrogenated. For instance, decahydro- or tetrahydro-naphthalene can be obtained via Raney-Nickel catalyst under appropriate reaction conditions.

❖ Hydroboration

The hydroboration may simply be defined as the addition of a hydrogen-boron bond to C-N double bonds, and C-O double bonds, and C-C double or triple bonds. Hydroboration is extremely valuable in synthetic organic chemistry. Herbert C. Brown, who developed the conceptual and technological framework of hydroboration, received Nobel Prize in Chemistry for his work.

Illustrative reaction: A typical organic reaction hydroboration-type addition to carbon-carbon multiple reactions is given below.

Mechanism Involved: The hydroboration reactions are primarily dictated by anti-Markovnikov's rule, which means that the H gets attached to the most substituted carbon of the multiple bond. This reversal of regiochemistry indicates a polar $B^{\delta+}-H^{\delta-}$ bond. The hydroboration reaction occurs via a four-membered transition state where the H and the B atoms are attached to the same double bond's face. Since the mechanism is of concerted-type, the C-H bond is slightly slower than the formation of the C-B bond. Therefore, the B atom develops a somewhat negative charge in the transition state; whereas the partially positive charge resides on the more substituted carbon, which can be compensated via the +I effect from substituting groups. In other words, we may say that it is a case of group transfer reaction. Nevertheless, orbitals' investigation showed that the transformation is 'pseudopericyclic'; and therefore, the Woodward-Hoffmann rule cannot be implemented strictly for reactivity rationalization.

The hydroboration generally gives rise to compounds beyond monoalkyl borane if the reagent used is BH_3 (particularly in the case of alkenes with very small sterically hindrance). The hydroboration of trisubstituted alkenes can give rise to dialkyl boranes; though any subsequent alkylation of the organoboranes is discouraged due to increased steric hindrance. The difference of rate in the generation of di- and tri-alkyl boranes can be employed in the synthesis of bulky boranes to fine-tune the regioselectivity.

❖ Michael Reaction

The Michael addition (or Michael reaction) may simply be defined as the addition of a nucleophile like a carbanion to an α, β-unsaturated carbonyl compound with an electron-withdrawing group. It is a type of conjugated addition, and this process is one of the most practical approaches for the formation of C-C bonds.

Illustrative Reaction: The typical organic reaction showing this type of addition is shown below.

Where R and R' on the nucleophile symbolize the electron-withdrawing groups such as cyano and acyl, making the nearby methylene H enough acidic to give rise to a carbanion when treated with a base (B). The R'' group on the activated olefin (Michael acceptor) is generally a ketone to makes the molecule an enone; nevertheless, it can also be a sulfonyl fluoride or nitro group.

Mechanism Involved: In the first step, the carbanion is formed due to the deprotonation of the substates, which is stabilized by electron-withdrawing groups. Three resonating structures (2A, 2B, and 2C) can be drawn for this hybrid species with two enolate ion types. The electrophilic alkene reacts with this nucleophile via conjugated addition mode. Finally, the abstraction of a proton by the enolate from solvent (or protonated base) gives rise to the final product.

It is also worthy to note that the Michael addition is primarily dominated by the orbital picture rather than the electrostatic interactions. The lowest unoccupied molecular orbital (LUMO) of α-, β- unsaturated carbonyl systems have a hefty magnitude of the coefficient for β-carbon, whereas the HOMO of resonance stabilized enolates have a large magnitude of the coefficient for carbon. Therefore, owing to the similar-energy polarized frontier orbitals and softens, they are suitable for generating a good C–C bond.

❖ Sharpless Asymmetric Epoxidation

The Sharpless asymmetric epoxidation may simply be defined as an enantioselective chemical reaction where primary and secondary allylic alcohols are converted into epoxy-alcohols using tert-butyl hydroperoxide (TBHP), chiral diethyl tartrate (DET), and titanium tetra(isopropoxide) as the catalyst.

Illustrative Reaction: The typical organic chemical reaction showing Sharpless asymmetric epoxidation is given below.

Mechanism Involved: The mechanism for asymmetric epoxidation starts with the substitution of the isopropoxide ligands in titanium tetra(isopropoxide) catalyst by the chiral diethyl tartrate, which is followed by the further displacement via TBHP in the resulting complex.

In the last, the allylic alcohol reagent displaces the fourth isopropoxide ligand (the only remaining). Although the resulting titanium complex is supposed to be a dimer, the monomer unit is much easier to tackle as far as the mechanism is concerned. After that, the olefin part gets oxidized byTBHP with the face of attack dictated by the chiral DET resulting in the final product i.e., stereoselective epoxy-alcohol.

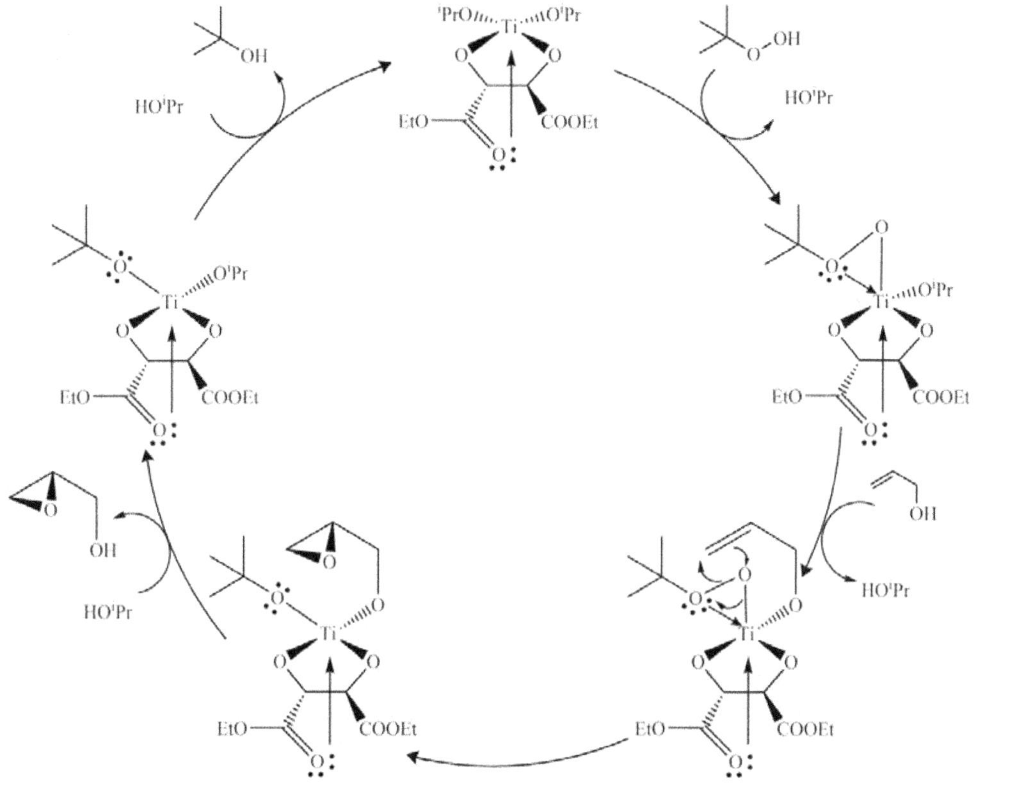

The Sharpless asymmetric epoxidation can be employed to synthesize different pheromones, leukotrienes, saccharides, terpenes, and antibiotics.

❖ Problems

Q 1. Discuss the mechanism and stereochemistry of electrophilic addition to carbon-carbon multiple bond.

Q 2. What is nucleophilic addition to carbon-carbon multiple bond? How is it different from free radical addition to carbon-carbon double bond?

Q 3. Define Markovnikov's rule.

Q 4. Discuss the chemoselectivity of electrophilic addition to carbon-carbon double bond.

Q 5. State and explain Michael reaction.

Q 6. Define hydroboration.

Q 7. Discuss the Sharpless asymmetric epoxidation.

Q 8. Write a short note the hydrogenation of aromatic rings.

Q 9. How does the hydrogenation of double bonds is different from the hydrogenation of carbon-carbon triple bonds?

❖ Bibliography

1. M.S. Singh, *Reactive Intermediates in Organic Chemistry*, John Wiley & Sons, Inc., New Jersey, USA, 2014.

2. D. Klein, *Organic Chemistry*, John Wiley & Sons, Inc., New Jersey, USA, 2015.

3. C. A. Coulson, B. O'Leary, R. B. Mallion, *Hückel Theory for Organic Chemists*, Academic Press, Massachusetts, USA, 1978.

4. H. Zimmerman, *Quantum Mechanics for Organic Chemists*, Academic Press, New York, USA, 1975.

5. J. Clayden, N. Greeves, S. Warren, *Organic Chemistry*, Oxford University Press, Oxford, UK, 2012.

6. R. L. Madan, *Organic Chemistry*, Tata McGraw Hill, New Delhi India, 2013.

7. M. B. Smith, *March's Advanced Organic Chemistry: Reactions, Mechanisms, and Structure*, John Wiley & Sons, Inc., New Jersey, USA, 2013.

CHAPTER 12

Addition to Carbon-Hetero Multiple Bonds

❖ Mechanism of Metal Hydride Reduction of Saturated and Unsaturated Carbonyl Compounds, Acids, Esters and Nitriles

Unlike the nucleophilic addition to carbon-carbon multiple bonds, the nucleophilic addition to carbon-heteroatom multiple bonds is much simpler to study as there is no regiochemical preference (i.e., the orientation of unsymmetrical addition) in most of the cases. For instance, consider $>C=O$, $-C\equiv N$, and $-C=N$ types of bonds which are extremely polar (great difference in electronegativity of participating atoms); so the carbon atoms bear a partial positive charge which makes the molecule an electrophile and the carbon atom as the electrophilic center; and therefore, the electrophilic species of the attacking reagent always goes to the oxygen or nitrogen whereas the nucleophilic part of the reagent attacks at the carbon.

The reaction given above is also called a 1, 2-type nucleophilic addition, and a racemic mixture will be obtained if the alkyl substituents are different.

Furthermore, it is obvious that the first step is just the same as the first step of nucleophilic displacement at a carbonyl's C atom; nevertheless, the latter case rarely takes place because R groups (H and carbon groups like alkyl aryl, etc.) are extremely poor leaving groups supporting the former case (i.e., addition). The acyl substitution dominates in the case of carboxylic acid derivatives (like amides or acid chlorides) because of the presence of 'relatively' good leaving groups like NH_2, OR, Cl, etc.

Hence, we can conclude that the nature of 'R' groups dictates whether the nucleophilic attack at carbon-heteroatom multiple bonds will give rise to the addition or substitution.

Until now we have discussed the basic ideas of addition to simple (i.e., non-conjugated) carbon-heteroatom multiple bonds; however, some systems do have unsaturation at least at α- and β-carbon. The nucleophilic addition in such cases is called conjugate addition and is different from ordinary nucleophilic additions to carbon-hetero bonds (i.e., 1, 2-nucleophilic additions) due to far separated attacking sites (i.e., 1, 4-nucleophilic additions). It is also important to recall the fact that normal alkenes neither show 1, 2- nor 1, 4-reactivity (possible only via activation by special substituents) due to lack of polarity.

The mechanism of conjugate addition can be understood by taking the example of an α, β-unsaturated carbonyl compound like cyclohexenone, where it can be showed that the β-position is an electrophilic site that can react with a nucleophile (from resonance structures). After the nucleophilic attack, the negative charge of the nucleophilic part of the attacking reagent is now distributed via resonance in α-carbon carbanion and alkoxide anion. Finally, the protonation results in a saturated carbonyl compound via keto-enol tautomerism (the equilibrium lies toward the keto form because it is more stable than the enol form due to a stronger C=O bond than C=C bond). Also, another electrophile will replace the proton if the reaction further proceeds via vicinal difunctionalization.

In this section, we will study the mechanism of a special type of addition to the carbon-heteroatom multiple bonds where the reduction of saturated and unsaturated carbonyl compounds (aldehydes, ketones, and acyl halides), acids, esters, and nitriles is carried out using metal hydrides.

➢ *Metal Hydride Reduction of Saturated Carbonyl Compounds*

Many metal hydrides can be used to reduce the saturated carbonyl compounds like aldehydes, ketones, and acid halides.

1. Reduction by Sodium Borohydride: The sodium borohydride ($NaBH_4$) is one of the most common sources of the hydride nucleophile. The hydride anion is produced in the course of reaction because of the polar nature of the metal-hydrogen bond.

Sodium Borohydride

The hydride anion's addition to carbonyl compound results in an alkoxide anion which in turn gives rise to a reduced product.

i) Reduction of aldehydes by NaBH₄:

The hydride anion's addition to aldehyde results in an alkoxide anion, which in turn, gives rise to primary (1°) alcohols on protonation.

The mechanism for the aldehydic reduction by metal hydride involves the nucleophilic addition of the hydride ion to the carbonyl carbon. In many cases, the Na^+ ion activates the carbonyl group by attaching itself to the oxygen atom, which in turn, will raise the electrophilic character of the C=O group.

ii) Reduction of ketones by $NaBH_4$:

The hydride anion's addition to ketone results in an alkoxide anion which in turn gives rise to secondary (2°) alcohols on protonation.

The mechanism for the ketone reduction by metal hydride involves the nucleophilic addition of the hydride ion to the carbonyl carbon. In many cases, the Na^+ activates the carbonyl group by attaching itself to the oxygen atom, which in turn, will raise the electrophilic character of the C=O group.

iii) Reduction of acyl halides by $NaBH_4$:

Two subsequent additions of hydride anions to acyl halides result in a sodium alkoxide anion which in turn gives rise to primary (1°) alcohols on protonation.

The mechanism for the acyl halides' reduction by metal hydrides involves the nucleophilic addition of the hydride ion to the carbonyl carbon. In many cases, the Na^+ activates the carbonyl group by attaching itself to the oxygen atom, which in turn, will raise the electrophilic character of the C=O group.

2. Reduction by Lithium aluminium hydride: The lithium aluminium hydride ($LiAlH_4$) is one of the most common sources of the hydride nucleophile. The hydride anion is produced in the course of reaction because of the polar nature of the metal-hydrogen bond.

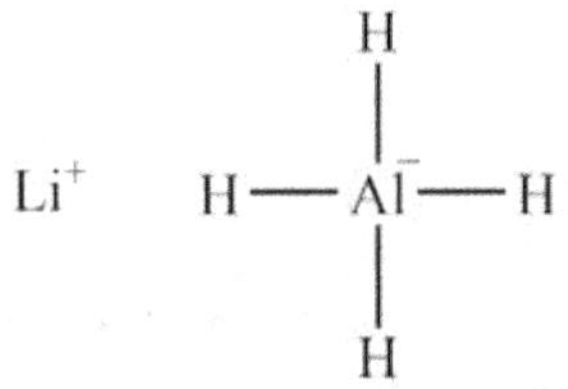

Lithium aluminium hydride

The hydride anion's addition to carbonyl compound results in a lithium alkoxide anion which in turn gives rise to a reduced product.

i) Reduction of aldehydes by LiAlH₄:

The hydride anion's addition to aldehyde results in an alkoxide anion which in turn gives rise to primary (1°) alcohols on protonation.

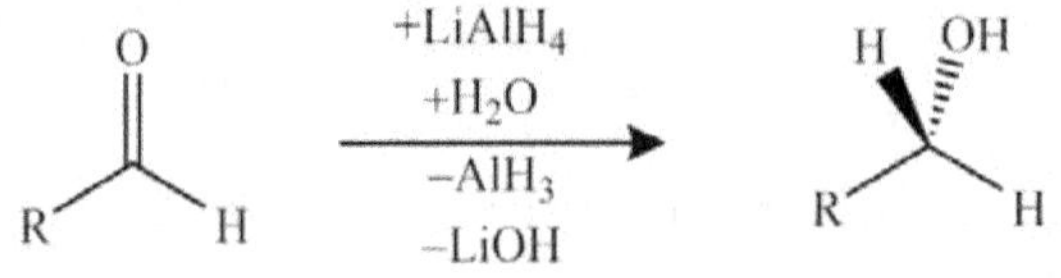

The mechanism for the aldehydic reduction by metal hydride involves the nucleophilic addition of the hydride ion to the carbonyl carbon. In many cases, the Li^+ activates the carbonyl group by attaching itself to the oxygen atom, which in turn, will raise the electrophilic character of the C=O group.

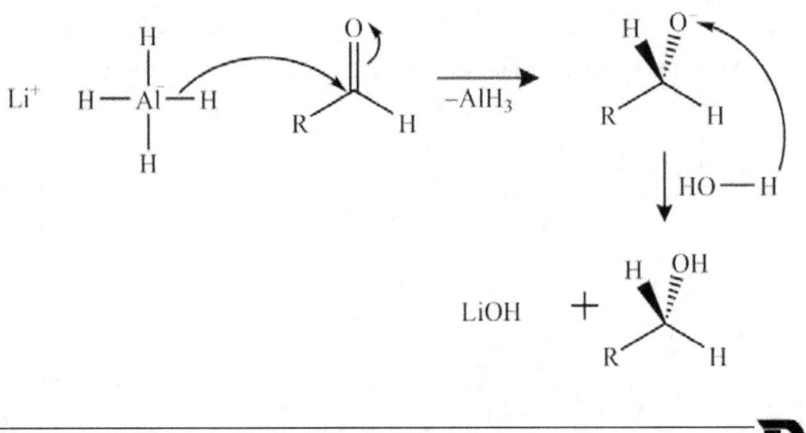

ii) Reduction of ketones by LiAlH₄:

The hydride anion's addition to ketone results in an alkoxide anion which in turn gives rise to secondary (2°) alcohols on protonation.

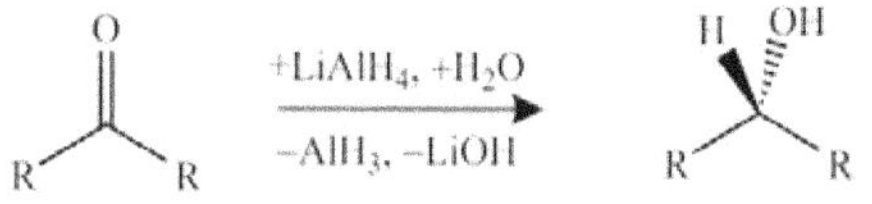

The mechanism for the ketone reduction by metal hydride involves the nucleophilic addition of the hydride ion to the carbonyl carbon. In many cases, the Li^+ activates the carbonyl group by attaching itself to the oxygen atom, which in turn, will raise the electrophilic character of the C=O group.

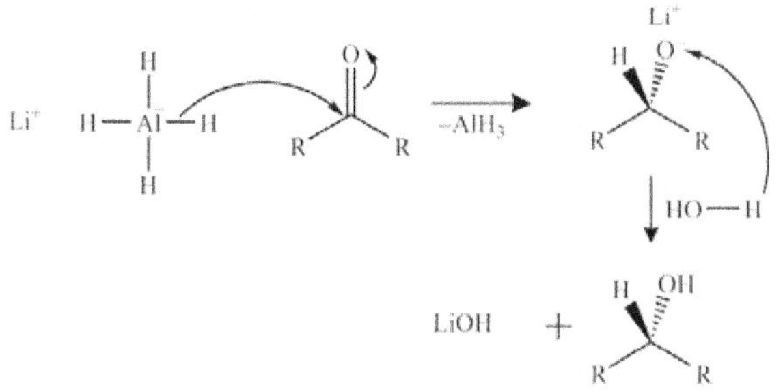

iii) Reduction of acyl halides by LiAlH₄:

Two subsequent additions of hydride anions to acyl halides results in an alkoxide anion which in turn gives rise to primary (1°) alcohols on protonation.

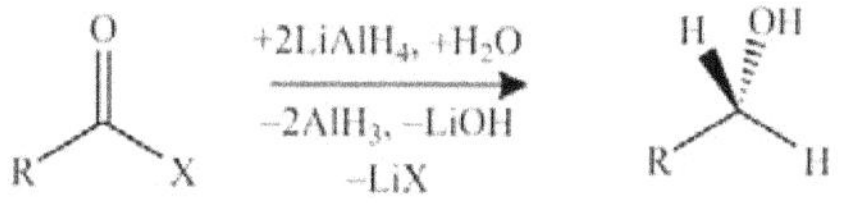

The mechanism for the acyl halides' reduction by metal hydride involves the nucleophilic addition of the hydride ion to the carbonyl carbon. In many cases, the Li^+ activates the carbonyl group by attaching itself to the oxygen atom, which in turn, will raise the electrophilic character of the C=O group.

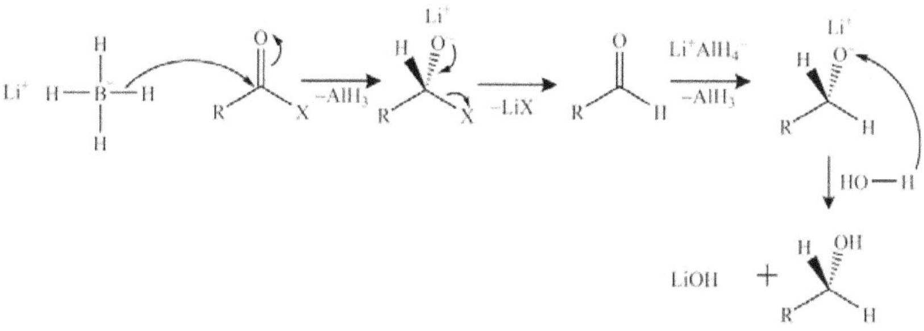

> ➤ *Metal Hydride Reduction of Unsaturated Carbonyl Compounds*

Many metal hydrides can be used to reduce the unsaturated carbonyl compounds like α, β-unsaturated aldehydes, ketones, and acid halides.

1. Reduction by Sodium Borohydride: It is well-known fact that isolated C=C bond cannot be reduced either by $LiAlH_4$ or by $NaBH_4$. However, if this C=C is in conjugation with C=O (i.e., conjugated aldehyde, ketones, or acid chlorides), the substrate can act as an electrophile via at the β-carbon or via the carbon of the carbonyl group. Now since the β-carbon is a "soft" electrophilic site, it will prefer to react with "soft" nucleophile like $NaBH_4$; whereas the carbonyl's carbon is a relatively "hard" electrophilic site, it will prefer to react with "hard" nucleophile like $LiAlH_4$ (HSAB Principle). Nevertheless, after the reduction of the C=C bond, the saturated aldehyde will also get reduced to the alcohol by $NaBH_4$ in the next step. Conversely, $LiAlH_4$ will show a 1, 2-addition, leaving an isolated C=C double bond unaltered.

The hydride anion's addition to unsaturated carbonyl compound results in an alkoxide anion which in turn gives rise to a reduced product.

i) Reduction of α, β-unsaturated aldehydes by $NaBH_4$:

The hydride anion's addition to the α, β-unsaturated aldehyde results in an alkoxide anion which in turn gives rise to primary (1°) alcohols on protonation.

The mechanism for the aldehydic reduction by metal hydride that involves the 1, 4-nucleophilic addition of the hydride ion followed by a 1-2-addition to the carbonyl carbon is given below.

ii) Reduction of α, β-unsaturated ketones by NaBH₄:

The hydride anion's addition to α, β-unsaturated ketone results in an alkoxide anion which in turn gives rise to secondary (2°) alcohols on protonation.

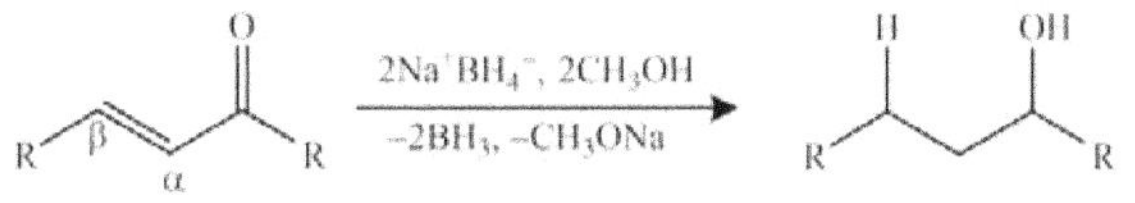

The mechanism for the ketone reduction by metal hydride that involves the 1, 4-nucleophilic addition of the hydride ion, followed by a 1, 2-addition to the carbonyl carbon is given below.

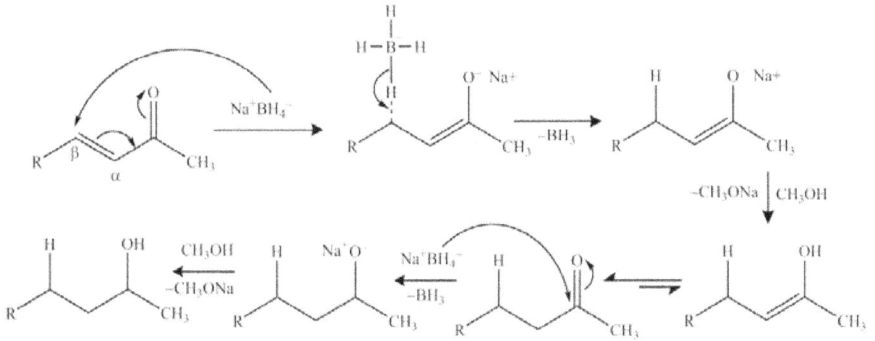

iii) Reduction of α, β-unsaturated acyl halides by NaBH₄:

Although it seems like an easy task, there are no reports of NaBH₄-catalyzed 1, 4-addition to acyl halides (with reasonable yield). Nevertheless, 1, 2-adducts for benzoyl chloride have been obtained.

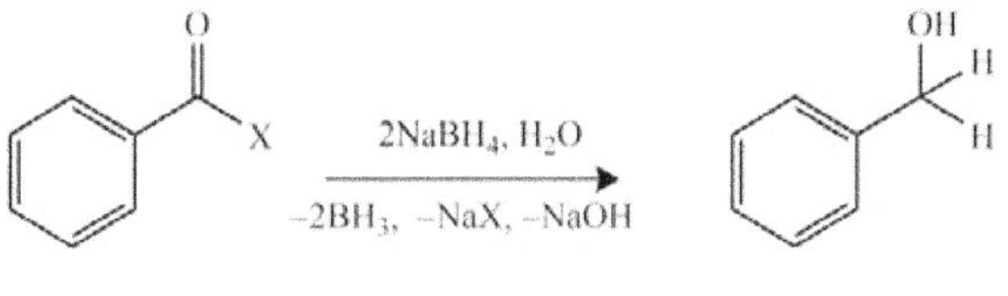

The mechanism for the transformation of benzoyl chloride to benzyl alcohol via NaBH₄ is given below.

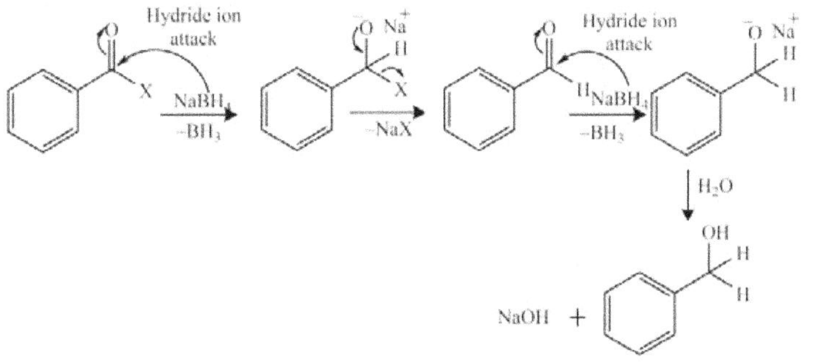

2. Reduction by Lithium aluminium hydride: Since the carbonyl's carbon is a relatively "hard" electrophilic site, it will prefer to react with "hard" nucleophile like $LiAlH_4$ (HSAB Principle). Hence, unlike $NaBH_4$, $LiAlH_4$ will show a 1, 2-addition, leaving an isolated double bond unaltered.

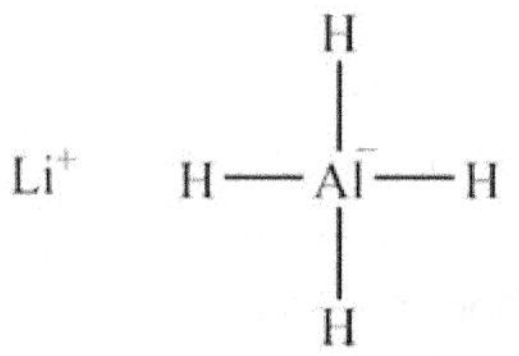

Lithium aluminium hydride

The hydride anion's addition to carbonyl compound results in a lithium alkoxide anion which in turn gives rise to a reduced product.

i) Reduction of α, β-unsaturated aldehydes by LiAlH₄:

The hydride anion's addition to α, β-unsaturated aldehyde results in an alkoxide anion which in turn gives rise to primary (1°) alcohols on protonation.

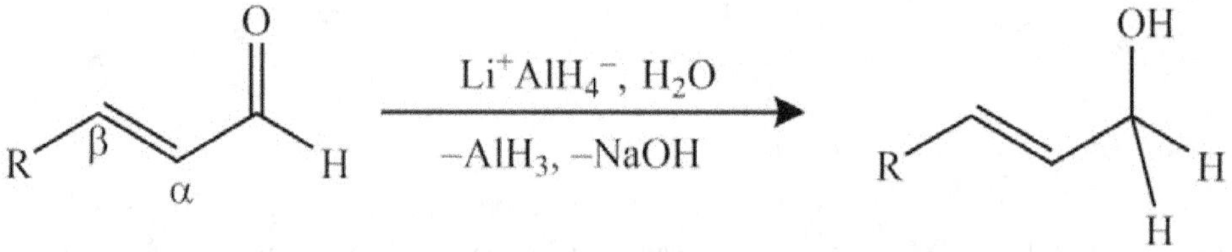

The mechanism for the α, β-unsaturated aldehydic reduction by metal hydride involves 1, 2-nucleophilic addition of the hydride ion to the carbonyl carbon as shown below. Furthermore, it is also worthy to note that conjugate addition in product might also be obtained alongside with very small yield.

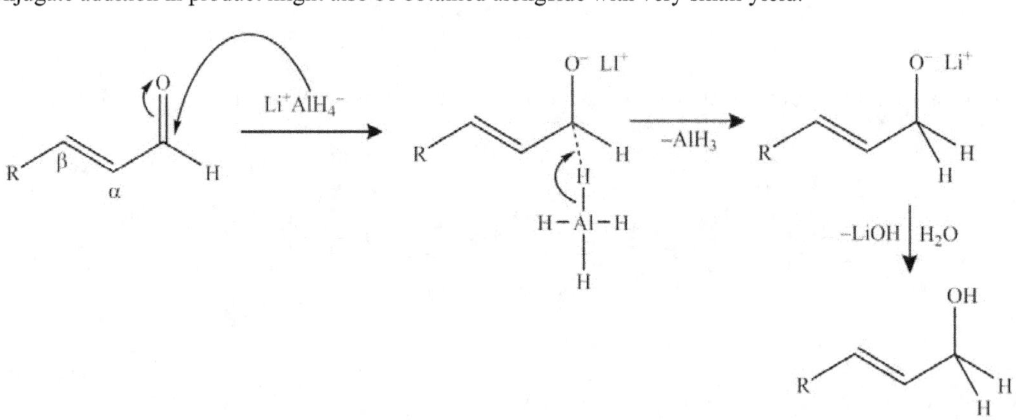

ii) Reduction of α, β-unsaturated ketones by LiAlH₄:

The hydride anion's addition to ketone results in an alkoxide anion which in turn gives rise to secondary (2°) alcohols on protonation.

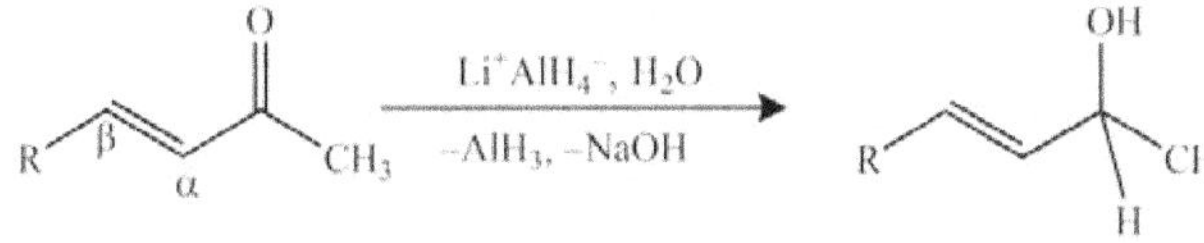

The mechanism for the α, β-unsaturated ketonic reduction by metal hydride involves 1, 2-nucleophilic addition of the hydride ion to the carbonyl carbon as shown below. Furthermore, it is also worthy to note that conjugate addition in product might also be obtained alongside with very small yield.

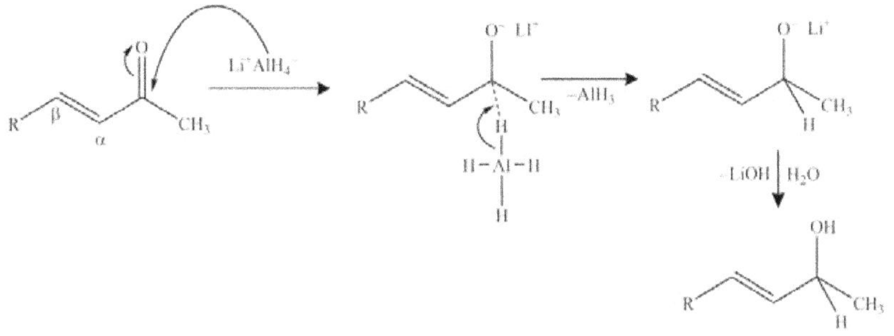

iii) Reduction of α, β-unsaturated acyl halides by LiAlH₄:

The hydride anion's addition to ketone or aldehyde results in an alkoxide anion which in turn gives rise to primary (1°) alcohols on protonation.

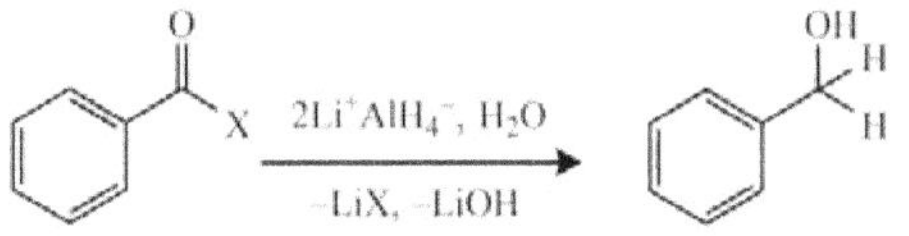

The mechanism for the transformation of benzoyl chloride to benzyl alcohol via LiAlH₄ is given below.

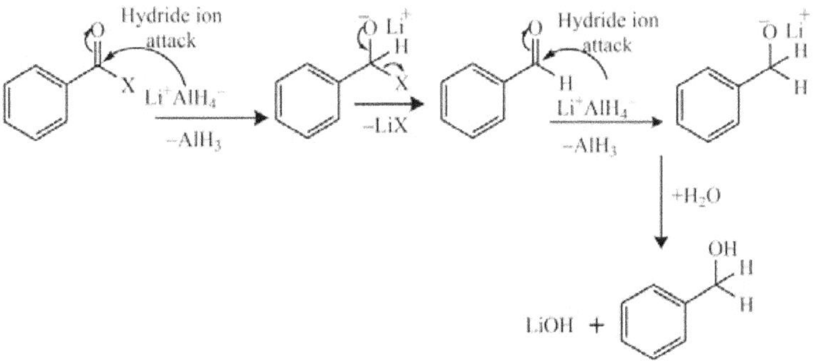

> ### *Metal Hydride Reduction of Acids*

Many metal hydrides can be used to reduce the saturated and unsaturated carbonyl compounds like carboxylic acid.

1. Reduction by lithium aluminium hydride: The lithium aluminium hydride ($LiAlH_4$) is one of the most common sources of the hydride nucleophile. The hydride anion is produced in the course of reaction because of the polar nature of the metal-hydrogen bond.

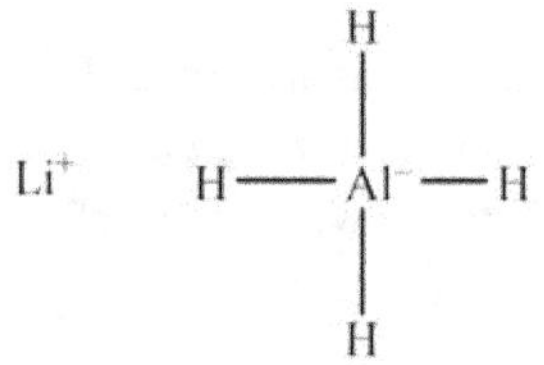

Lithium aluminium hydride

The hydride anion's addition to carbonyl compound results in an alkoxide anion which in turn gives rise to a reduced product.

The hydride anion's attack on carboxylic acid results in a carboxylate anion, which in turn, is attacked by AlH_3 to yield aldehyde. This aldehyde then gives rise to primary (1°) alcohols after 1, 2-addition.

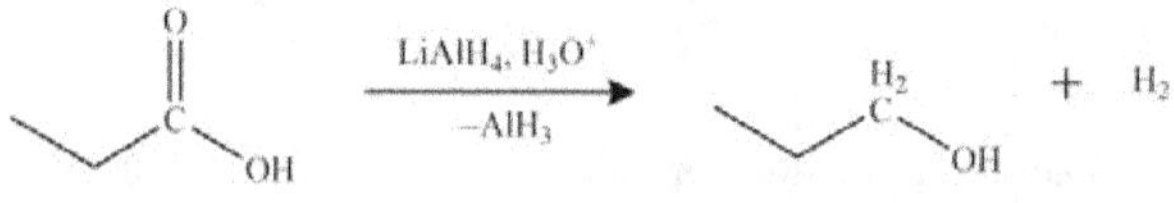

The mechanism for the carboxylic acid's reduction by metal hydride (by LiAlH4 in this case) to give primary alcohols is given below.

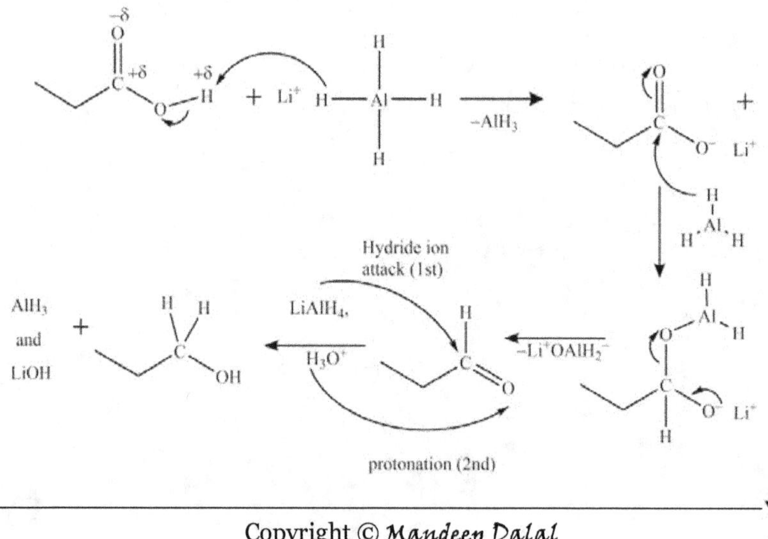

2. Reduction by aluminium hydride: The aluminum hydride, i.e., AlH_3, is one of the most common types of electrophilic addition for the reduction of carboxylic acid because simple borohydride cannot reduce carboxylic acids.

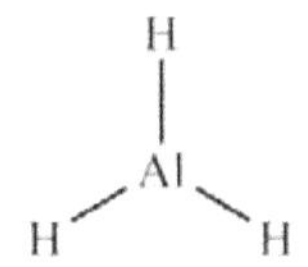

Aluminium hydride

In some cases, the reactivity of aluminium hydride is like lithium aluminium hydride; whereas sometimes it acts as borane (BH_3).

The hydride anion's addition to carboxylic acid results in a complex series of transition states which in turn gives rise to primary ($1°$) alcohols on protonation.

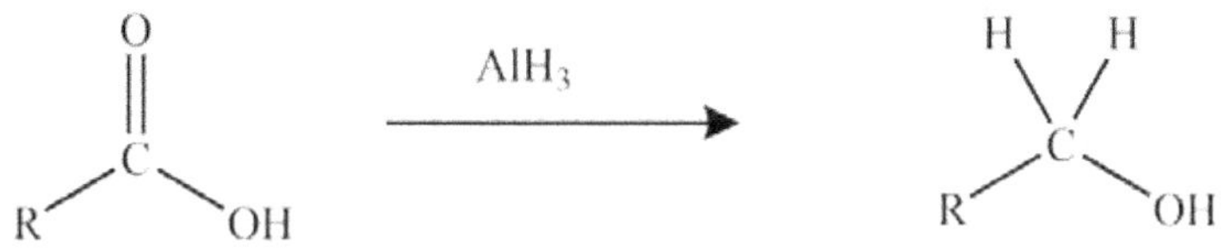

The mechanism responsible for the reduction of carboxylic acids by aluminium hydride involves the following steps.

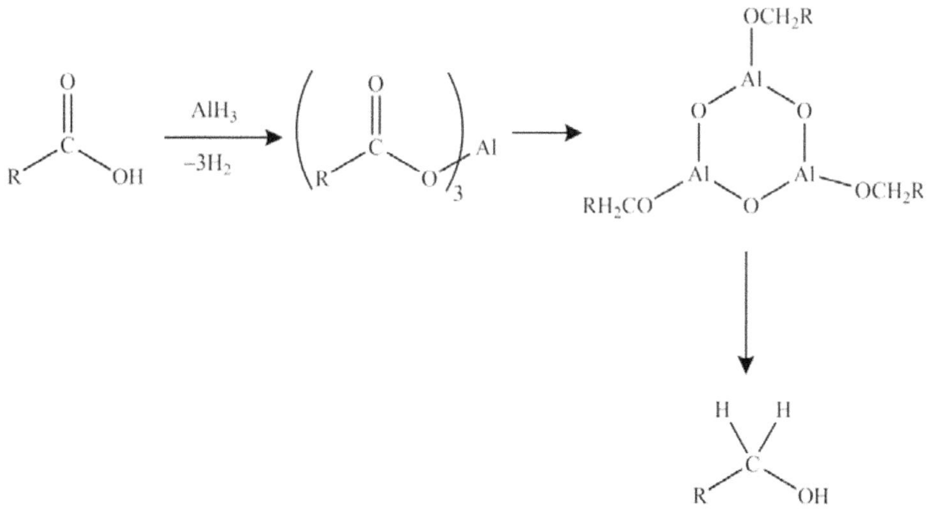

> ➤ *Metal Hydride Reduction of Esters*

Metal hydrides like lithium borohydride and lithium aluminium hydride can be used to reduce the carbonyl compounds like esters.

1. Reduction by lithium Borohydride: The lithium borohydride ($LiBH_4$) is one of the most common sources of the hydride nucleophile. The hydride anion is produced during reaction because of the polar nature of the metal-hydrogen bond.

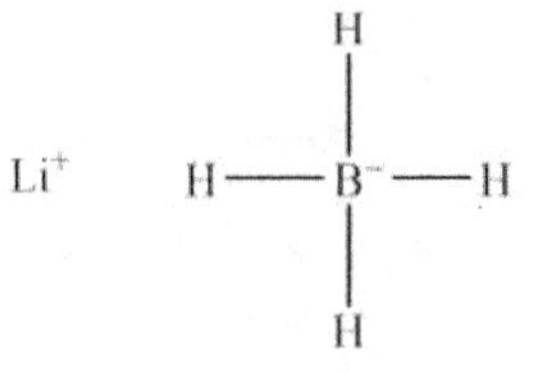

Lithium borohydride

The hydride anion's addition to carbonyl compound results in an alkoxide anion which in turn gives rise to the reduced product.

Two subsequent hydride anions' additions to ester result in an alkoxide anion which in turn gives rise to primary (1°) alcohols on protonation.

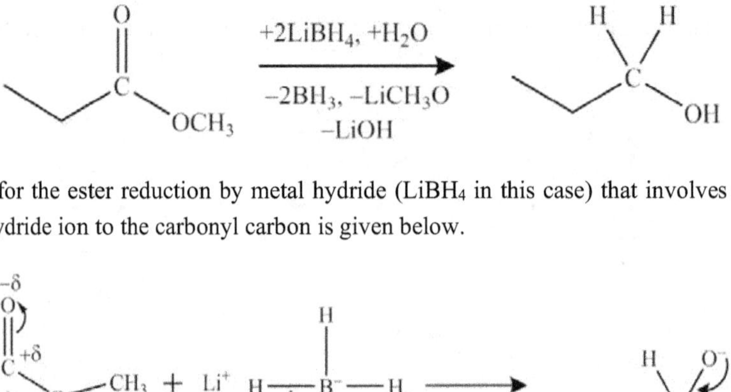

The mechanism for the ester reduction by metal hydride ($LiBH_4$ in this case) that involves the nucleophilic addition of the hydride ion to the carbonyl carbon is given below.

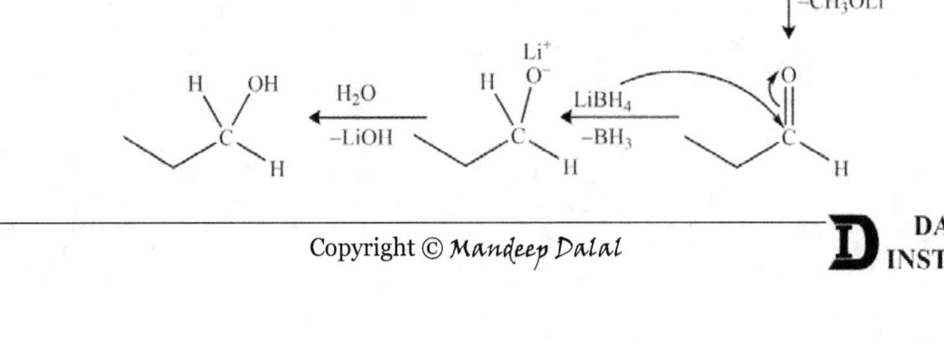

2. Reduction by Lithium aluminium hydride: The lithium aluminium hydride ($LiAlH_4$) is one of the most common sources of the hydride nucleophile. The hydride anion is produced during reaction because of the polar nature of the metal-hydrogen bond.

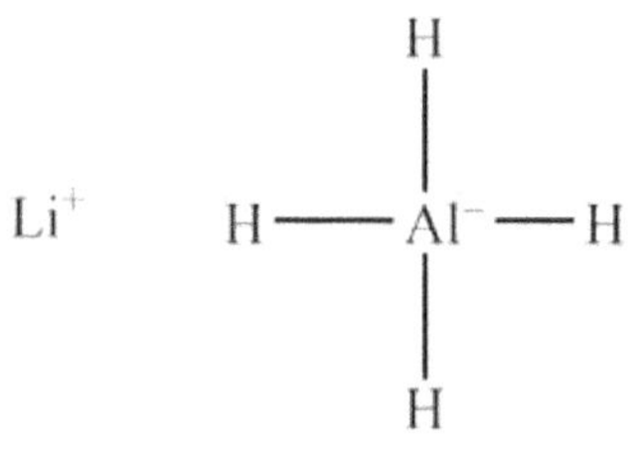

$$Li^+ \qquad H-Al^--H$$

Lithium aluminium hydride

The hydride anion's addition to carbonyl compound results in an alkoxide anion which in turn gives rise to the reduced product.

Two subsequent hydride anions' additions to ester result in an alkoxide anion which in turn gives rise to primary (1°) alcohols on protonation.

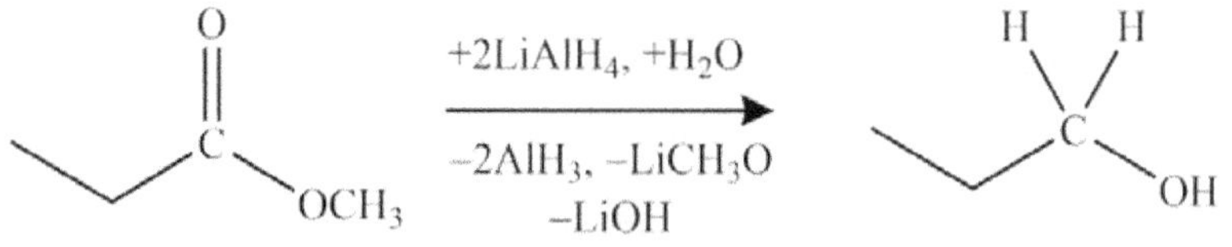

The mechanism for the ester reduction by metal hydride ($LiAlH_4$ in this case) that involves the nucleophilic addition of the hydride ion to the carbonyl carbon is given below.

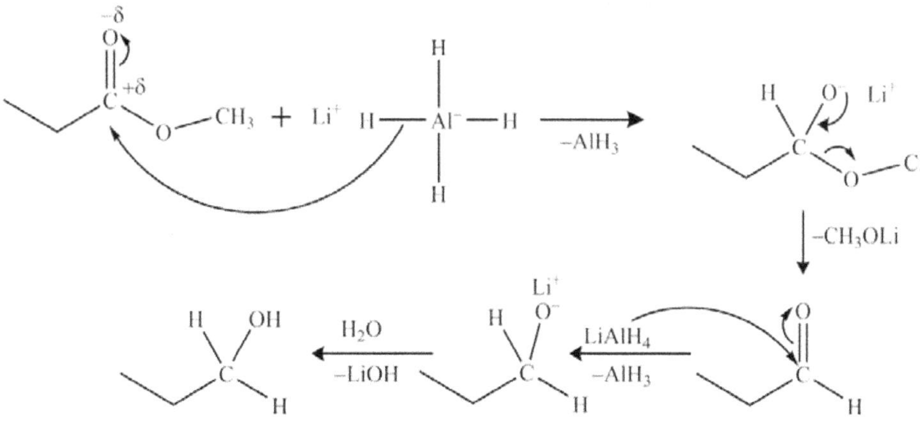

> ### *Metal Hydride Reduction of Nitriles*

The nitriles' reduction may simply be defined as the chemical transformation in which a nitrile is reduced to either an aldehyde or an amine by the use of a suitable reagent. Many metal hydrides can be used to reduce the nitrile compounds to amines but $LiBH_4$ and $LiAlH_4$ are most common.

1. Reduction by lithium borohydride: The lithium borohydride ($LiBH_4$) is one of the most common sources of the hydride nucleophile. The hydride anion is produced during the reaction because of the polar nature of the metal-hydrogen bond.

$$Li^+ \qquad H-\overset{\overset{\displaystyle H}{|}}{\underset{\underset{\displaystyle H}{|}}{B^-}}-H$$

Lithium borohydride

The hydride anion's addition to nitrile compounds results in an anion which in turn gives rise to a reduced product.

Two subsequent additions of hydride anions to the carbon-nitrogen bond result in a lithium salt which in turn gives rise to primary (1°) amines on protonation.

$$R-C\equiv N \quad \xrightarrow[-BH_3]{LiBH_4/\ H_2O/\ H^+} \quad R-\overset{\overset{\displaystyle H}{|}}{\underset{\underset{\displaystyle H}{|}}{C}}-NH_2 \quad + \quad LiOH$$

The mechanism for the nitriles' reduction by metal hydride ($LiBH_4$ in this case) that involves the nucleophilic addition of the hydride ion to the carbon-nitrogen bond is given below.

2. Reduction by Lithium aluminium hydride: The lithium aluminium hydride (LiAlH$_4$) is one of the most common sources of the hydride nucleophile. The hydride anion is produced during reaction because of the polar nature of the metal-hydrogen bond.

Lithium aluminium hydride

The hydride anion's addition to nitrile compounds results in an anion which in turn gives rise to a reduced product.

Two subsequent additions of hydride anions to the carbon-nitrogen bond result in a lithium salt which in turn gives rise to primary (1°) amines on protonation.

The mechanism for the nitriles' reduction by metal hydride (LiAlH$_4$ in this case) that involves the nucleophilic addition of the hydride ion to the carbon-nitrogen bond is given below.

DALAL INSTITUTE

❖ Addition of Grignard Reagents, Organozinc and Organolithium Reagents to Carbonyl and Unsaturated Carbonyl Compounds

As far as the addition to carbon-heteroatom multiple bonds is concerned, three organometallic reagents are much more important than the others; organomagnesium (Grignard reagents), organozinc, and oreganolithium compounds. In this section, we will discuss the addition of these three types of reagents to carbonyl and unsaturated carbonyl compounds.

➤ *Addition of Grignard Reagents to Carbonyl and Unsaturated Carbonyl Compounds*

The typical addition mode of organomagnesium compounds (Grignard reagent) to common carbonyl compounds like ketone and aldehydes is given below.

Halomagnisium alkoxide

The halomagnesium alkoxide thus formed can react with H_2O (when an HX type mineral acid is available) to result in alcohols; which in turn, could be susceptible to dehydration (acid-catalyzed) if it is tertiary-type. To stop the alcohol's dehydration, we need to add some ammonium chloride (NH_4Cl) to the water so that because its acidic character can be employed to transform ROMgX to ROH.

or

It is also worthy to note that bulky groups in the keto group, or in the reagent itself, greatly affect the nucleophilic addition in a negative way, or gets completely prohibited in some cases.

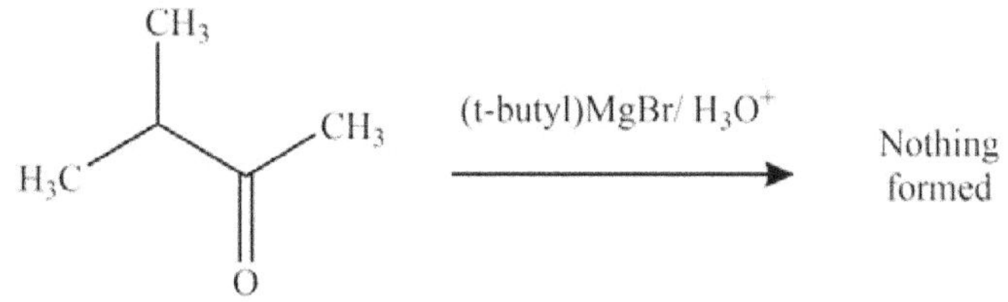

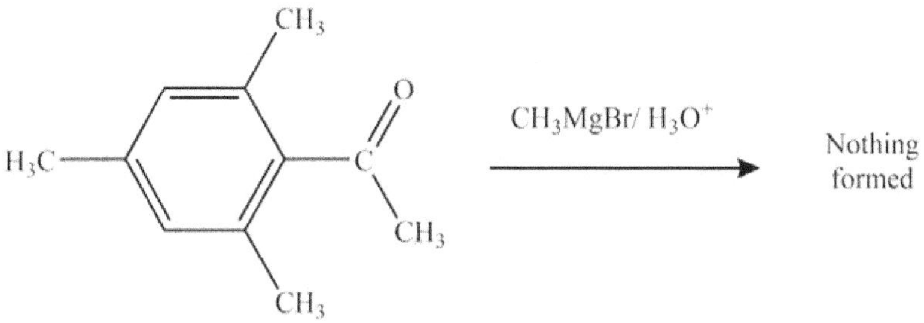

However, if a β-hydrogen is present in the bulky group of Grignard reagent, the transfer of hydride ion can result in the reduction of even highly hindered ketones. Moreover, if this is carried out via chiral Grignard reagent, the resulting product will also become optically active proving that the transfer of hydride ion does take place via the generation of cyclic transition state having six members.

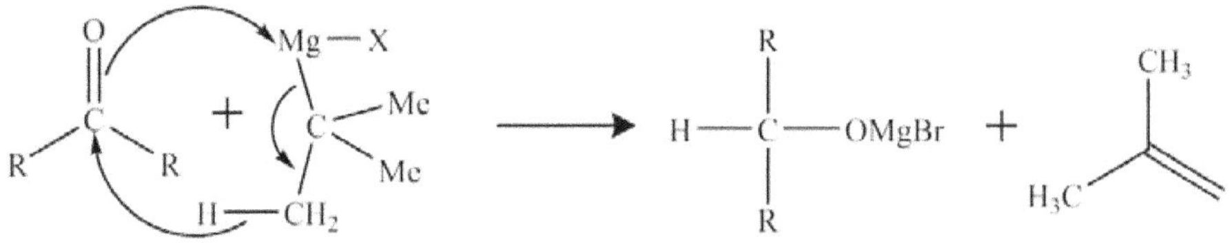

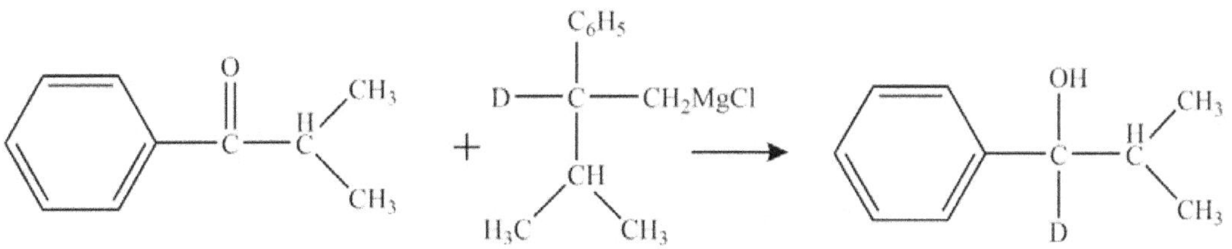

Also, if the Grignard addition takes place in cyclic ketones, the nucleophilic attack will happen from the carbonyl's face with less steric hindrance.

Since the organomagnesium reagent moves towards the substrate from a less hindered end, the final products' nature is also a function of the Grignard reagent used as it will push the hydroxy group to the axial site. On the other hand, a less bulky R (in comparison to OH) will make the hydroxy group occupy an equatorial site.

On a final note, if the aldehyde or the ketone used is α-, β-unsaturated, the nucleophilic addition of Grignard reagents becomes must faster and effective than normal carbonyl compounds, yielding 1, 4- and 1, 2-adddition products simultaneously. Moreover, these α-, β-unsaturated ketone and aldehydes give 1, 4- and 1, 2- adducts as the major products, respectively.

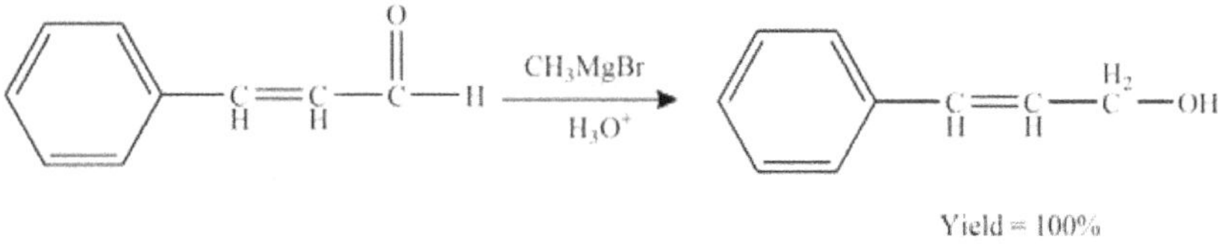

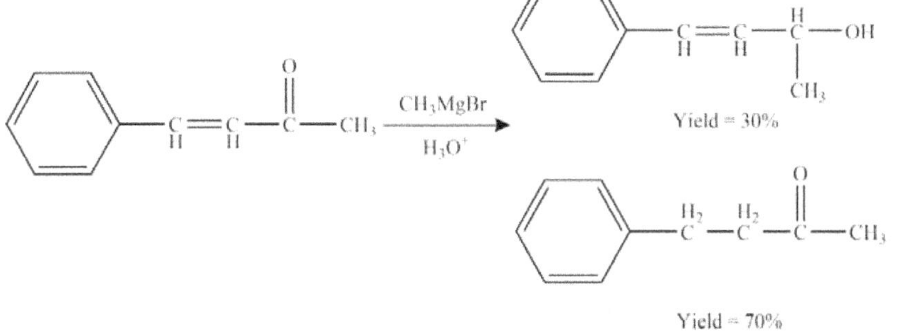

Yield = 100%

Yield = 30%

Yield = 70%

Nevertheless, it should also be noted that only 1, 4-adduct will be obtained if the addition over ketone is carried out in the presence of Cu_2Br_2.

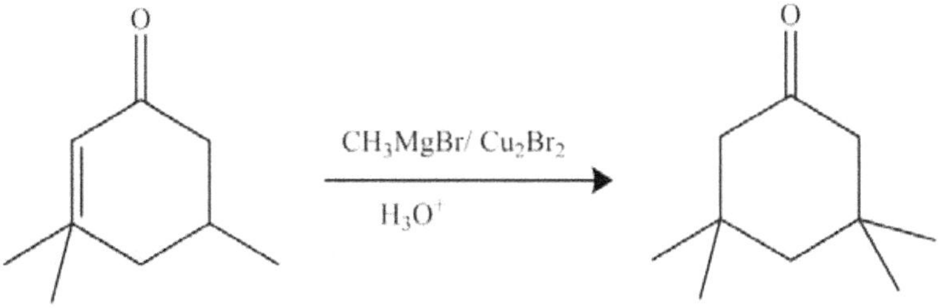

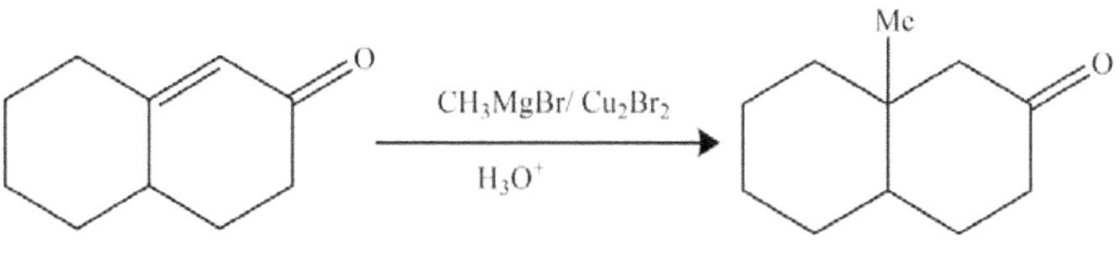

Since the Grignard reagents are very strong bases, they are not suitable to act as nucleophiles with substrate containing acidic hydrogens. In other words, Grignard reagents will act as a base and will abstract the acidic hydrogen instead of participating as a nucleophile to attack the carbonyl group.

> ➤ *Addition of Organozinc Reagents to Carbonyl and Unsaturated Carbonyl Compounds*

The rate of reaction for carbonyl compounds with dialkylzinc reagents is quite slow. It has also been observed that the rate for higher dialkylzinc is even lesser than lower dialkylzinc reagents. For instance, the reaction of diethylzinc with acetaldehyde takes hours for completion whereas the higher homologs may even take weeks. Nevertheless, allylzinc reagents show greater reactivity towards nucleophilic addition than normal dialkylzinc systems. Furthermore, the metal halide Lewis acids have been shown to enhance the rate of addition via dialkylzinc reagents.

$$H_3C-\overset{\overset{\textstyle O}{\|}}{C}-H \;+\; (C_2H_5)_2Zn \;\;\xrightarrow[\text{ether}]{MgBr_2}\;\; H_3C-\overset{\overset{\textstyle OH}{|}}{\underset{H}{C}}-C_2H_5$$

Yield = 60%

The heteroatom's presence at the α-site (relative to CO group) has also been found to be supportive of nucleophilic addition organozinc reagents.

$$+ \;\; (C_2H_5)_2Zn \;\;\xrightarrow{\text{ether}}$$

On a final note, it has also been proved that many titanium catalysts are supportive of the reactivity of organozinc reagents, specially $TiCl_4$ and $Ti(O^iPr)_4$.

$$R-\overset{}{\underset{H}{C}}=\overset{\overset{\textstyle Br}{|}}{C}-CHO \;\;\xrightarrow[\;Ti(O^iPr)_4\;]{(C_2H_5)_2Zn\;/\;H_3C-C_6H_5}\;\; R-\overset{}{\underset{H}{C}}=\overset{\overset{\textstyle Br}{|}}{C}-\overset{\overset{\textstyle OH}{|}}{\underset{H}{C}}-C_2H_5$$

$$Bu_3Sn-\overset{}{\underset{H}{C}}=\overset{\overset{\textstyle Br}{|}}{C}-CHO \;\;\xrightarrow[\;Ti(O^iPr)_4\;]{(ClC_4H_8)_2Zn\;/\;H_3C-C_6H_5}\;\; Bu_3Sn-\overset{}{\underset{H}{C}}=\overset{}{\underset{H}{C}}-\overset{\overset{\textstyle OH}{|}}{\underset{H}{C}}-(C_2H_4)_4-Cl$$

> ➤ *Addition of Organolithium Reagents to Carbonyl and Unsaturated Carbonyl Compounds*

Organolithium reagents react with organic carbonyl derivatives to generate lithium alkoxide, which in turn gives rise to alcohols upon hydrolysis.

Sometimes a bioproduct via the α-deprotonation can also be obtained because besides being a nucleophilic attacker, organolithium is a powerful base too.

Furthermore, it is also worthy to note that organolithium reagents are better than their organomagnesium counterparts; and therefore, some highly hindered carbonyls (who were unable to react at all with Grignard reagents) can also be used as a substrate to produces quite stable products.

On a final note, the conjugated addition doesn't happen in the case of organolithium, leaving 1, 2-adducts as the only products.

❖ Wittig Reaction

The Wittig olefination (or Wittig reaction) may simply be defined as a chemical transformation where a ketone or aldehyde reacts with a triphenyl phosphonium ylide (Wittig reagent).

The conversion of aldehydes and ketones to alkenes is one of the most common uses of Wittig reactions. Usually, the Wittig reaction is employed to add a methylene group using $Ph_3P=CH_2$ (methylenetriphenylphosphorane or Wittig reagent). The importance of Wittig reaction can be imagined by the fact that George Wittig, who invented this reaction, was awarded the Nobel prize in 1979 for the same work.

With help of Wittig reagent, a camphor-like ketone, which has a very much sterically hindered carbon, can also be transformed into its methylene derivative. Now before we proceed further to study different aspects of Wittig reaction, we need to first know what a Wittig reagent actually is and how does it behave around different kinds of substrates.

➢ *The Wittig Reagent (An Organophosphorus Ylide)*

The Wittig reagent is a ylide, and a ylide may be defined as a compound with opposite charges on adjacent atoms both of which have complete octets. These ylides are obtained as the zwitterionic conjugate bases of the cationic part of phosphonium salts.

An Ylide

Since these ylides are stabilized by *pπ-dπ* bonding, the carbanions adjacent to the phosphonium centers also get stability benefits from the same. It is also obvious that the phosphorus's ability to hold more than eight valence electrons permits for a resonance structure with double-bonded; and therefore, enhances the stability.

> ➤ *Mechanism of Wittig Reaction*

The NMR studies have confirmed the formation of two intermediates after the generation of the first carbon-carbon bond during the Wittig reaction, the betaine (a dipolar species) and oxaphosphatane (a four-membered heterocyclic structure). The final product will be obtained by the cleavage of oxaphosphatane to alkene and phosphine oxide which is irreversible and exothermic in nature. Precisely, the mechanism can primarily be divided into three steps as given below.

1. Nucleophilic attack on the carbonyl:

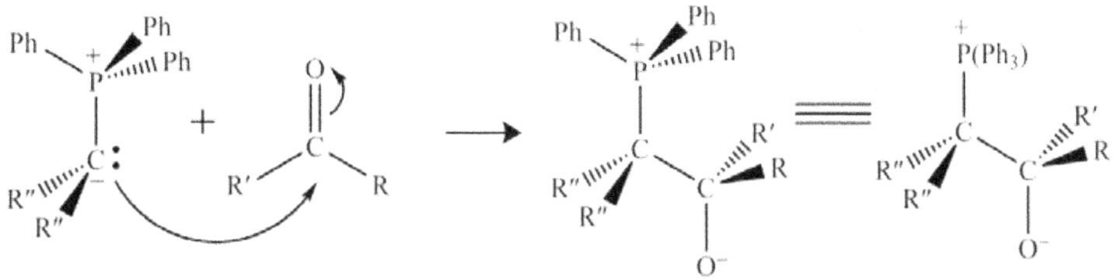

Betain

2. Formation of four-membered ring:

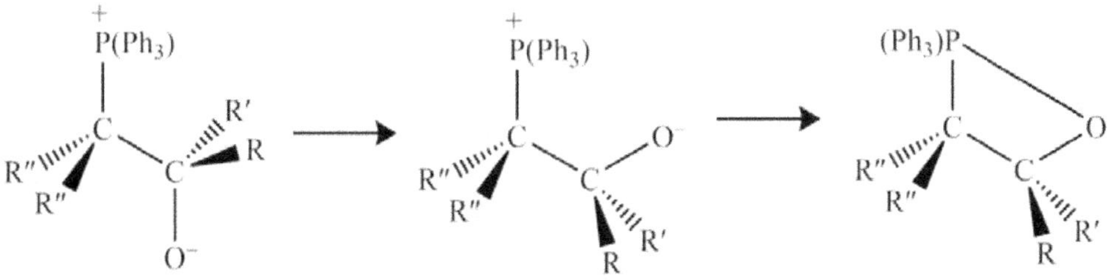

Oxaphosphatane

3. Generation of the alkene:

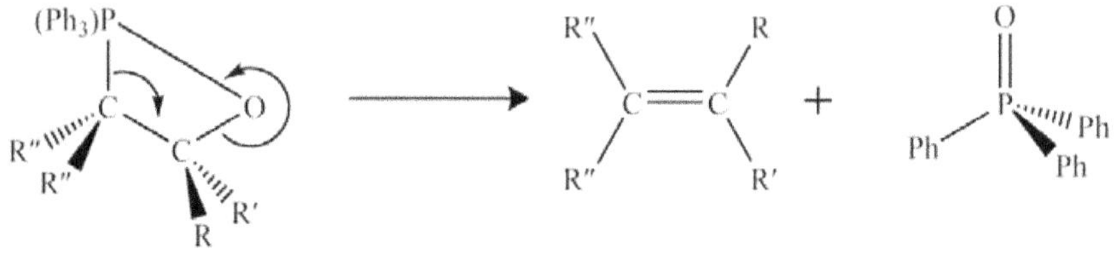

Oxaphosphatane

A major benefit of the alkene synthesis via Wittig's route is that, unlike alcohol dehydration, the site of the double bond is fixed absolutely.

> *Stereochemistry of Wittig Reaction*

In the case of aldehydes, the geometry around double bonds can easily be predicted by analyzing the ylide's nature. With unstable ylides (R_3 = alkyl), (Z)-alkenes are formed with reasonable to very high selectivity. With stable ylides (R_3 = ester or ketone), (E)-alkenes are formed with a very high magnitude of selectivity. The selectivity ratio (E/Z) is usually very poor with semi-stabilized ylides (R_3 = aryl).

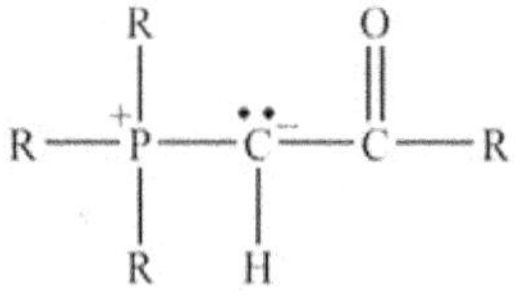

will give E-alkene 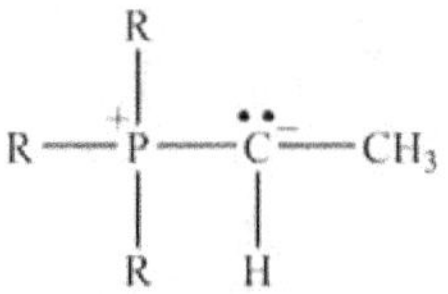   will give Z-alkene

If we want to get (E)-alkene but from a destabilized ylide, the Schlosser modification of the Wittig reaction can be employed. Otherwise, the (E)-alkene selectively can also be obtained via Julia olefination and its different variants. Since the (E)-enoate (α, β-unsaturated ester) are prepared via Horner-Wadsworth-Emmons reaction, the same can be used as a substitute for the Wittig reaction. On a final note, the Still-Gennari modification of the Horner-Wadsworth-Emmons reaction can be used to get (Z)-enoate.

> *Examples of Wittig Reaction*

Some of the most common examples of organic chemical transformation via Wittig reagent are given below.

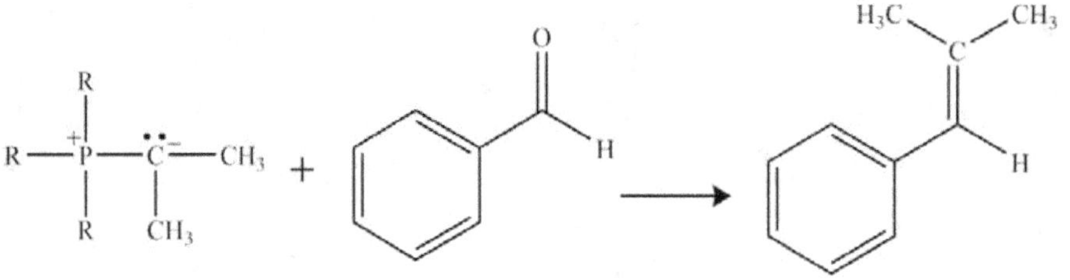

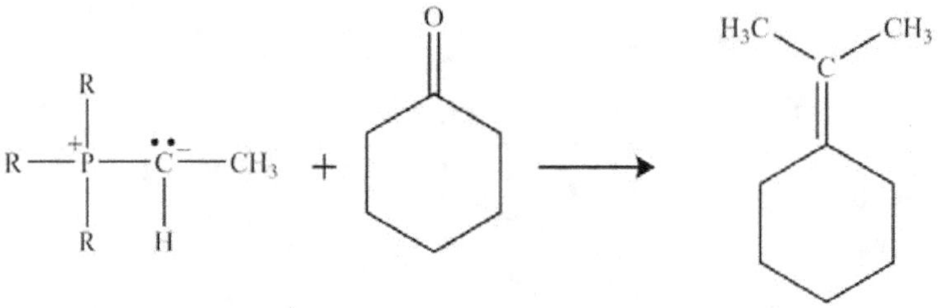

It has been observed that the Wittig reagents usually tolerate carbonyl compounds with numerous types of functional groups like OH, OR, epoxide, aromatic nitro, and ester groups.

> ➢ *Applications of Wittig Reaction*

Some of the most common applications of organic chemical transformation via Wittig reagent are given below.

1. The conversion of aldehydes and ketones to alkenes is one of the most common uses of Wittig reactions.

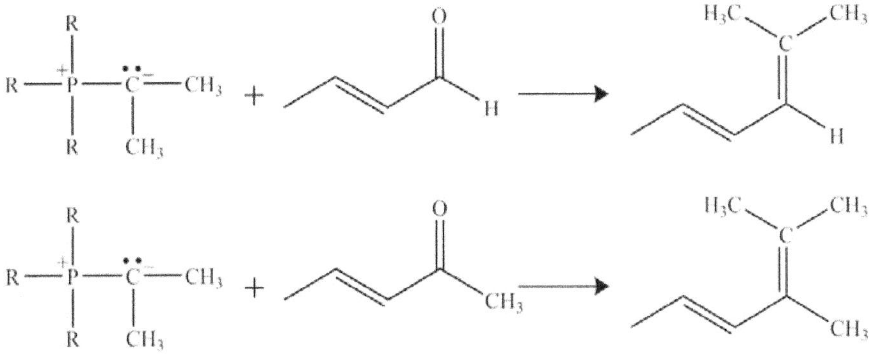

2. The Schlosser modification Wittig reaction can be used to get allylic alcohols by the reaction of the betaine ylide with a secondary aldehyde.

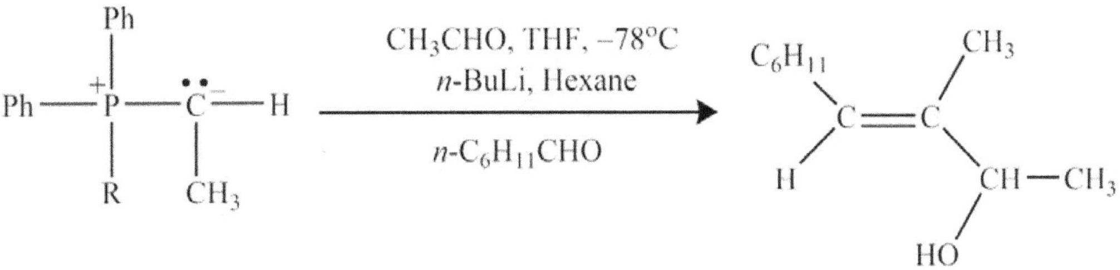

3. Even a sterically hindered ketone such as camphor can be converted to its methylene derivative.

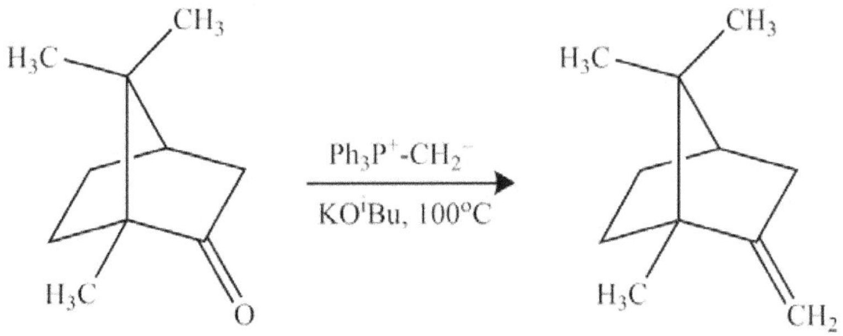

➢ *Limitations of the Wittig reaction*

Some of the most common limitations of organic chemical transformation via Wittig reagent are given below.

1. The Wittig reaction proceeds mainly via the erythro betaine intermediate that gives rise to the Z-alkene, which is problematic if the E-isomer is the desired product. This limitation can be overcome by converting the erythro betaine into threo betaine using phenyllithium at low temperature which can afford to yield the E-alkene (Schlosser modification).

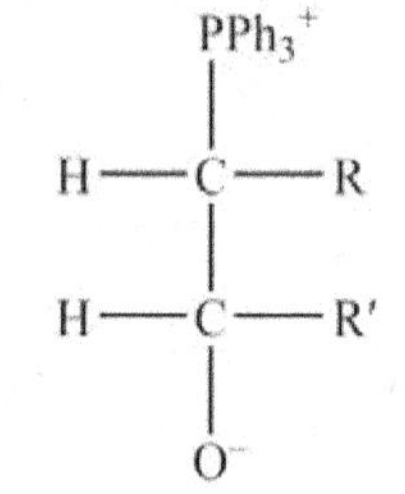

erythro betain
(gives Z-alkene)

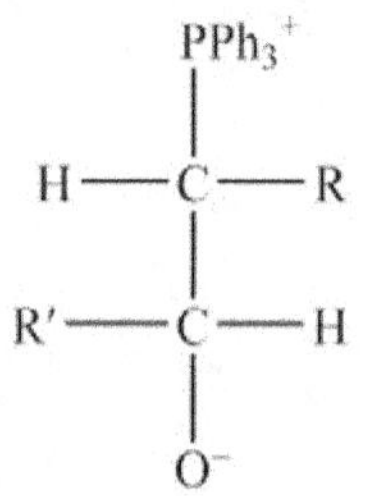

threo betain
(gives E-alkene)

2. The yield given by conventional Wittig reaction is very low when a sterically hindered ketone is used, and the rate of transformation was also found to be very small. This is especially true for stabilized ylides. This limitation can be overcome by using a phosphonate ester (Horner-Wadsworth-Emmons reaction).

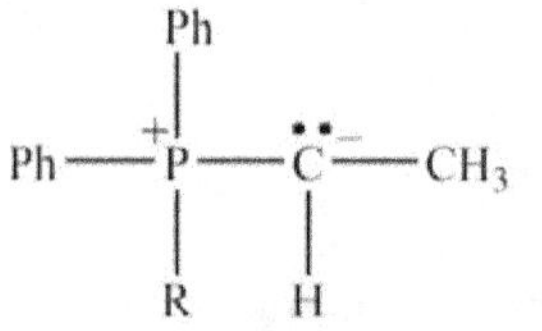

Non-stablized ylide

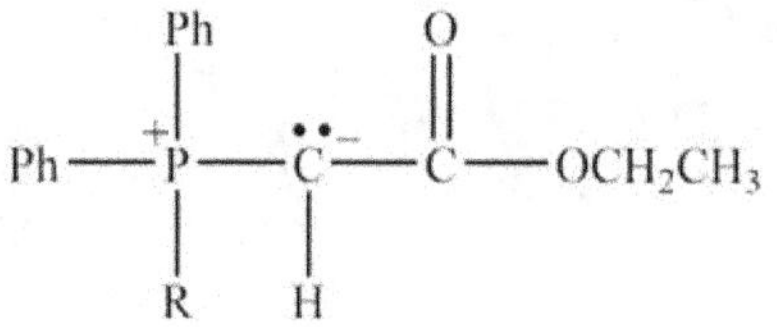

Stablized ylide

❖ Mechanism of Condensation Reactions Involving Enolates: Aldol, Knoevenagel, Claisen, Mannich, Benzoin, Perkin and Stobbe Reactions

So far we have discussed the structure and reactivity of addition reactions of carbon-heteroatom multiple bonds, now we need to study some of the most important name reactions (Aldol, Knoevenagel, Claisen, Mannich, Benzoin, Perkin, and Stobbe condensations) involving this type of mechanism.

➢ *Aldol Condensation*

The aldol condensation may simply be defined as a condensation reaction in organic chemistry where an enol or an enolate ion reacts with a carbonyl compound to give a β-hydroxyaldehyde or β-hydroxyketone (an aldol addition), followed by the dehydration to form a conjugated enone.

The aldol condensation was invented in 1872 by a French chemist, Charles Wurtz, who first synthesized the β-hydroxy aldehyde using acetaldehyde.

Aldol addition (first part)

Dehydration of Aldol adduct (second part)

Mechanism of aldol condensation: The mechanism of aldol condensation can be fragmented into two parts; the first part is a simple aldol reaction (including 3 steps), whereas the second part includes the dehydration reaction (1 step i.e., elimination of an alcohol or H_2O molecule).

The dehydration of aldol product can occur via two pathways; a strong base like NaOH deprotonates the aldol product to an enolate, which then eliminates via the E_1CB route, whereas the second pathway for dehydration needs acid catalyzation for the enol mechanism. The dehydration part may also take place by decarboxylation if an activated carboxyl group is available. All four steps for aldol condensation are given below.

i) Step 1:

Enolate ion

ii) Step 2:

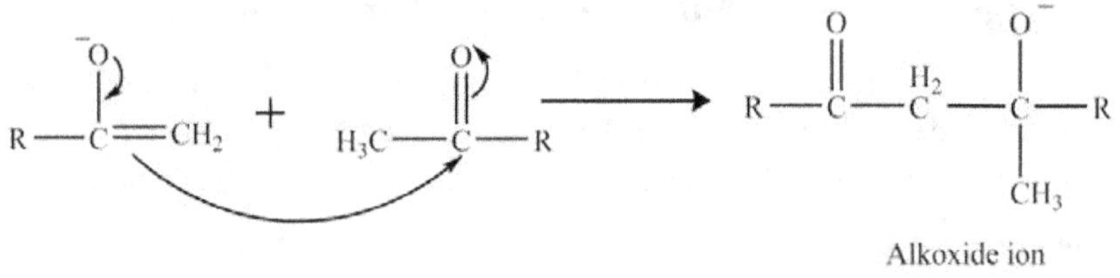

Alkoxide ion

iii) Step 3:

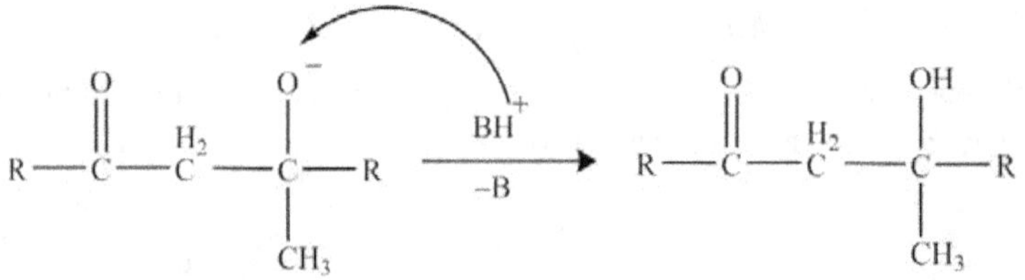

iii) Step 4:

Aldol condensation product

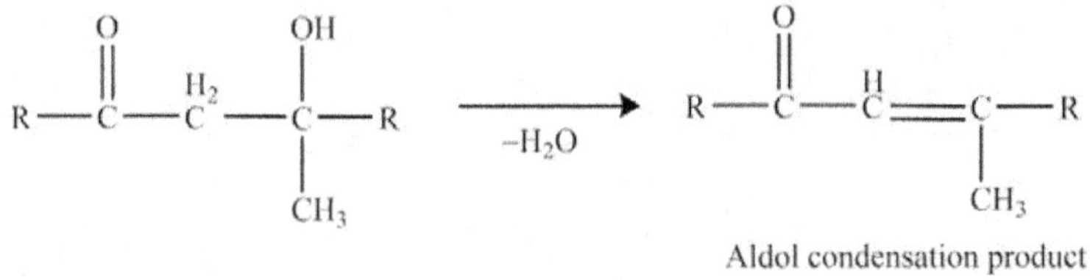

Furthermore, the aldol condensation may also be fine-tuned under kinetic or thermodynamic control for the desired product.

For the acid-catalyzed mechanism, the enol form acts as the nucleophile rather than electrophile, which is obviously triggered by the protonation of oxygen.

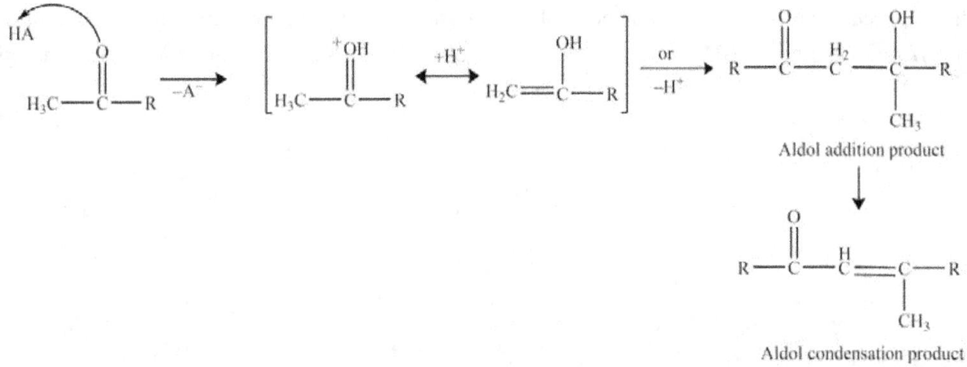

Aldol addition product

Aldol condensation product

Stereochemistry of aldol condensation: Two stereoisomers will be obtained if enolate-yielding ketones are unsymmetrical. Consequently, syn- and anti-isomers will be formed from the condensation between aldehyde and ketonic enolate. It has also been observed that syn-isomer is generally the major product and ant is minor. Furthermore, the substituted enolate can exist either as Z- or E configuration, the corresponding treatment of aldehyde will give rise to syn- and anti-product.

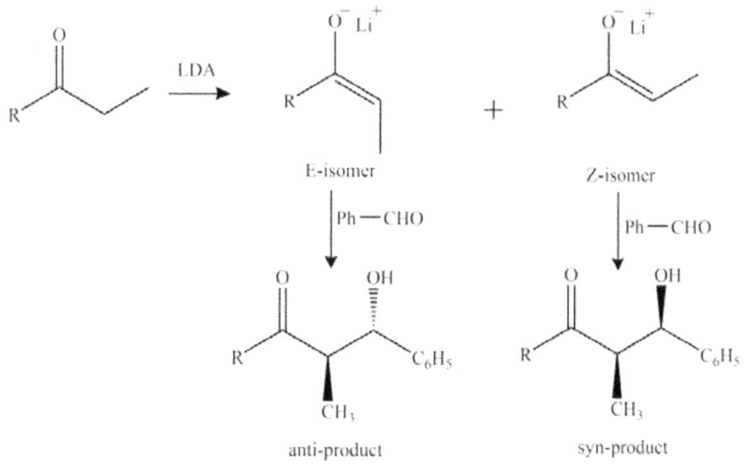

Since some enolate can only exist as E-configuration, they will exclusively result in the anti-product; for instance, consider the following reaction.

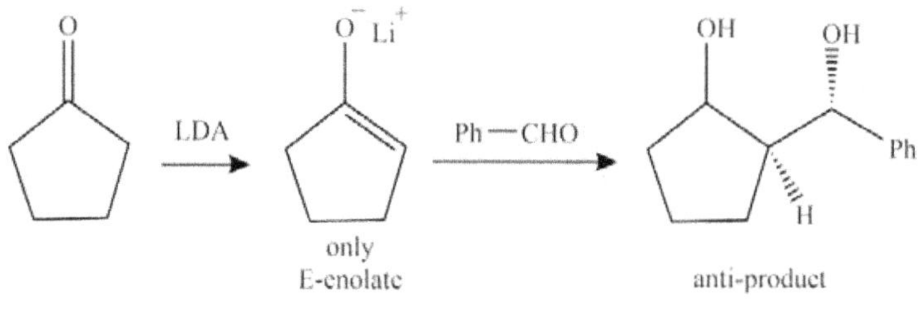

Ketones with very bulky group can only react Z-enolate giving exclusively syn product; consider the reaction between tertiary butyl ketone and benzaldehyde.

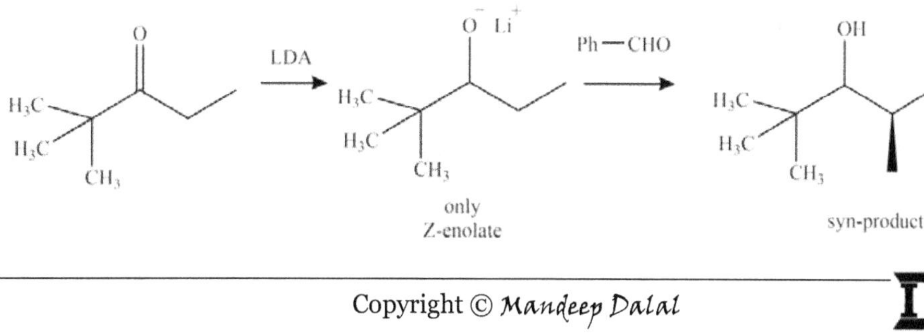

Examples of aldol condensation: Some of the most common examples of organic chemical transformation involving aldol condensation are given below.

i) The aldol addition of acetaldehydes followed by dehydration.

$$H_3C-CHO \;+\; H_3C-CHO \xrightarrow{\text{Dil. NaOH}} H_3C-\underset{H}{\overset{OH}{C}}-\overset{H_2}{C}-CHO$$

Aldol addition (first part)

$$H_3C-\underset{H}{\overset{OH}{C}}-\overset{H_2}{C}-CHO \xrightarrow[-H_2O]{OH^-/\,\Delta} H_3C-\underset{H}{C}=\underset{H}{C}-CHO$$

Dehydration of Aldol adduct (second part)

ii) The aldol addition of aromatic aldehydes and ketones followed by dehydration.

$$\underset{Ar}{\overset{O}{\|}} \;+\; \underset{H \quad Ar'}{\overset{O}{\|}} \xrightarrow[-H_2O]{\text{Dil. NaOH}} \underset{Ar \qquad \qquad Ar'}{\overset{O}{\|}}$$

Applications of aldol condensation: Some of the most common applications of organic chemical transformation involving aldol condensation are given below.

i) The Aldol condensation reactions are vital to the synthesis of many organic compounds as they give a good way to form C−C bonds. For instance, the Robinson annulation involves an aldol condensation; the Wieland-Miescher ketone is a chief reactant for many organic reactions.

ii) Aldol condensation reactions are also taught in organic chemistry at the university-level as they can illustrate important reaction mechanisms. In other words, it includes the nucleophilic addition of an enolate to an aldehyde to give or "aldol" (aldehyde + alcohol) or a β-hydroxy ketone.

iii) The aldol condensation also finds its applications in the field of biochemistry. Nevertheless, the aldol condensation reactions in such cases are not officially condensation reactions because they don't involve the small molecule's loss.

> ### *Knoevenagel Condensation*

The Knoevenagel condensation may simply be defined as a nucleophilic addition of an active methylene compound to a carbonyl group followed by the elimination of a water molecule (i.e., dehydration), resulting in an α, β-unsaturated ketone generally.

This reaction is a modification to aldol condensation and was invented by a German chemist Emil Knoevenagel; and therefore, is also named after him.

Where the carbonyl group is a ketone or an aldehyde and $Z–CH_2–Z$ is the active methylene group. The catalyst used in this reaction is typically a weakly basic amine. Furthermore, it is also worthy to note that the active hydrogen component can also be of $Z–CHR–Z$ or $Z–CHR_1R_2$ form, where Z is an electron-withdrawing functional group.

Mechanism of Knoevenagel condensation: The mechanism of Knoevenagel condensation includes two steps where the step includes the deprotonation of methylene by a base to result in a resonance stabilized carbanion. The carbanion formed during the first step acts as a nucleophile, and attacks at the carbon of carbonyl group of the ketone to yield to give aldol addition product followed by dehydration (second step). Both the steps for clear depiction are illustrated below.

i) Step 1:

i) Step 2:

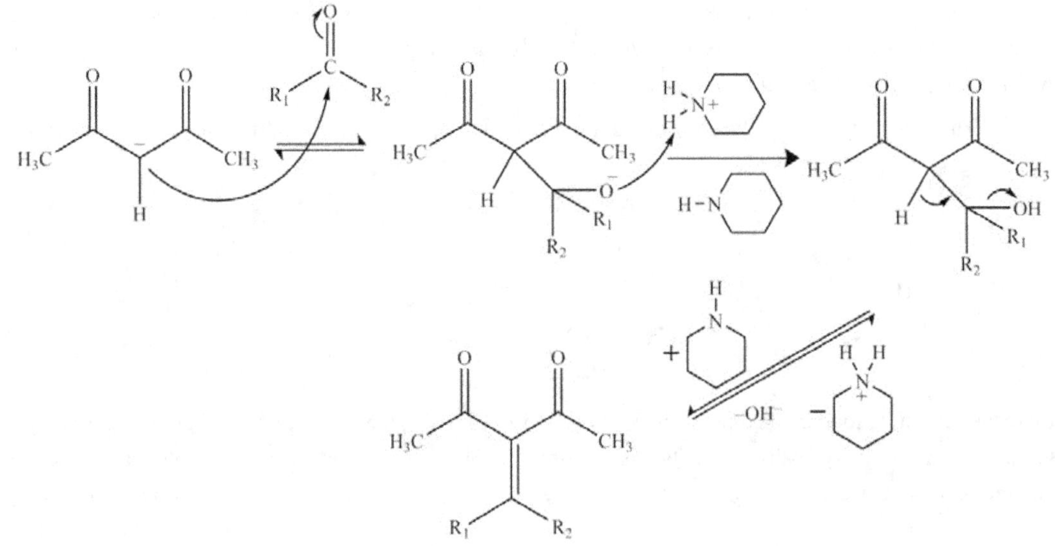

Stereochemistry of Knoevenagel condensation: In most of the Knoevenagel condensation reactions, both cis and trans isomers are formed guided by mainly steric and mode of attack.

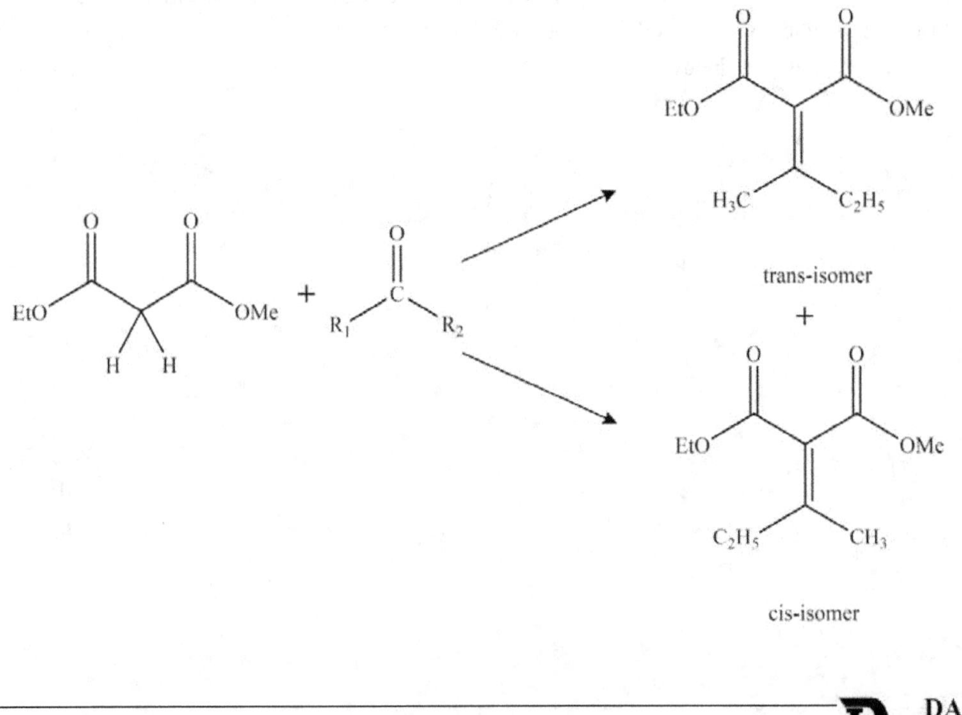

Examples of Knoevenagel condensation: Some of the most common examples of organic chemical transformation Knoevenagel condensation are given below.

1. The reaction between benzaldehyde and acetylacetone.

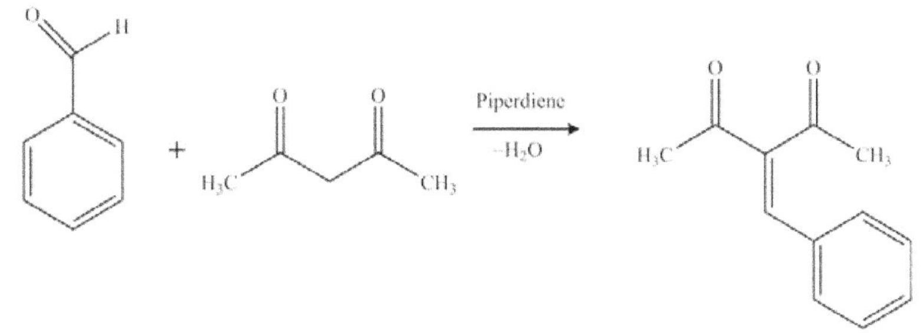

2. The reaction between benzaldehyde and malonic acid.

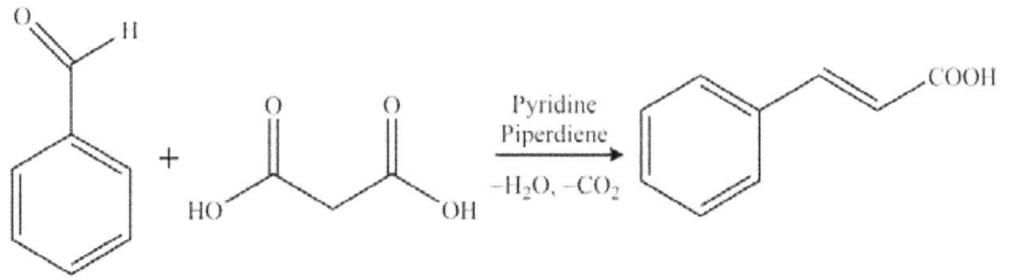

3. The reaction between 2-hydroxybenzaldehyde and diethyl malonate.

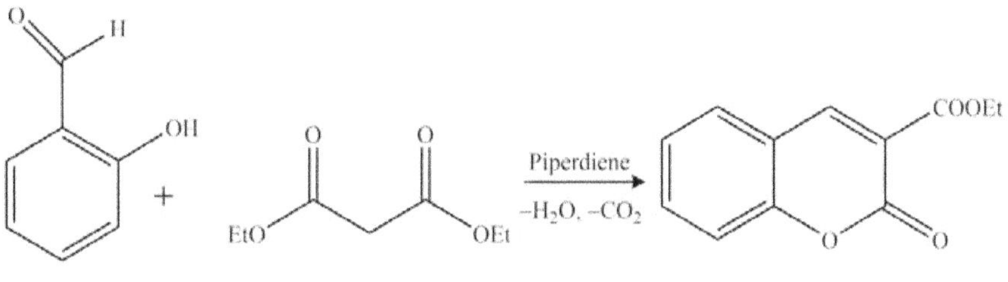

Applications of Knoevenagel condensation: Some of the most common applications of organic chemical transformation involving Knoevenagel condensation are given below.

1. The Knoevenagel condensation is the main step in the commercial production of antimalarial drug lumefantrine (a Coartem's component).

2. A Knoevenagel condensation reaction is confirmed in the reaction of thiobarbituric acid with 2-methoxybenzaldehyde in C_2H_5OH using piperidine as a basic assistant, yielding subsequent formation of a charge-transfer complex molecule.

3. The Knoevenagel condensation is also found in a multicomponent reaction with microwave-assisted synthesis with cyclohexanone, malononitrile, and 3-amino-1,2,4-triazole.

> ***Claisen Condensation***

The Claisen condensation may simply be defined as a chemical reaction giving carbon-carbon bond between two esters or one ester and another carbonyl compound in the availability of a strong base, resulting in a β-diketone or a β-keto ester.

This reaction is a modification to aldol condensation and was invented by a German chemist Rainer Ludwig Claisen in 1887; and therefore, it is also named after him. The primary condition for Claisen condensation is that one reagent (at least) must have α-hydrogen so that it can be enolized via deprotonation. A typical Claisen condensation is shown below.

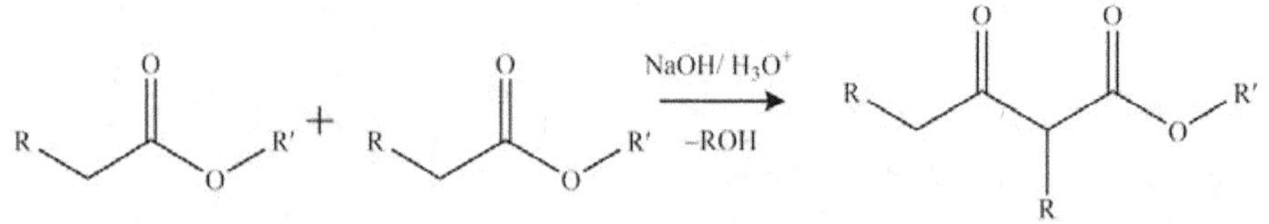

It is obvious that the molecule eliminated in this 'modified aldol condensation' is not water but alcohol. Now depending upon various enolizable and nonenolizable carbonyl compounds, many types of Claisen condensations can be obtained.

Mechanism of Claisen condensation: The mechanism starts with the detachment of α-proton by a strong base giving a resonance stabilized enolate anion. After that, the carbonyl carbon of the second ester is attacked by the enolate anion. The alkoxy anion is then eliminated, and reattached, followed by the elimination of the alcohol molecule. Finally, a proton from aqueous acid is added to neutralize the enolate to give rise to the final product (β-diketone in this case).

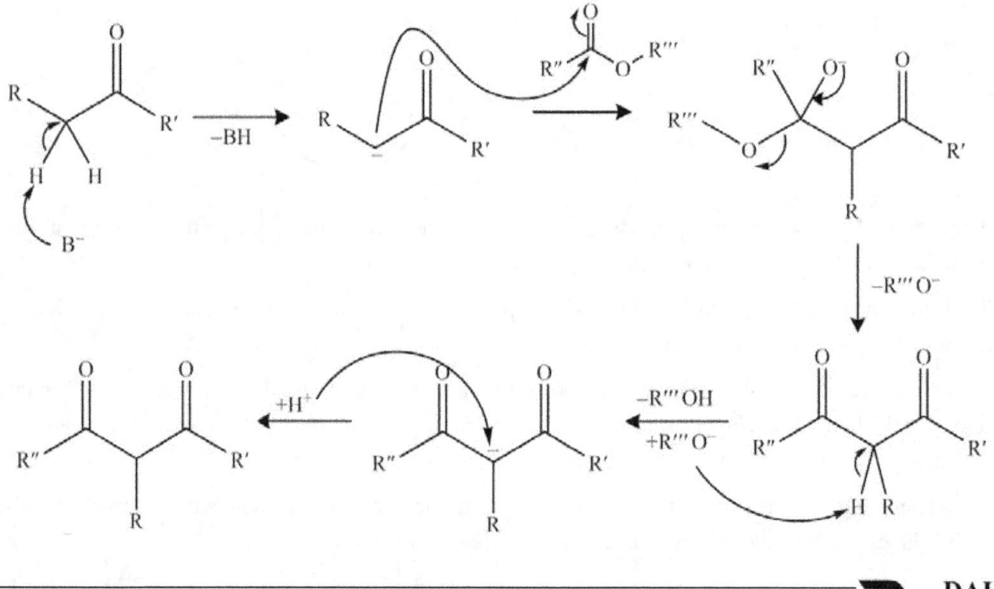

Stereochemistry of Claisen condensation: In the case of different R groups (R''' $\neq$ R'), the Claisen condensation will give rise to chiral β-diketones or β-diketoesters as shown below.

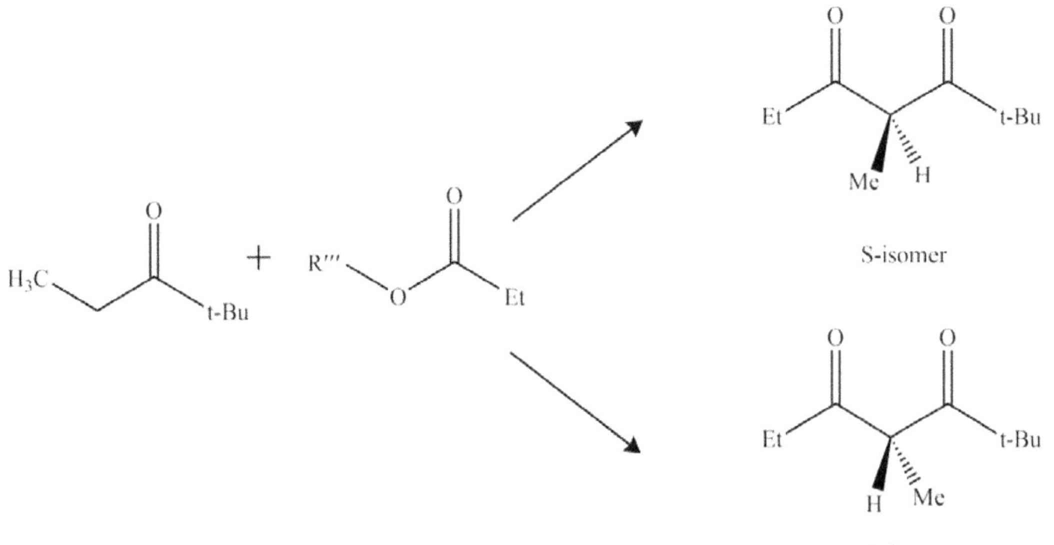

S-isomer

R-isomer

Examples of Claisen condensation: Some of the most common examples of organic chemical transformations involving Claisen condensation are given below.

1. The condensation reaction of ethyl acetate.

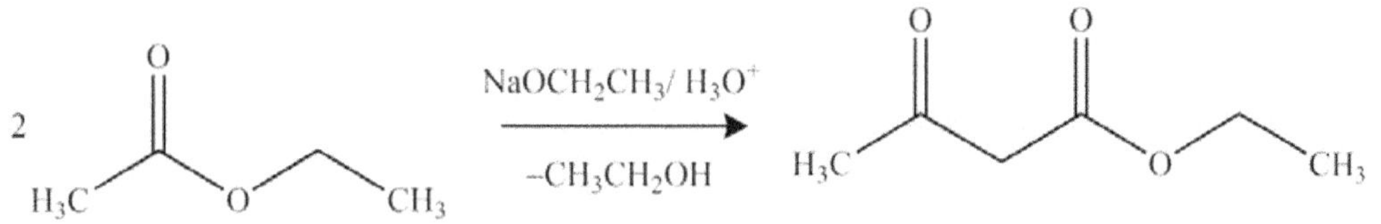

2. The condensation reaction of diethyl adipate.

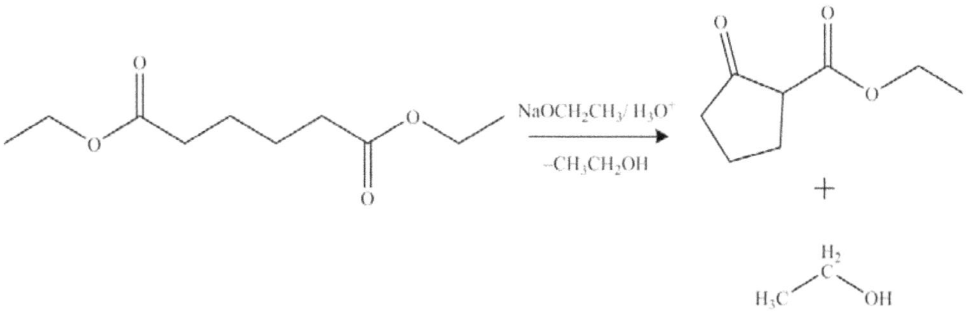

3. The condensation reaction of ethyl benzoate.

3. The reaction between 2,2-dimethylpentan-3-one and ethyl propionate.

Applications of Claisen condensation: Some of the most common applications of organic chemical transformation involving Claisen condensation are given below.

1. Crossed and simple Claisen condensations have been widely used in the preparation of a huge range of organic compounds, like terpenes, vitamins, flavones, alkaloids, etc.

2. Crossed Claisen condensation reaction between two different esters (both are having α-H) have little to no synthetic impoartance, and we will obtain an a mixture of four products. Nevertheless, if no α-hydrogen are available in one of the esters, it will act as a acceptor for carbanion and the self-condensation of the other ester is diminished. Most popularly used esters which have zero α-hydrogen are ethyl formate, ethyl benzoate, ethyl carbonate, ethyl oxalate, etc.

3. Since the esters are generally less acidic than ketones, the rate of their base-catalyzed condensation reaction (aldol-type) is very small; and therefore, the ketone can act as nucleophiles in crossed Claisen condensation reactions to give rise to a huge number of different kinds of products.

> ➢ ***Mannich Condensation***

The Mannich condensation may simply be defined as an organic chemical transformation where a carbonyl functional group's neighboring proton (acidic in nature) undergoes amino alkylation by formaldehyde and ammonia (or a secondary, or primary amine), giving rise to a β-amino-carbonyl compound called Mannich base.

This reaction was invented by an eminent German chemist Carl Mannich in 1912; and therefore, is also named after him.

The Mannich condensation is a case of nucleophilic addition of an amine to a carbonyl group trailed by the dehydration to yield a Schiff base, which in turn, reacts in an electrophilic addition mode with a compound containing acidic hydrogen (next step).

Mechanism of Mannich condensation: The Mannich condensation's mechanism begins with the generation of an iminium ion from the formaldehyde and the amine used. The protonated oxygen is highly acidic with a pKa value of −2. The reaction will be stopped when the carbonyl gets deprotonated by amine base; and therefore, it is necessary to perform at a pH of about 5. Hence, the right pathway should begin with a nucleophilic attack at carbonyl's carbon by the nitrogen atom.

iminium ion

The carbonyl compound like ketone will undergo tautomerization to yield enol form, which can attack the iminium ion afterward. It is also important to note that the enolization and Mannich addition can occur twice with methyl ketones, trailed by an β-elimination to give rise to β-amino enones.

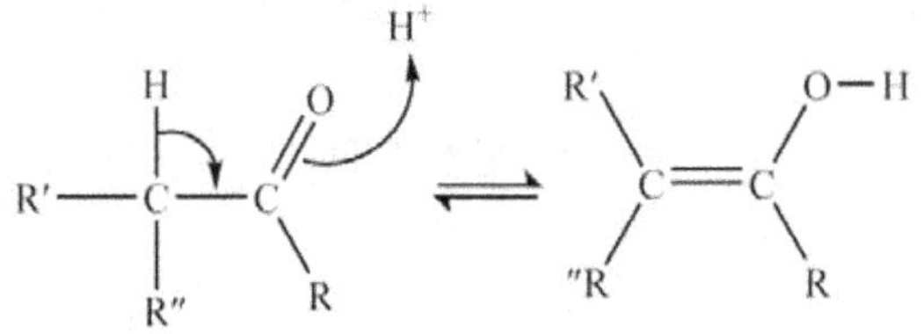

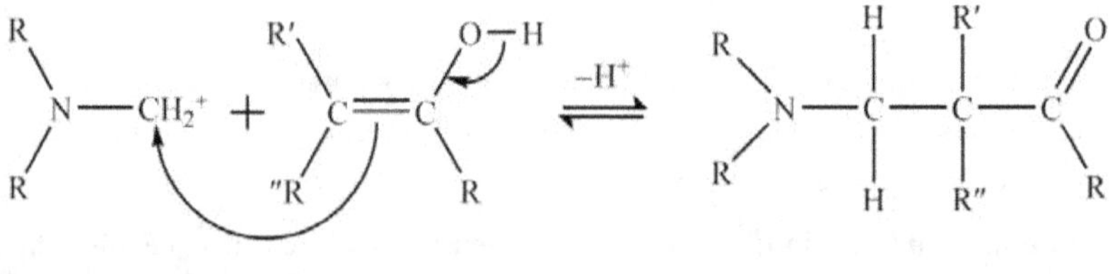

Stereochemistry of Mannich condensation: Asymmetric Mannich reactions have also been studied in the recent era. It has been observed that if two prochiral centers are present in an appropriately functionalized ethylene bridge of Mannich adduct, two diastereomeric pairs of enantiomers are obtained. One of the most commonly reported examples (also first) of asymmetric Mannich reaction was performed using (S)-proline as a naturally occurring chiral catalyst.

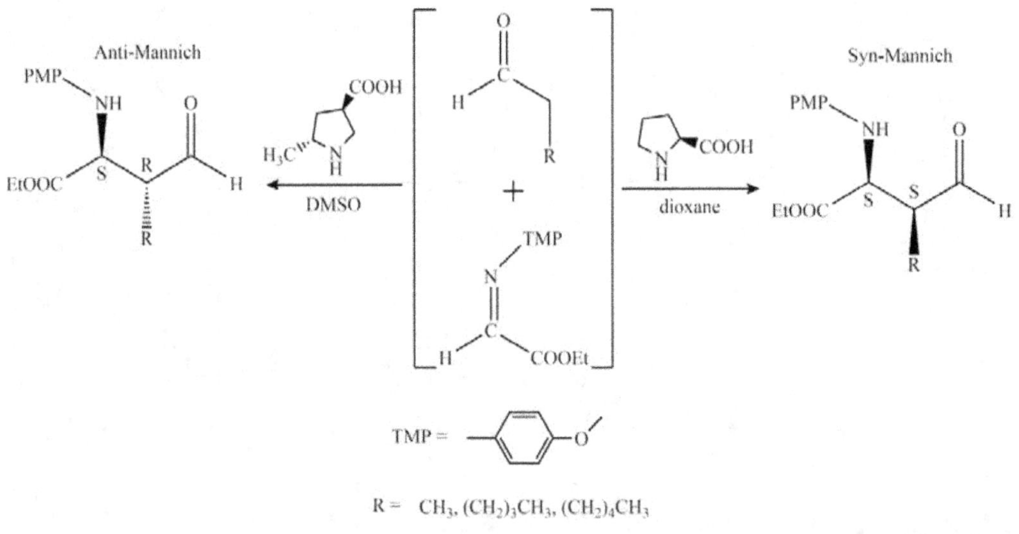

$R = CH_3, (CH_2)_3CH_3, (CH_2)_4CH_3$

Examples of Mannich condensation: Some of the most common examples of organic chemical transformation Mannich condensation are given below.

1. The reaction between aniline, benzaldehyde, and an aromatic ketone.

2. The reaction between amide, benzaldehyde, and an aromatic ketone.

3. Magnesium(II)-Binaphtholate-catalysed condensation reaction with malonates

Applications of Mannich condensation: Some of the most common applications of organic chemical transformation involving Mannich condensation are given below.

1. The Mannich condensation is used in the synthesis of peptides, alkyl amines, antibiotics, nucleotides, alkaloids like tropinone, and many important agrochemicals.

2. Mnay polymers, Formaldehyde tissue crosslinking, catalysts, Pharmaceutical drugs like rolitetracycline (fluoxetine (antidepressant), tolmetin (anti-inflammatory drug), and tramadol are formed via Mannich condensation.

3. Many detergents and soaps are synathesized via Mannich condensation which find applications in cleaning industry, epoxy coatings, and automotive fuel treatments.

4. The thermal decay of Mannich reaction products gives rise to α, β-unsaturated ketones by (e.g. methyl vinyl ketone through 1-diethylamino-butan-3-one).

➢ *Benzoin Condensation*

The benzoin condensation may simply be defined as an addition reaction involving two aldehydes (generally aromatic aldehydes or glyoxals) to give rise to an acyloin.

This reaction was invented by two German chemists Justus von Liebig and Friedrich Wohler, and its classic case is the conversion of benzaldehyde to benzoin.

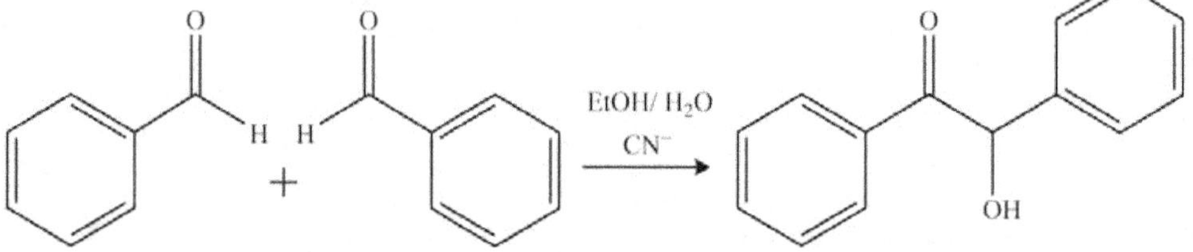

Mechanism of benzoin condensation: The benzoin condensation is catalyzed by nucleophiles like N-heterocyclic carbene or cyanides. A. J. Lapworth proposed a mechanism in 1903 which says that the cyanide anion from sodium cyanide reacts with the given aldehyde via nucleophilic addition (first step); followed by the rearrangement of the intermediate imparting polarity reversal of the carbonyl group, which in turn, attacks another carbonyl group via nucleophilic addition (second step). The benzoin as the final product will be obtained by the proton transfer and cyanide ion's elimination happening afterward. Also, being a reversible transformation, the products' distribution is governed by the comparative thermodynamic stability of the reactants and products.

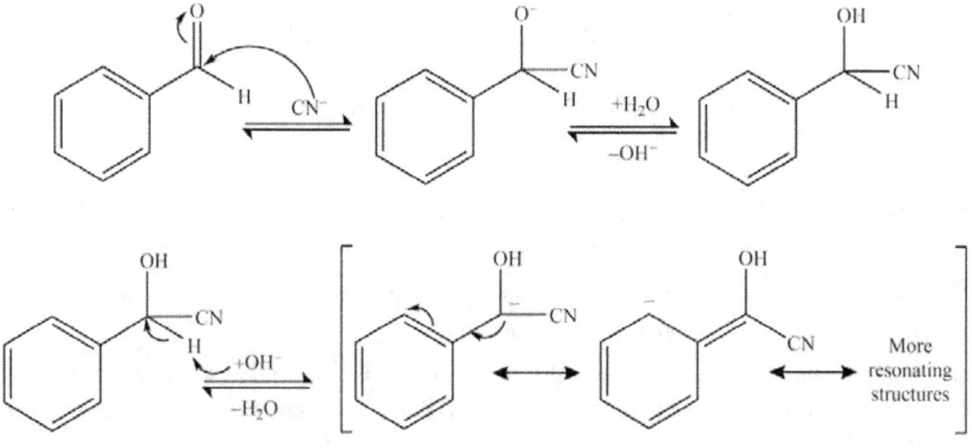

By looking at the mechanism, it can clearly be seen that one aldehyde accepts a proton whereas the other one donates a proton. Although most of the aldehydes are capable of donating as well accepting proton (like benzaldehyde); some aldehydes like 4-dimethylaminobenzaldehyde can only donate protons.

Exploiting this possibility, mixed benzoins can easily be synthesized. Nevertheless, the homo-dimerization should be sidestepped by careful matching of proton donating-accepting aldehydes.

Stereochemistry of benzoin condensation: If participating aldehydes are different in benzoin condensation reactions, the final product will exist as an enantiomeric pair.

S-isomer

R-isomer

Examples of benzoin condensation: Some of the most common examples of organic chemical transformations benzoin condensation are given below.

1. The *n*-heterocyclic carbene-catalyzed cross-benzoin reactions with chemoselective behavior.

2. A regiospecific catalyzed cross silyl benzoin reaction

Applications of benzoin condensation: Some of the most common applications of organic chemical transformation involving benzoin condensation are given below.

1. The benzoin condensation can be extended to aliphatic aldehydes if thiazolium salts are used. The resulting compounds are vital in the heterocyclic synthesis. Also, the 1,4-addition of an aldehyde analogous to an enone is labeled as the Stetter reaction.

2. In biochemical systems, the coenzyme thiamine is accountable for the biosynthesis of acyloin-like compounds via benzoin condensation; and this coenzyme has a thiazolium moiety, which becomes a nucleophilic carbene after deprotonation.

3. The asymmetric version of benzoin condensation has been carried out by using chiral triazolium and thiazolium salts; triazolium salts yielded higher enantiomeric excess in comparison to thiazolium salts.

4. Owing to the thermodynamical control, retro benzoin condensation can be very valuable. If acyloin or benzoin can be prepared by another route, then they can be transformed into ketones via cyanide or thiazolium catalytic use. The mechanism will almost be the same except that it takes place in the backward direction; which in turn, allows ketonic access.

> *Perkin Condensation*

The Perkin condensation may simply be defined as an organic transformation where an α, β-unsaturated aromatic acid is obtained by the aldol condensation of an acid anhydride and an aromatic aldehyde, in the availability of an alkali salt (acting as a base catalyst) of the acid.

This reaction was invented by an English chemist William Henry Perkin to make cinnamic acids; and therefore, is also named after him.

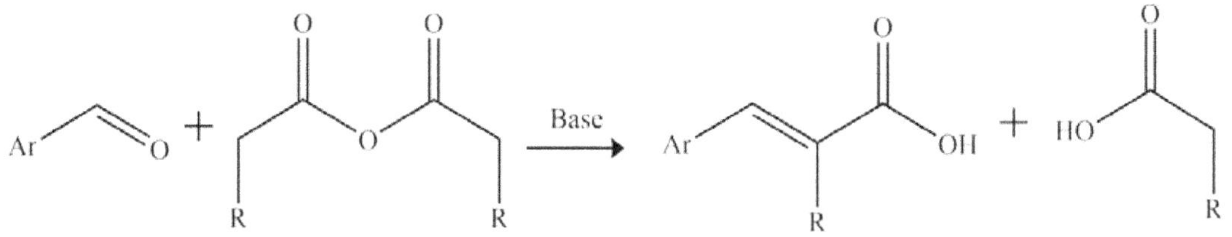

The relative arrangement of the aromatic ring and carboxylic acid in the end product of Perkin condensation can either be Z or E.

Mechanism of Perkin condensation: The most widely accepted mechanism for the Perkin condensation is given below.

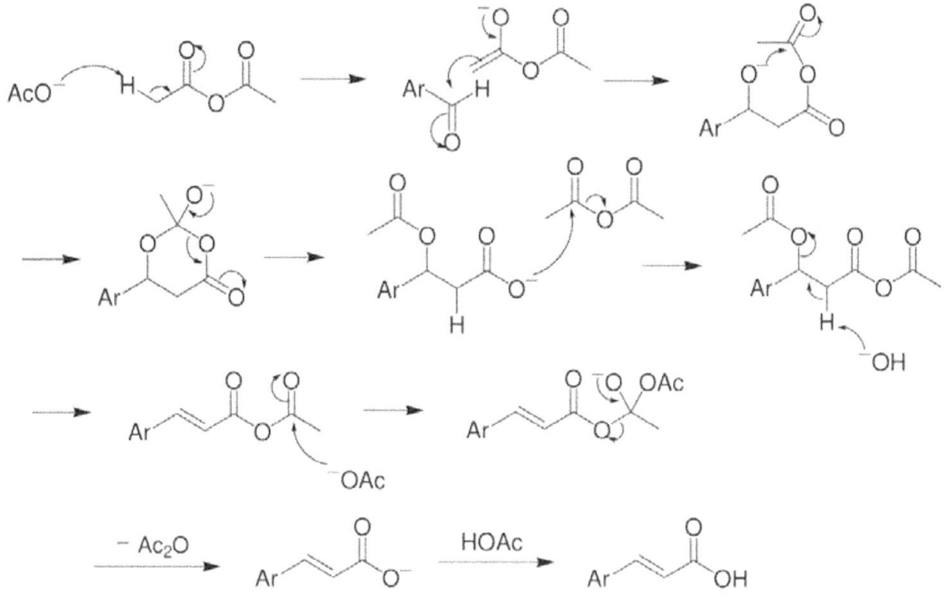

It is also worthy to note that the mechanism given above is not accepted by all of the scientific community; and therefore, many other sorts can also be found in different texts. One of such versions differs in the aspect of decarboxylation without transfer of acetic group.

Stereochemistry of Perkin condensation: In the Perkin condensation, the geometrical arrangement of the aromatic ring and carboxylic acid in the ending product can either be Z- or E-type (although the amount of major and minor will be different).

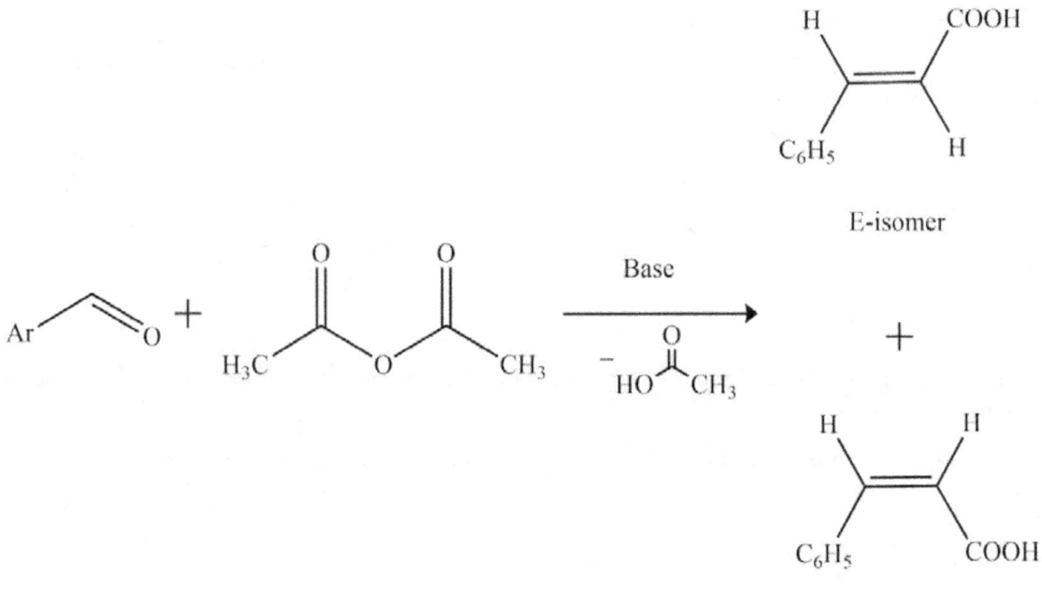

Examples of Perkin condensation: Some of the most common examples of organic chemical transformations involving Perkin condensation are given below.

1. The most popular example of Perkin condensation is the reaction between benzaldehyde and acetic anhydride to give cinnamic acid.

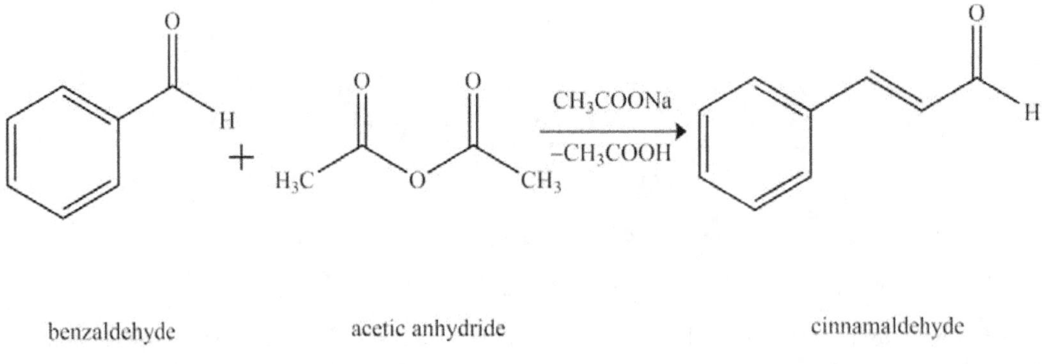

benzaldehyde   acetic anhydride   cinnamaldehyde

2. The reaction between sodium salt of salicylaldehyde and acetic anhydride to give coumarin is also an example of Perkin condensation.

2-formylphenolate

+

acetic anhydride

$\xrightarrow[-CH_3COOH]{CH_3COONa}$

2*H*-chromen-2-one

3. One more example of Perkin condensation includes the generation of coumarin from 2-hydroxybenzaldehyde.

2-hydroxybenzaldehyde

$\xrightarrow[H_2O]{Ac_2O,\ AcONa}$

2*H*-chromen-2-one

Applications of Perkin condensation: Some of the most common applications of organic chemical transformation involving Perkin condensation are given below.

1. One of the most important applications of Perkin condensation is in the laboratory preparation of the phytoestrogenic stilbene resveratrol.

2. Perkin condensation is used to synthesis of 'coumarin' which finds uses in medicine, rodenticide precursor, laser dyes, aromatizers, and perfumes.

3. Perkin condensation is the most popular route for the synthesis of cinnamic acid which is an extremely important compound for synthetic indigo, flavorings, and pharmaceuticals industry.

> *Stobbe Condensation*

The Stobbe condensation may simply be defined as a modification to Claisen condensation where the diethylesters of succinic acid react with aldehydes (or ketones) to give rise to alkylidene succinic acids or their monoesters in presence of a relatively less strong base.

This reaction is a modification to Claisen condensation and was invented by a German chemist Hans Stobbe; and therefore, is also named after him. The initial reaction was observed in 1893 when H. Stobbe observed that the reaction between acetone and diethyl succinate (in the presence of C_2H_5ONa) yielded an α-, β-unsaturated ester (tetraconic acid) and its monoethyl ester, instead of a 1-, 3-diketone product via normal Claisen condensation.

In the later years, Stobbe and his co-workers observed that this is quite common when succinic acid's esters are treated with aldehyde or ketones.

Mechanism of Stobbe condensation: The most widely accepted mechanism for Stobbe condensation that can explain the generation of an ester group, as well as the formation of a carboxylic acid group is a function of a lactone intermediate as shown below.

The carbonyl component isn't restricted in Stobbe condensation; and therefore, it even can have α-hydrogens. Nevertheless, if α-hydrogens are present in the carbonyl component, the double bond migration can trigger the formation of many types of final products.

Stereochemistry of Stobbe condensation: Only one alkene stereoisomer will be obtained if symmetrical ketones are used; nevertheless, unsymmetrical ketones will give rise to a mixture of alkene stereoisomers.

Z-isomer

+

E-isomer

Assuming that the R″ has higher priority than R′

Examples of Stobbe condensation: Some of the most common examples of organic chemical transformation Stobbe condensation are given below.

1. One of the most popular examples of Stobbe condensation is the reaction between acetone and diethyl succinate to give tetraconic acid and its monoethyl ester.

acetone

diethyl succinate

Tetraconic acid

+

Tetraconic acid monoethylester

2. The reaction between benzophenone and diethyl succinate to give corresponding monoethyl ester is also an example of Stobbe condensation.

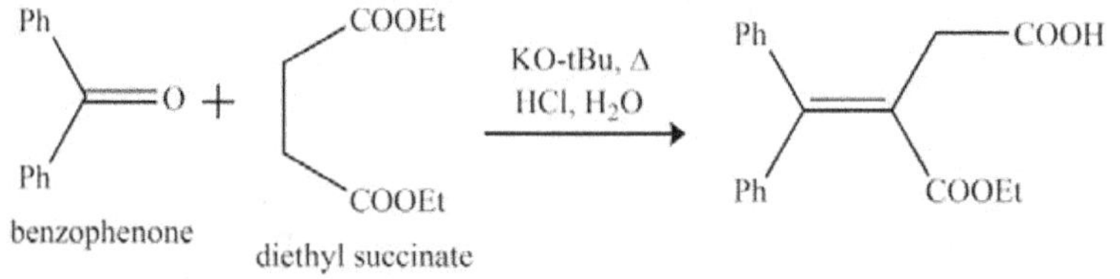

3. One more example of Stobbe condensation includes the generation of acids and monoethyl esters from the reaction between alkyl aryl ketone with diethyl succinate.

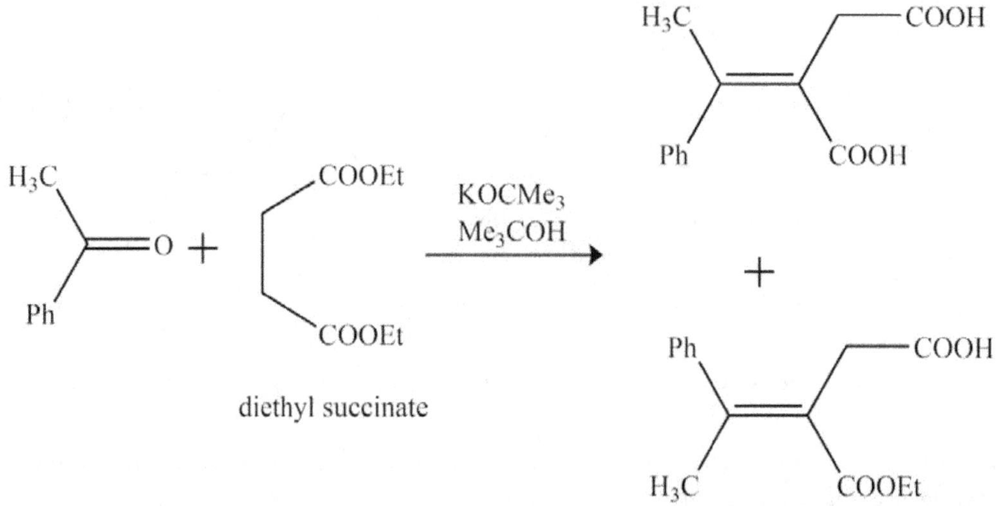

Applications of Stobbe condensation: Some of the most common applications of organic chemical transformation involving Stobbe condensation are given below.

1. Stobbe condensation is widely used to synthesize different types of organic acids.

2. one of the major applications of Stobbe condensation is the synthesis of polycyclic ring systems. For instance, the Stobbe products from aryl ketones can give rise to naphthol or indenone derivatives when undergoes dehydration route.

3. Tetralone and phenenthren derivatives can also be obtained using Stobbe condensation.

4. Reinhard Sarges' synthesis of tametraline and synthesis of dimefadane are also based upon the employment of Stobbe condensation in the first step.

❖ Hydrolysis of Esters and Amides

In this section, we will discuss the mechanism of acid- and base-catalyzed hydrolysis of esters and amides (both are the derivatives of carboxylic acid) in detail.

➤ *Hydrolysis of Esters*

Although the esters are derived from acids, they are generally neutral compounds. In an archetypal ester reaction, the OR group (i.e., alkoxy) of the ester is swapped by another group. One such type of reaction is the ester hydrolysis where the OH group (generated by the water-splitting) replaces the alkoxy group of esters under consideration. The ester hydrolysis can either be catalyzed by an acid or by a base.

Illustrative Reaction: The typical organic chemical reaction depicting acid hydrolysis of esters is shown below.

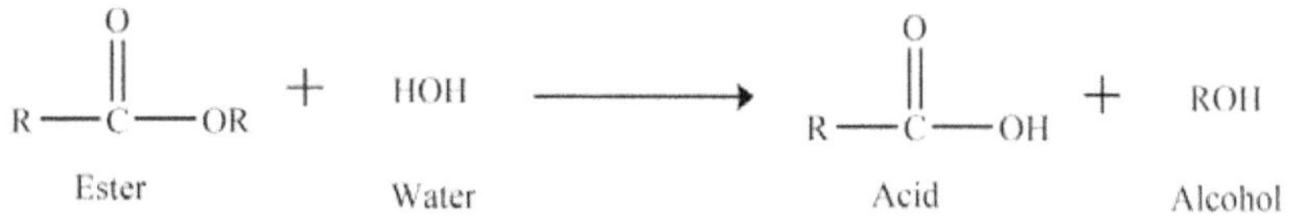

$$R-\underset{\substack{\| \\ O}}{C}-OR \;+\; HOH \longrightarrow R-\underset{\substack{\| \\ O}}{C}-OH \;+\; ROH$$

Ester   Water   Acid   Alcohol

Mechanism involved: Since the ester hydrolysis can either be catalyzed by an acid or by a base; a brief overview for both kinds must be understood for a better understanding.

i) Acid-catalyzed mechanism of ester hydrolysis:

The mechanism for acid-catalyzed ester hydrolysis is a case of 'less reactive system type', and all the steps involved are shown below.

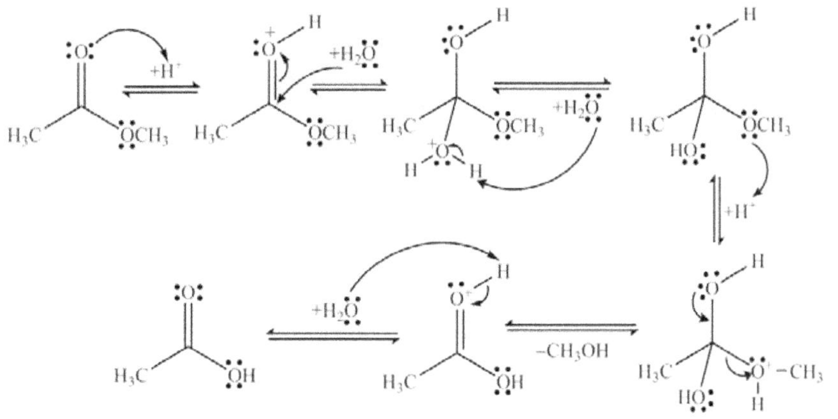

Furthermore, it is also worthy to note that the acidic hydrolysis of esters is just the reverse of esterification where an ester is heated with a large amount of water in the presence of a strongly acidic catalyst. Also, acidic ester hydrolysis is a reversible process and does not complete with 100% yield (like esterification).

ii) Base catalyzed mechanism of ester hydrolysis:

The mechanism for base-catalyzed ester hydrolysis is a case of 'reactive system type', and all the steps involved are shown below.

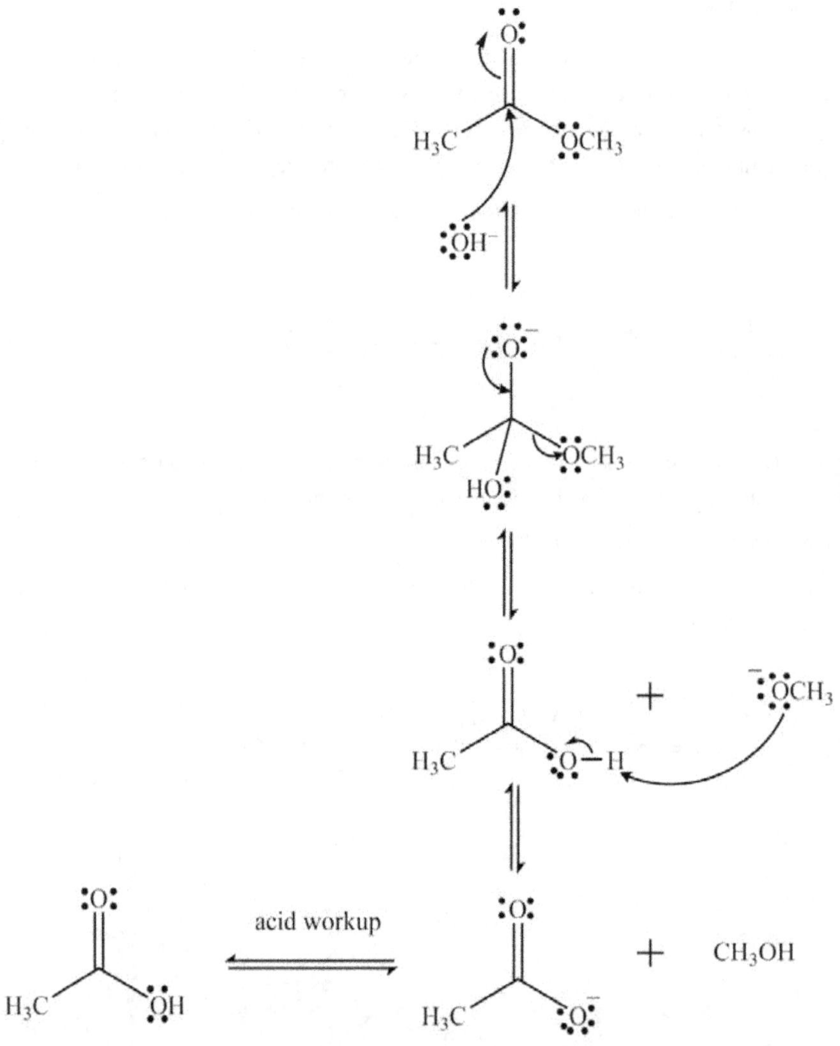

The mechanism given above gives rise to the breakage of the acyl-oxygen bond (second step); and is supported by experimental pieces of evidence through if the compound is isotopically labeled (i.e., ^{18}O). A similar conclusion was drawn if esters of chiral alcohols were used. The base-catalyzed ester hydrolysis is popularly known as the "saponification" process due to its use of soap-synthesis.

> ***Hydrolysis of Amides***

Amides are derivatives of carboxylic acid where the OH group has been substituted by NR_2, NH_2, NHR, or amine. Since the reaction between an amine and a carboxylic acid giving amide occurs via the release of the water molecule (condensation reaction), the amides' hydrolysis can be labeled as the reverse of condensation reaction as the amine and acid are being reproduced. The amides' hydrolysis isn't easy and requires conditions like the heating of amide with aqueous acid for a long interval of time. Like the hydrolysis of esters, the amide hydrolysis can either be catalyzed by an acid or by a base.

Illustrative Reaction: The typical organic chemical reaction depicting acid hydrolysis of amides is shown below.

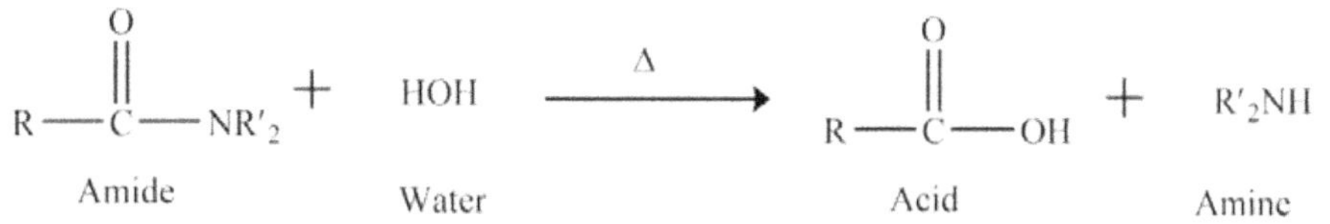

$$R - \overset{\overset{\textstyle O}{\|}}{C} - NR'_2 \;+\; HOH \;\xrightarrow{\;\Delta\;}\; R - \overset{\overset{\textstyle O}{\|}}{C} - OH \;+\; R'_2NH$$

Amide   Water   Acid   Amine

Mechanism involved: Since the amide hydrolysis can either be catalyzed by an acid or by a base; a brief overview for both kinds must be discussed for a better understanding.

i) Acid-catalyzed mechanism of amide hydrolysis:

The mechanism for acid catalyzed amide hydrolysis is a case of 'less reactive system type', and all the steps involved are shown below.

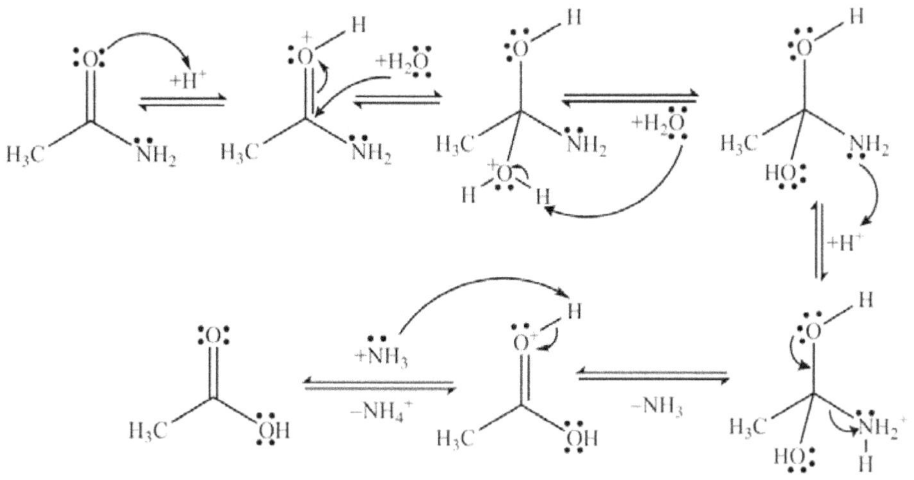

It is obvious from the mechanism given above that the acid catalysed amide hydrolysis is quite analogous to the acid catalysed esters' hydrolysis; and proceed via the protonation of the carbonyl group and not the amide one.

ii) Base catalyzed mechanism of amide hydrolysis:

The base-catalyzed amide hydrolysis is extremely difficult to carry out but possible if the amide is heated for a very long span of time. All the steps involved in the base-catalyzed hydrolysis of amide are shown below.

It is obvious that the major problem in the way of substitution to happen is the need for a good leaving group; however, the deprotonated amine so strongly basic that it is almost the opposite of a good leaving group. Consequently, the breaking of amide is proved to be extremely difficult even if we couple very high temperatures with a base like KOH.

❖ Ammonolysis of Esters

Before we study the ammonolysis of esters, we need to distinguish the term 'ammonolysis' from the term 'aminolysis' first. The precise definition of 'ammonolysis' includes the chemical reactions in which a compound is split into two parts by its reaction with ammonia; however, in broader terms, admins can also be used. On the other hand, the precise definition of 'aminolysis' includes the chemical reactions in which a compound is split into two parts by its reaction with amine; nevertheless, in broader terms, ammonia can also be used. Hence, we can conclude that the terms 'ammonolysis' and 'aminolysis' are pretty much similar not only w.r.t names but also in their approach; and therefore, are used in an exchangeable manner in different textbooks.

> ### Definition and Examples Reactions of Ammonolysis of Esters

Now we come to the 'ammonolysis' of esters, which popularly means that the esters can be converted into primary, secondary, and tertiary amides (along with alcohols) by treating them with ammonia, primary amines, and secondary amines respectively.

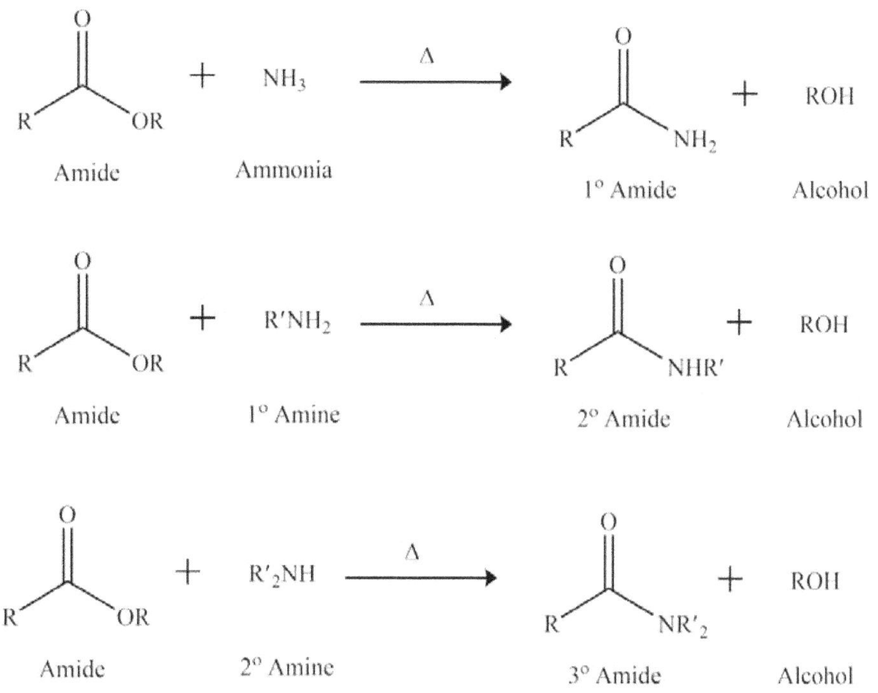

Since the RO^- is a very poor leaving group, the conventional nucleophilic addition-elimination pathway will not be useful as far the practicality is concerned. Hence, unlike the reaction of acyl chlorides with amines, the corresponding nucleophilic addition-elimination in case esters requires much stronger conditions.

> ### *Mechanism of Ammonolysis of Esters*

Before we discuss the mechanism of ammonolysis of esters, we understand different outcomes first. Initially, an ammonia molecule (or amine) attacks the carbonyl via nucleophilic addition; whilst a large amount of ammonia still present in the reaction solution, followed by the formation of an anionic tetrahedral intermediate due to deprotonation.

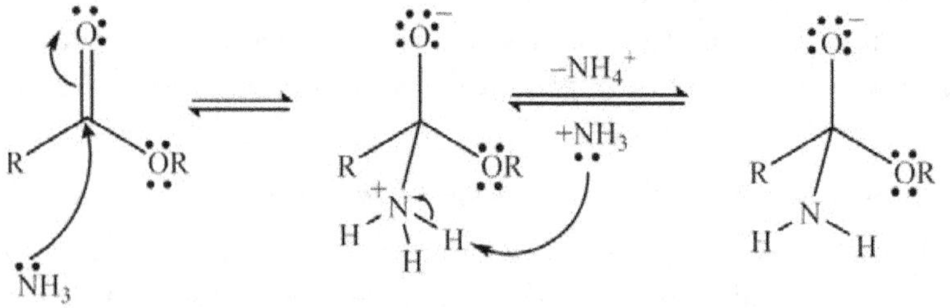

At this point, C=O double bond can only be restored only if either the alkoxy (RO^-) or the amide (NH_2^-) group is detached. Now although both are very poor leaving groups; the pKa values of alcohol and ammonia suggested that alkoxy groups (RO^-) are much weaker bases than ammonia's conjugate base (i.e., NH_2^-), and therefore, is a better leaving group. Consequently, the reassortment of the C=O bond will happen via loss of alkoxy group giving rise to an amide product.

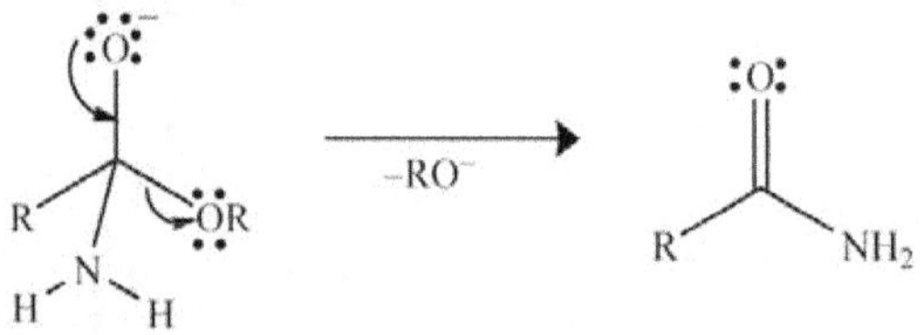

However, it is also worthy to note that the relative betterment of alkoxy as a leaving group doesn't make it a good leave group on the absolute scale; and therefore, the ammonolysis of esters isn't a very effective route for the amides' synthesis, indicating acyl chlorides as more suitable substrates.

❖ Problems

Q 1. Discuss the basic mechanism of addition to carbon-heteroatom multiple bonds.

Q 2. What is carbonyl reduction? How $LiAlH_4$ acts differently than $NaBH_4$ in such transformations?

Q 3. Give the mechanism involved in the metal hydride reduction of nitriles.

Q 4. What are Grignard reagents? Explain with a suitable example.

Q 5. State and illustrate the mechanism of the Wittig reaction.

Q 6. Write down the mechanism involved in the Claisen condensation.

Q 7. Illustrate the mechanism of hydrolysis of ester and amides.

Q 8. Define the process of ammonolysis.

❖ Bibliography

1. M.S. Singh, *Reactive Intermediates in Organic Chemistry*, John Wiley & Sons, Inc., New Jersey, USA, 2014.

2. H. Zimmerman, *Quantum Mechanics for Organic Chemists*, Academic Press, New York, USA, 1975.

3. J. Clayden, N. Greeves, S. Warren, *Organic Chemistry*, Oxford University Press, Oxford, UK, 2012.

4. R. L. Madan, *Organic Chemistry*, Tata McGraw Hill, New Delhi India, 2013.

5. M. B. Smith, *March's Advanced Organic Chemistry: Reactions, Mechanisms, and Structure*, John Wiley & Sons, Inc., New Jersey, USA, 2013.

6. D. Klein, *Organic Chemistry*, John Wiley & Sons, Inc., New Jersey, USA, 2015.

7. C. A. Coulson, B. O'Leary, R. B. Mallion, *Hückel Theory for Organic Chemists*, Academic Press, Massachusetts, USA, 1978.

INDEX

D

E

DALAL
INSTITUTE